AF776171

This is an authorized collector edition only if signed and numbered by the author. #98, signed John J Brown

DAN CALKIN and his ELFs

(He never called them ELVES)

John J. Brown

John J. Brown
Model Aviation Books
Santa Ana, CA

1999

DAN CALKIN and his ELFS
He never called them ELVES

by John J. Brown
edited by Cecelia Calkin Jackson

John J. Brown

(714) 544-3358 voice
(714) 544-3486 fax
e-mail jjrbrown@earthlink.net

This book was produced on a Macintosh Performa Model 6320CD using QuarkEXpress 3.32, Photoshop 4.0 and a Umax Scanner Model 1200S. Pre-press printing was done on an Apple Laserwriter Pro 630 postscript printer.

Our thanks to Jim Hastings at The Enlarger in Santa Ana, California, for his help with the photographic work. Also a special thanks to Doug Lee, owner of Stilley's Photo Shop in Santa Ana, California, for his advice and one day black and white film processing. Steve Martin gets a lot of credit for technical assistance with Photoshop and QuarkExpress and for his design of the dust jacket. Model Aviation logo by Bill Hannan of Hannan's Runway.

Published by:

Model Aviation Books
John J. Brown
SPECIALIST IN MODEL AIRCRAFT LITERATURE
13362 FAIRMONT WAY • SANTA ANA, CA. 92705

Library of Congress Catalog Card Number: 99-95426
ISBN 0-9673666-0-7

Printed and bound in the United States of America

DEDICATION

It would be impossible to embark on a project of this magnitude on a whim without the total support of the woman in my life. My wife, Jean, encourages me to do anything that interests me. She knows that whatever makes me happy will somehow make her happy.

A second dedication goes to Mrs. Edwin Daniel Calkin, also known as Josephine Jane Calkin. The first time I talked to her on the telephone, I was struck by the enthusiasm in her voice. That enthusiasm was contagious. Add to that her willingness to let me copy the Calkin ELF scrap book, her gift of prototype and other early ELF engines, her loan of the existing ELF records, and her willingness to answer innumerable questions has changed the basis of this book from educated guesses to factual information. Following that were many photographs, pieces of literature, tooling and odds and ends of parts and it can be seen who should really get the credit for this book.

Finally, my hat is off to Cecelia, the daughter of Dan and Josephine. I had originally set out merely to document the various models of ELF engines. After numerous conversations with Cecelia, she suggested that I write a book. At first I was shocked. I told her that I didn't have enough information for a book and I had never written a book before, but, after a while I realized that a book was indeed possible and I thought I would try it. Without her encouragement and assistance, this book would not have happened.

Foreword
by Alex Imrie

The story of Dan Calkin and his diminutive ELF engines is one of great perseverance. Naturally, trial and error entered into their development but the feat of producing such a small engine as the Model 1 ELF in 1935 was all the more remarkable when one realises that this was true pioneering work that required the highest skill and a firm understanding of the principles involved. There was no great song and dance of advertising when this engine first became available, it relied entirely on its quality to forge its reputation.

One of the first users overseas was England's pioneer modeller C E Bowden, whose ambition at the time was to build a power model 'small enough to carry in an attache case with lunch and a bottle of beer'. Apart from the 'astonishingly' small size he was greatly impressed with the fine workmanship and the engine's easy starting and smooth running - 'it keeps on running like a little clock'. He used his two ELF's to power many of his small models over a number of years and later owned an early ELF "Four" Model 40 that fascinated him with its excellent power and 'exciting exhaust note'. When noted modeller Paul Plecan wrote:- 'Within that rough sand-casting was a heart of pure gold', he was writing of the first ELF production engine, (he owned #105) but his comment can be applied equally to all of Dan Calkin's engines.

John Brown's narrative puts this into perspective and enables today's engine enthusiasts to learn much of the activities behind the scenes that resulted in the high quality ELF products. Readers should be aware that over a period of years John has devoted himself to the history of our hobby. He owns what must be without doubt the finest collection of model aviation reference material in the world and in article form has written about his findings in *The Engine Collectors' Journal* when one of his main interests concerned compressed-air motors. Some thirty years ago, in presenting such an engine history, John wrote: 'When you try to follow a trail that is forty years old, you must either be darned good or very lucky to come up with a story containing anything that approaches factual information.'

The following excellent compilation shows that the author is obviously 'darned good' and while one could say that chance brought him to the ELF story, John's 'luck' in ferreting out information is the direct result of his tenacious approach at finding the truth and he tells something of his search for the facts in these pages. What is not so obvious is his absolute dedication to the task in hand where he leaves no stone un-turned in his quest for more information.

It gives me great pleasure to be able to write this Foreword to John's book. I seriously doubt if such a detailed accurate account of the development of any other model company's engines of the period can be done.

This book sets the highest standards and all enthusiasts will be grateful to John Brown for his fine portrayal of such an important engine manufacturer and his products.

TABLE OF CONTENTS

List of Illustrations

List of Illustrations, continued

List of Illustrations, continued

List of Illustrations, continued

List of Illustrations, continued

List of Illustrations, continued

List of Illustrations, continued

List of Illustrations, continued

INTRODUCTION

The 1930s were a time for great advances in aviation that was still riding the crest of excitement over Lindbergh's successful New York to Paris flight. The advances in aviation were matched by advances in model aviation. While model airplanes were being built for everything from toys to scientific research vehicles, most were being built by an air minded youth captivated by the excitement and romance of flight. A boy launching a rubber powered free flying model airplane imagined himself as another aviation pioneer like Lindbergh. The dreams of working as an aviation researcher were mixed with thoughts of dashing winged knights, barnstorming dare devils, and speed thrill seekers such as Rickenbacker, the Red Baron, Frank Hawks or Al Williams.

Though a few internal combustion engines had been built for model airplanes by home craftsmen, most were too heavy, cranky and cumbersome for a practical model airplane. By the early 1930s though, a few men were building successful model power plants that could be made available for general use. Names like Bill Brown, Bill Atwood, Irwin Ohlsson, Dan Bunch and Dan Calkin come to mind. This book is about the quest of one man—Dan Calkin—to build a better and smaller model airplane engine.

Dan Calkin and his contemporary, Bill Brown, were two of the real pioneers in the development and production of model airplane engines. To make an engine or two is one thing, to manufacture them in quantity over a 25 year period is quite another. Calkin may be the greater pioneer because he accepted the challenge of making very small engines. The early ELF single had a .138 cubic inch displacement versus .60 cubic inch for the Brown Jr. I don't mean to minimize the accomplishments of Bill Brown. I do mean to emphasize the accomplishments of Dan Calkin. For a one-man shop to compete with the large manufacturers with automatic equipment and high volume production is a success story in its own right. Many manufacturers of model airplane engines couldn't survive, but ELF did. How many other manufacturers who started before WWII survived to the middle 1950s?

Very little factual information about the company, the man or his engines has been available up to now. Much of the information that has been passed around about Calkin and the ELF engines has been in the nature of opinions, speculation, and deliberate misrepresentation. I hope to clear up much of the misinformation and shed new light on the man and his engines. This has only been possible through the help and support of Dan's wife Josephine, daughter Cecelia, and former ELF employees Philip Foster, Carl VanCourt, Dick Rothrock, and Leonard "Lucky" Lucas. In my search I learned that former ELF employees/associates Rufus Rothrock, Fred "Red" Burkhart, Dan Totten, and Harry Fosbury were all deceased. Oddly, I must list Allan R. Shively as a valuable source of information. Even though Allan died several years ago, he was among the first ELF engine collectors and he diligently pursued ELFs and information about them for more than 20 years. He put together much of the information and mis-information about the various models that we still use today.

My search for information caused me to track down another early ELF researcher, Doctor Richard E. Nichol, commonly known as Doc Nichol. From his articles in the *Engine Collector's Journal* I knew that Doc had studied the ELF line extensively in the 1960s and had interviewed Dan Calkin on several occasions. Through another stroke of luck, I found Doc. When I telephoned him he told me that at one time he had planned to write a book about the ELF story and, though he had accumulated a lot of information, he no longer had the time to do so. He was pleased that someone was doing it and he very generously sent me a copy of his thick notebook with all the information he had accumulated. It was fascinating reading his notebook. He not only confirmed many things I had learned, giving me more confidence in my work, but he had some interesting insights and new information which really inspired me. Thanks a lot, Doc.

Ted Enticknap has been my mentor through all of this. During the course of this project, I telephoned Ted hundreds of times to discuss various details with him. He really helped me make some sense of very confusing and conflicting information. Ted is a long-time ELF expert and he knew a former ELF employee—Phil Foster. Ted put me in touch with Phil.

Alex Imrie, who writes the "'Vintage Corner" column for *Aeromodeller* magazine put me in contact with another former ELF employee, Carl VanCourt. Later, I found Dick Rothrock and Leonard Lucas simply by using the telephone directory.

Another major source of information has been the help of about 150 ELF owners who took the time to fill out and return a long survey sheet. The survey sheets were printed through the generosity of Tim Dannels, editor of *The Engine Collectors' Journal*. So many people helped in so many ways that I can't possibly single everyone out for special thanks, but it all helped, and it all contributed to the story. All the people, whose names I can remember, who contributed to the story, are listed in Appendix E. I refer to them as ELF Insiders. Let's give them all a big hand (applause).

As a long time model airplane builder I have always admired the ELF engines of Dan Calkin. These engines were different. ELFs were not mass produced. They were a labor of love. Every engine bore Dan's skilled touch, his pride of workmanship, his best engineering genius. Maybe this is why the ELF engines of Dan Calkin are perhaps the most highly prized among those of us who treasure the model airplane engines of the past.

As Irwin Ohlsson told me, "They were made like a watch." Dan ran every engine himself and any engine that didn't meet his standards did not go out the door. They were crafted in a one-man shop during most of 21 years. It was also a one-woman shop. The woman was Dan's wife Josephine.

In the spring of 1994, a pharmacist friend of mine, Fred Stewart, introduced me to one of his customers, Dave DeMilt, who was a model builder. I arranged for him to visit my home. We were exchanging model airplane stories when he told me about a house painter who, on January 5, 1979, had painted the Huntington Beach, California, home of a woman who had a large twin engine model airplane in her garage. She wanted to give the model away because it took up too much room in her garage. She also told him that at one time her husband had manufactured model airplane engines in Portland, Oregon. As soon as I heard that I got excited, because the only engine ever manufactured in Portland was the ELF.

I set out to find this woman. The painter now lived in Las Vegas, but after two weeks of searching, I had the woman's phone number. I called and asked the woman who answered the phone if she was the Mrs. Calkin. She said, "That's me."

I visited Mrs. Calkin and she showed me the Calkin scrap book and told me a little about Dan Calkin and his ELFs; she said that he never called them ELVES.

Because I knew of several people who had tried to put together the story of the ELF engines and had failed, I wanted to try. I was allowed to copy the entire scrap book and I started looking for more information. Mrs. Calkin put me in touch with her daughter, Cecelia, who spent her childhood in the ELF factories, helping out in any way she could. When she was in high school, Cecelia went to the shop after school every day and made the castings, did machining and whatever else was required.

Many, many people have given me a lot of help, but the one who was absolutely indispensable is Mrs. Calkin, who I always called Josephine.

Dan Calkin before the first flight of the Model 10 ELF in October 1939.

Chapter 1 BEFORE ELF

THE FORMATIVE YEARS

What sort of man was Dan Calkin? In 1928 he built the world's smallest gas engine. In 1933 he was credited by Ripley's "BELIEVE IT OR NOT" with building the world's smallest gas engine powered model airplane. He manufactured high quality, reliable engines for model airplanes, mostly multi-cylinder engines for over two decades, then became a rocket scientist. He is an enigma. He accomplished great things but, until recently, almost nothing was known about him. Here is the story, derived from records and countless interviews with family, friends, and co-workers, of the man and his ELF model airplane engines.

Figure 1-1 A 1996 photo of Dan's boyhood home in Kansas City, Missouri. (Chuck Welman photo)

Edwin Daniel Calkin, known as Dan to nearly everyone, was born in Prosser, Washington, on the 25th of February, 1911. He was the only child of Ralph Aretus Calkin and Etta Wise Calkin.

Ralph Calkin's profession was dentistry. He provided dental services for the federal government, mostly in the northwestern part of the country, later for the US Army, then the Veterans Administration, eventually becoming the Chief Dental Surgeon. Because of his occupation he and his family moved frequently.

SCHOOL DAYS

Dan had an inquisitive mind. He was very interested in engineering, mathematics, and science.

During his high school days, Dan lived with his parents in several different houses in the Kansas City, Missouri, area. Through most of his teens, they lived at 5906 Charlotte Street. (Figure 1-1) Before moving to Charlotte Street he had accumulated a large stock of chemicals for his chemistry experiments. Not wanting to risk moving his chemicals again, his parents capitalized on his interest in engineering and mechanics and bought him a lathe to replace the chemicals.

DAN CALKIN

"A youth to fame and to fortune unknown."

Student Council 4
Engineers, Vice-President 4
Orchestra 3, 4

Figure 1-2 Dan Calkin from his 1927 Southwest High School yearbook.

Dan first attended Westport High School (now a junior high school). When a high school was built for the kids in the newer part of town where Dan lived, he transferred to the school known as Southwest High School. He was active in student government and the school newspaper. He also enjoyed playing the clarinet in the school band and orchestra, and was a leader in the Faraday Club, an engineering club. School records show that he excelled in math and science and was an honor student. Graduation was on June 10, 1927. Figure 1-2 shows Dan Calkin as he looked in the 1927 Southwest High School Yearbook.

During his high school years he developed asthma and chronic allergy problems. These and other severe health problems were factors he had to contend with for the remainder of his life.

Figure 1-3 A small electric motor built by Dan Calkin.

Figure 1-4 Calkin-built brass steam engine.

Figure 1-5 Dan machining a project on his lathe from a 1928 newspaper article. (Calkin family scrap-book)

Figure 1-6 Dan's Aunt Lillian holding the 40 inch boat. (Note: The dry cell batteries and the Model T Ford spark coil had been removed for this photo). (Calkin family photo)

The 1928 Boat Engine

Dan taught himself how to use the lathe. Some early projects include a little electric motor and a small brass steam engine. (Figures 1-3 and 1-4). He made three gasoline engines while in high school and junior college, doing both the casting and machining at home. Figure 1-5 shows a photo of Dan Calkin featured in the *Kansas City Journal-Post* on Sunday, April 22, 1928.

The *University of Washington Daily*, October 31, 1930, newspaper article quotes Dan as saying:

> The first (engine) failed to run at all, the second exploded in a tangled mess at the first trial. Engine number three was successful.

Dan was only 17 years old when he made his first successful engine which he planned to use in a boat. The boat engine had a bore and stroke of $\frac{3}{4}$ inch giving it a displacement of 0.33 cubic inches (5.43 cc). The crankcase was sand cast of aluminum and the cylinder was machined from a solid piece of steel. The piston incorporated two rings, one at the top, the other at the bottom of the piston. The engine was 4 inches tall overall and was used to power a 40 inch long boat. See Figure 1-6.

Ignition was provided by a modified Model T Ford spark coil, with two large dry cell batteries. Dan made the spark plug of stacked layers of mica, the standard practice of the aircraft industry of that time. It used a $\frac{5}{16}$ inch x 32 thread size.

The steel flywheel was almost three inches in diameter. The carburetor featured a mechanism to prevent back-flow through the intake tube in case the engine backfired. The name CALKIN was cast into the cover of the intake port. The level of workmanship on this engine was amazingly good for a young person who had taught himself pattern making, foundry work and machining.

An analysis of the literature of that era suggests that Dan's engine was inspired by a design for a single cylinder gasoline engine appearing in the

Figure 1-7 A 1928 photo of the boat engine in Dan's hand. (Calkin family photo)

Figure 1-8 A 1997 photo of the boat engine. (Note: The original carburetor has been updated).

Figure 1-9 Dan and Josephine on a date during their Kansas City Junior College days. (Calkin family photo)

Figure 1-10 Dan Calkin, from his Junior College yearbook. (Calkin family photo)

Figure 1-11 Josephine Peck, from her Junior College yearbook. (Calkin family photo)

1919 book *Model Making* by Raymond F. Yates. This was confirmed by conversations with former employees. The Yates engine had a bore and stroke of one inch. There is some evidence that suggests that the first two engines Dan attempted (the unsuccessful ones) were Yates engines. When they wouldn't run, he made numerous changes and, wanting to make a smaller engine, Dan reduced the displacement to about one-third the displacement of the Yates engine and incorporated other modifications and several improvements into his engine. This is the engine in Figures 1-7 and 1-8. The Yates book also explained how to make a spark plug and gave basic information about pattern making and foundry work, soldering, and use of a lathe.

KANSAS CITY JUNIOR COLLEGE

After graduation from high school in June of 1927, Dan attended Kansas City Junior College, again excelling in mathematics and science. While attending junior college, he met the love of his life, a girl named Josephine Jane Peck. Josephine was also a student at the junior college. The photo in Figure 1-9 shows Dan and Josephine on a date during their junior college days. They were standing in front of the 1929 Buick belonging to Dan's father. Dan graduated from Kansas City Junior College on June 7, 1929. Before graduating, Dan and Josephine became engaged to be married. They had known each other for a year. The couple exchanged their photographs taken for the college yearbook. (Figures 1-10 and 1-11)

Semi Annual Joint Meeting of
Western Washington Section
and
University of Washington
Student Branch
AMERICAN SOCIETY OF
MECHANICAL ENGINEERS
University of Washington Commons
April 15 1931
—PAPERS—

Private Ownership of Airplanes
Don Euler

High Temperature Insulation
Donn Barthalet

Merrill Movable Wing
Maynard Pennell

Design and Construction of
Model Gas Engines
E.D. Calkin

Piano Selections - Wiley Shumm
Morry Hatch's Orchestra

Figure 1-12 Announcement of the presentation that Dan Calkin and others gave regarding their senior theses. (Calkin family scrap-book)

THE UNIVERSITY OF WASHINGTON

Dan enrolled at the Massachusetts Institute of Technology in Cambridge, Massachusetts. He attended MIT for one semester starting in September 1929, then transferred to the University of Washington because they would allow him to work on his small gas engines as a senior project.

It was about that time that Dan's father and mother moved from Kansas City due to the transfer of Dr. Calkin to the government hospital at American Lake, Washington.

Figure 1-13 The University of Washington engine on its display stand. (Calkin family photo)

During summer breaks from college, Dan learned to play golf at the American Lake country club where his father was a member.

Figure 1-14 A recent photo of the University of Washington engine.

Dan earned his Bachelor of Science degree in mechanical engineering at the University of Washington, graduating on June 15, 1931. During his senior year, Dan was listed as one of four people presenting a thesis at an April 15, 1931, ceremony of the American Society of Mechanical Engineers at the University of Washington. (Figure 1-12). He titled his thesis the *Design and Construction of Model Gas Engines*. The engine design for his senior project would become Dan's first airplane engine. During the summer of 1931, Dan did the casting and began machining the crankcase and the cylinder for this engine.

University of Washington Engine

The U of W engine was a 0.23 cubic inch (3.77 cc) displacement sideport engine having an $^{11}/_{16}$ inch bore and a stroke of $^{5}/_{8}$ inch. In Figures 1-13 and 1-14, notice that it had the vertically split crankcase that characterized all of Dan's engines. Other features were a sand cast aluminum seven fin cylinder, a piston fitted with a single ring plus an oil groove, and a float type carburetor. The

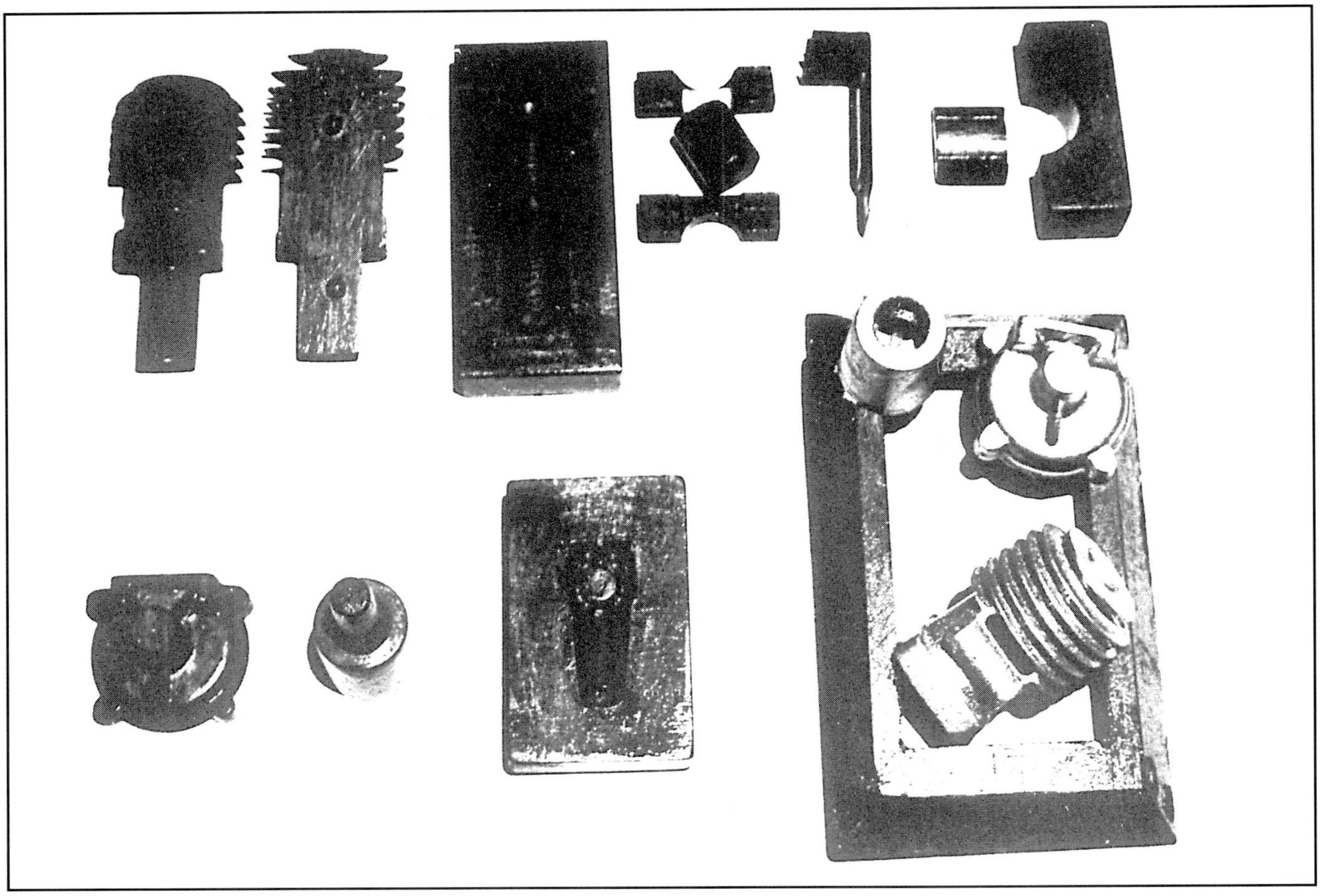

Figure 1-15 (Above) Patterns and some castings for the U of W engines. (Calkin family photo)

Figure 1-16 (Below) Machined parts for the U of W engine. (Calkin family photo)

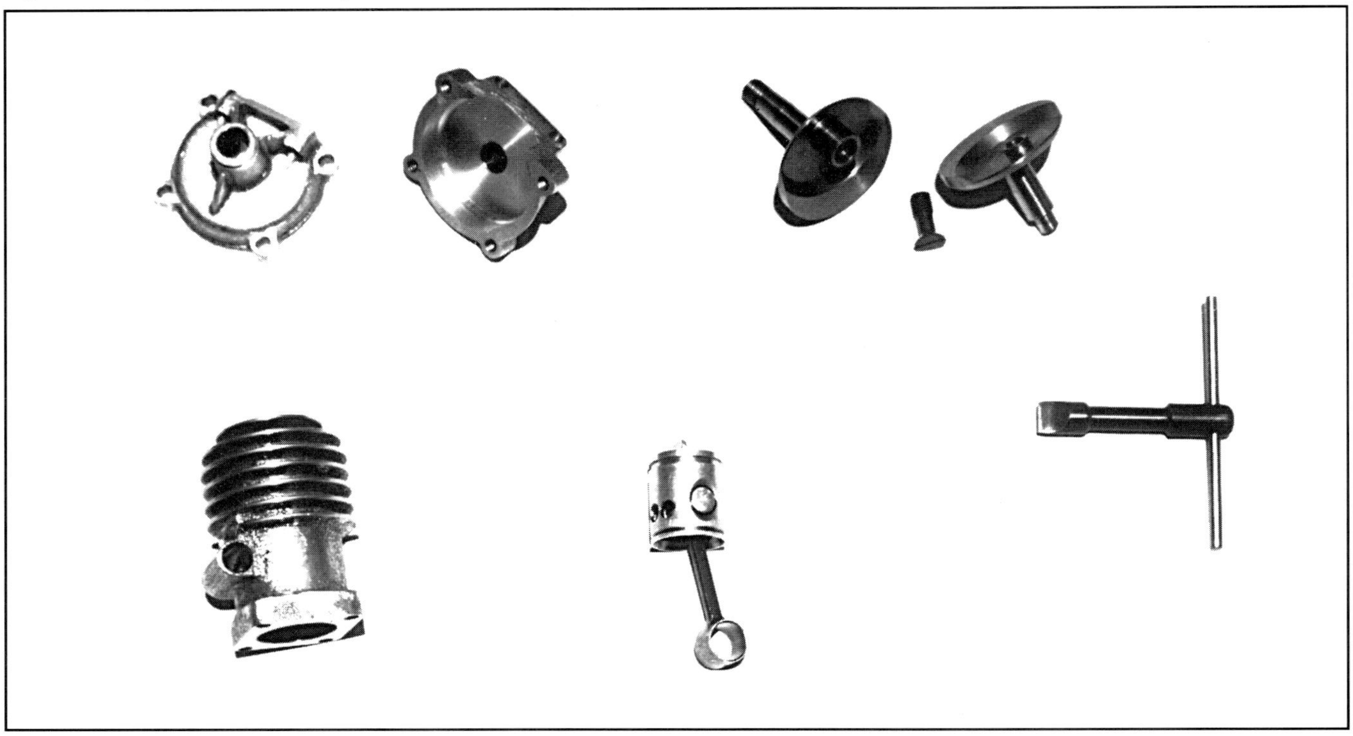

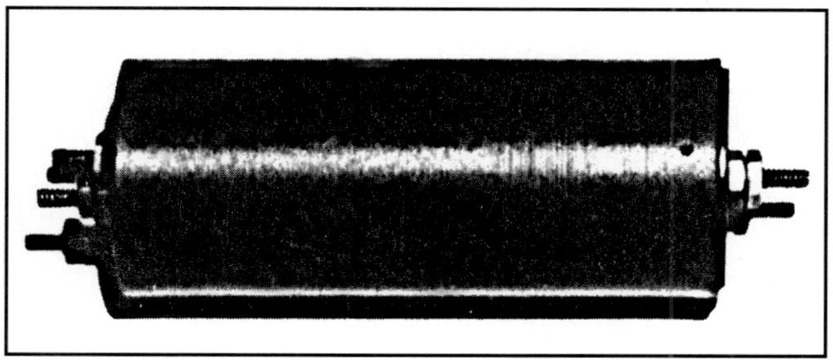

Figure 1-17 Dan Calkin's first ignition coil. (Calkin family photo)

spark plug Calkin made for this engine had a very fine $^{5}/_{16}$ inch x 48 pitch thread.

The propeller drive washers and front prop washers had eight screws through the hub in a circular pattern surrounding the central crankshaft. The timer was very similar to the one used later on the Model 1 ELF production engines, only larger. The engine was mounted radially at the front of the crankcase.

Additional details of its construction can be seen in Figures 1-15 and 1-16 showing the wood patterns and the machined castings for the engine.

MASSACHUSETTS INSTITUTE OF TECHNOLOGY

During the fall of 1931, Dan returned to MIT to work on his Masters degree and there, in early 1932, he completed work on the engine he had designed at the University of Washington. In a letter to his parents dated March 13, 1932, Dan wrote:

> When I was showing the graduate students the engine, some one asked Prof. Taylor if I was taking the engine design class. The Prof. said I wasn't and the fellow evidently suggested that I should be, for Prof. Taylor said, "...taking it, he should be teaching it."

The coil made for this engine (Figure 1-17) was Dan's first attempt at making an ignition coil. Dan is known to have consulted with one of the professors in the electrical department about coil design.

While he was studying at MIT, his father had been transferred to Portland, Oregon. Dan graduated with a Masters of Science degree from MIT on June 3, 1933, and, with his formal education completed, took the train from Boston, Massachusetts, to again live with his parents. Dan joined his parents at their rented house at 309 SE 33rd Street.

Dan had hoped to go to work for a company such as Boeing Aircraft but, with the country mired in a depression, jobs were nearly impossible to find. In a 1976 letter to his granddaughter, Allanna, Dan wrote:

> When I graduated from college in 1931 there were no jobs. That is what brought the ELF engine into being. It was made in a small machine shop that I ran for 25 years until 1956 when I joined Douglas Aircraft as a controls engineer.

ENGINE DEVELOPMENT

After moving to Portland, Dan devoted himself full time to developing his engines. His parents provided him with an allowance saying that it was better to give him the money then when he needed it rather than waiting to give it to him as an inheritance when he would have much less need for the money.

Dan enlisted the help of his cousin, Dan Totten, who was a farm boy with no background at all in metal working, but he was a willing worker and learned quickly.

It was through his cousin, Dan Totten, that Dan first met Carl VanCourt, a close friend. Carl had purchased a small one cylinder *Airequine* engine from the Dynamic Manufacturing Company of Chicago. It had a one inch bore and stroke, a three-ringed piston and an 18 inch cast aluminum propeller. The ignition system and carburetor were both from a motorcycle engine. Carl was never able to get the engine to run and all it would do was "pop." Dan Totten told his cousin, Dan Calkin, about the engine and both visited Carl's house to see it. Dan Calkin looked at it and remarked, "It's a piece

Figure 1-18 The only photo of the 1932 experimental engine. (Calkin family photo)

Figure 1-19 Before the first flight of the first ELF powered five foot airplane. (Calkin family photo)

of junk, I could make you one half that size and it would run." Carl replied, "O.K. you build it and I'll fly it." That is just what happened. Dan built engines and Carl flew them with great success. This was the beginning of a long friendship.

Figure 1-20 A view of the power plant in the Calkin 5 foot airplane. Note the engine's exhaust pipe. (Calkin family photo)

The 1932 Experimental Engine

The two Dans set up a small shop in the basement of the 33rd Street house and worked to develop a model airplane engine as a commercial project. After months of work, a small engine was produced that was close to the desired product. One problem Dan encountered at first was that the engine would only "four cycle," meaning that it would fire on every other revolution. Finally the problem was solved and the engine would "two-stroke," that is, fire on every revolution of the engine as it should. The only surviving photo of this engine is shown in Figure 1-18. The cylinder looks like the Model 1 production cylinder, but the crankcase and carburetor are different.

The First ELF Powered airplane

A low wing model plane with a 5 foot span was built to test fly the engine and, when the plane was finished, Dan and his parents went to the Rose City Municipal Golf Course to fly it. The manager there gave him permission to fly in the late afternoon after the golfers had finished. In a 1934 letter to Josephine, this is how Dan described the first attempts:

> There is a fairway one hundred yards wide with an adjoining fairway nearly as wide, both over 400 yards long. We took the plane 130 yards from the car and started it out. The first time it cut little circles on the grass without going anywhere at all. The next time it took out right across the fairway, went in a dead straight line hopping along the ground until it hit a tree. This wrenched the wing and crushed a plane (wing panel) on one tip and broke the ignition wire. We tried it again after moving the weight further back, and it rose in the air 8 feet after take off but immediately came down again and ran along the ground. We changed the prop, but had no better luck. The gas line and ignition wires kept breaking.

From this inauspicious start things slowly improved and successful flights finally became commonplace. Dan and the airplane are shown in Figures 1-19 and 1-20.

Figure 1-21 The newly married Mr. and Mrs. Edwin Daniel Calkin. (Calkin family photo)

Later, in a 1941 advertising brochure, he said :

> In 1932 Dan Calkin built his first class "A" motor of record. After many successful flights with it, he named it the "ELF." (Figure 1-18)

By June of 1934 Dan and his cousin set about making castings for three more engines of the same type as the engine they were testing. About this time Dr. Calkin purchased a house just a few blocks away from their current home. They moved into the home at 3055 NE Everett Street on June 13, 1934.

WEDDING BELLS

In early August of 1934, after a five year engagement, Dan traveled to Kansas City to marry his college sweetheart. Dan and Josephine were married on August 15, 1934. Their wedding picture is shown in Figure 1-21. Josephine had graduated with honors a year earlier from Lindenwood College in St. Charles, Missouri. Following the wedding, Mr. and Mrs. Edwin Daniel Calkin took the train for the three day trip to Portland, Oregon. One thing that Josephine would not leave behind was her medium grand piano. That was later shipped to Portland.

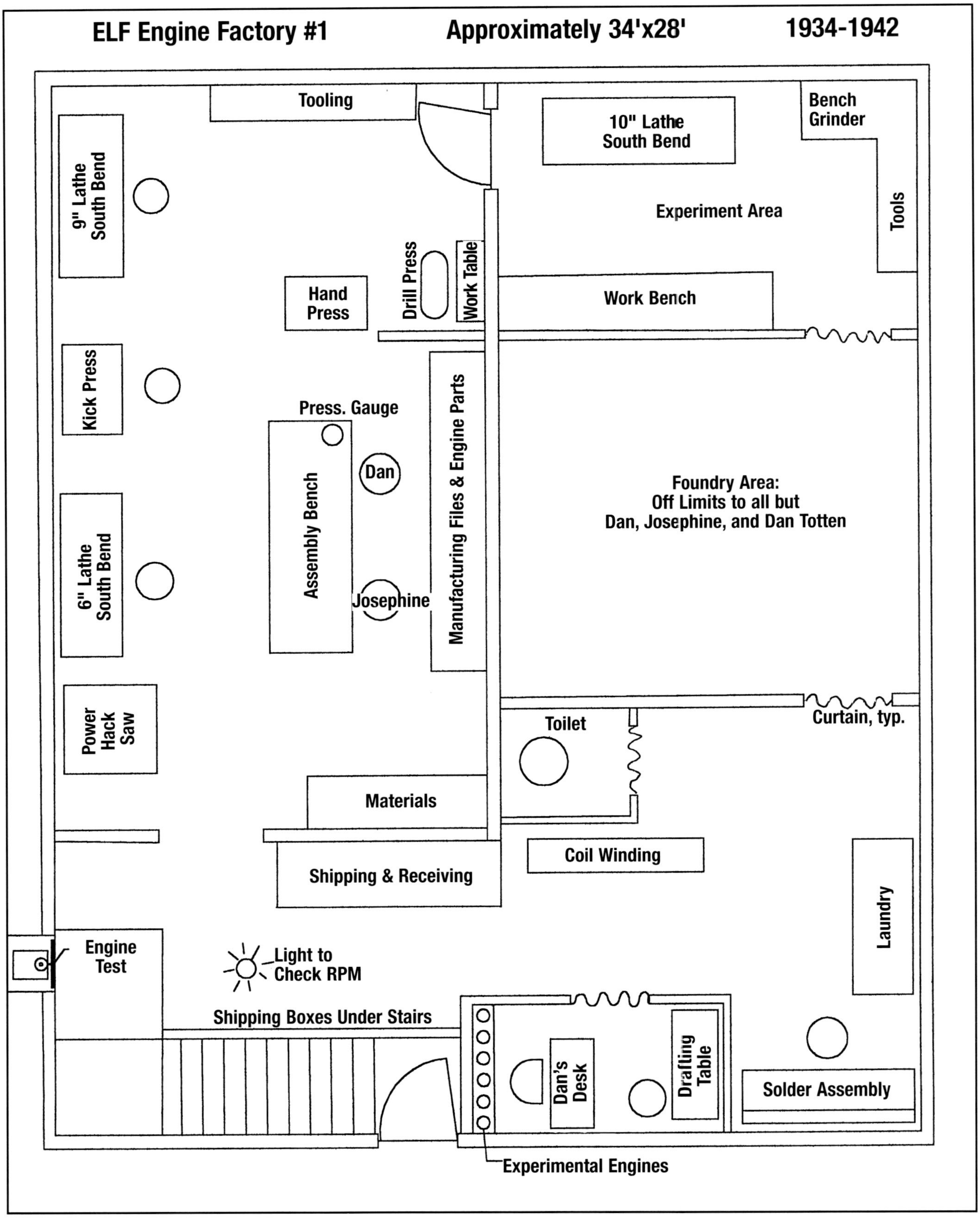

Figure 2-2 The floor plan of Dan Calkin's ELF factory in the basement of his home at 3055 Everett Street, Portland, Oregon. (A Phil Foster sketch redrawn on a computer by John J. Brown, Jr.)

ELF

A Four Ounce Gasoline Engine for Model Airplanes
ELF ENGINE COMPANY
DANIEL CALKIN
3055 N. E. Everett Street
Portland, Oregon

Chapter 2 THE EARLY ELF YEARS

EVERETT STREET

After arriving in Portland, Dan and Josephine moved into the two story house (plus an attic and basement) at 3055 NE Everett Street with Dan's parents. See Figure 2-1. The house was built in 1905, but even in 1996 it was still in very good condition and was one of the nicest homes in the area. When I visited the house in 1996, the current owner allowed me to look around the house, including the basement. I was startled to see that, under the basement stairs, there were shelves and drawers of exactly the same type Dan later built into his Huntington Beach, California, home.

Figure 2-1 The home of the Calkin family and the ELF Engine Company at 3055 Everett Street. (Calkin family photo)

The two Dans, Calkin and Totten, with help from Dr. Calkin, set about putting together a workshop in the basement of the Everett Street house. Dan set up an office in the attic.

Dan's Everett Street factory consisted of a 930 square foot basement area. A sketch of the work area is shown in Figure 2-2. Notice the engine test stand under the porch but accessible from the stair area. A 60 Hz light bulb provided the stroboscopic capability to check engine speed for 3600 RPM.

Dan had some clever ideas for making castings and wanted to keep these as trade secrets, so the foundry area was off limits to all visitors. One thing Dan Calkin did was to experiment with adding extra copper to the casting alloy. He found that using an aluminum alloy with 5-10% copper in the casting alloy greatly improved the machinability of the castings, while enhancing their appearance. Dan would weigh the aluminum to be melted and then added a number of copper pennies to the molten aluminum to get the desired percentage. At the Everett Street shop Dan usually made all the castings himself although he sometimes shared the job with Dan Totten. All castings used in all ELF engines were sand cast.

1934 Prototype Engines

The three sets of castings, made previously at the 33rd Street house, were machined and, one by one, Dan worked through and solved dozens of problems. These engines still had provisions only for radial mounting. The carburetor and timers were different from the later production engines but otherwise very similar. (No photo available).

1935 Prototype Engines

The machining and testing of the three engines convinced Dan that the basic engine design was what he wanted. In the first half of 1935 he built another engine with minor external changes, most notably the addition of beam mounts for easier installation of the engine in an airplane. Except for a few small changes, this engine was the basis for the Model 1 ELF production engines. Obvious differences are: the prop washers had six screws through them; the timer arm casting differed from the production engines; and there was no ELF logo on the right side of the engine. The 1935 prototype engine is shown in Figures 2-3 and 2-4. Probably two more of this type were made without the ELF logo, one of which was bought by Earl Vivell.

Dan needed a simple distinctive logo for his engines. The name ELF inside the two offset half circles was suggested by his father, Dr. Calkin.

FIRST ELF PRODUCTION ENGINE

The first production model ELF was a vertical plug single cylinder engine. See the line drawing in Figure 2-5. This engine continued the evolution of Dan's miniature engine design.

Model 1 Design Considerations

Recall that in those days gas engines were competing for duration against rubber powered model airplanes. Dan designed an engine that was perfectly suited for the task. Since the primary consideration was fuel economy he designed an engine that produced its maximum torque at a relatively low speed and matched it to a rather large 12 inch diameter x $6\frac{1}{2}$ inch pitch propeller of his own design. With this combination, the engine would run for about 45 minutes on one ounce of fuel at 3600 RPM. A suitable airplane would have about a three foot wing span (not unlike a conventional rubber powered model). This was the perfect engine for the times. To

Figure 2-3 The 1935 ELF Model 1 prototype engine.

Figure 2-4 The other side of the 1935 ELF Model 1 prototype.

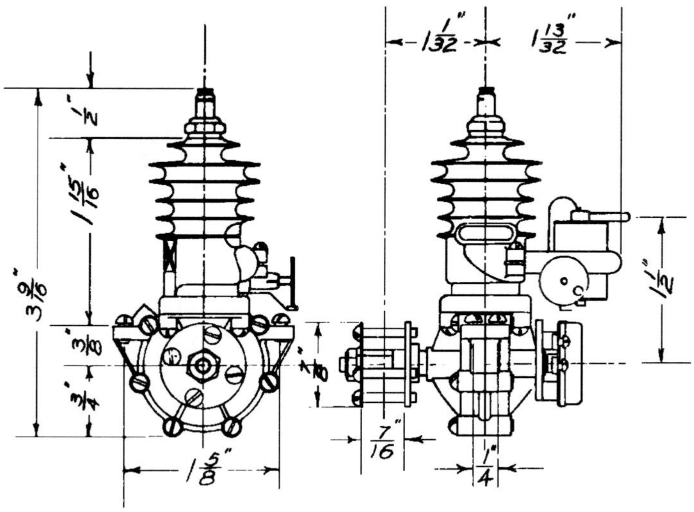

Figure 2-5 **ELF Engine Company drawing of Model 1 ELF, shown at 50% size.**

and crankcase sand cast of aluminum, cylinder fitted with a thin steel liner. Two bearing crankshaft fitted with steel counter weights. Engine weighs 4 oz., can run either upright or inverted. First SMALL model gasoline engine made. Was pictured in a three foot airplane in Ripley's BELIEVE IT OR NOT 12/27/36. (See Figure 2-6).

The aluminum cylinder casting had cored-in porting and housed a thin (0.010 inch thickness) steel cylinder liner. The piston was fitted with three rings, two at the top and one at the bottom. After making a couple of engines the ELF logo was added to the bypass area (right side) of the cylinder casting.

The cast aluminum crankcase was vertically split and machined from two identical castings with no

Figure 2-6 Dan Calkin's 3 foot gasoline powered model was featured by RIPLEY. (Calkin scrap-book)

Dan's knowledge, the only other engines available then were the *Brown Jr.* and the *Loutrel*, but they were both much larger than the ELF. Many modelers didn't own cars and had to transport their models via bus, trolley, subway or bicycle to reach a flying site. A three foot wing span airplane was much better suited for this purpose than a six foot span model.

Throughout his career Dan always had a personal goal of keeping the weight of every part of his engines and his airplanes to a minimum, as long as it was consistent with good engineering and a reliable product.

A further important consideration was that the engine be resistant to crash damage. To accomplish this objective, Dan developed some unique ways of mounting the engine to cause the engine mounts to fail before bending the crankshaft or breaking the crankcase of the engine. Dan also believed in using props made of soft wood that would break before bending the crankshaft.

Model 1 Design

One sheet in the Calkin scrapbook describes this engine as:

> Single .136 cubic inch, 2 cycle, four port, one cylinder, air cooled, bore 0.542", stroke $^{19}/_{32}$", piston displacement $2^{1}/_{4}$cc cylinder

gasket between them. Thus, each end of the crankshaft was supported by a bearing surface with a thrust washer on each side. This arrangement made possible the use of the engine with either pusher or tractor propellers. The rear crankcase was machined to accept the timer assembly. Dan machined the crankshaft from solid steel drill rod stock with pressed-on counterweights. The propeller drive washer used four screws in a circular pattern to assure that the propeller did not slip on the shaft. This followed full size aircraft practice.

The engines were designed for either upright or inverted operation but they were initially shipped with a carburetor installed for upright use. The carburetor housed a float which would rise when the bowl filled, thus shutting off the flow of fuel coming into it. Fuel flowed by gravity from an elevated external tank to the float bowl, then to the engine. This gave the advantage of allowing the fuel tank to be located near the airplane's center of gravity (CG). The carburetor could be turned upside down for inverted operation, but then it would be somewhat in the way of the exhaust and made mounting the engine more difficult. At the suggestion of Carl VanCourt, Dan designed a reversed carburetor for inverted operation and gave the first inverted unit to Carl, owner of one of the early engines. Carl had asked for engine number 13, but it had already been sold, so Dan took another engine from the first fifty and numbered it 113 for Carl, instead. Carl's #113 engine, as shown in Figure 2-7, was typical of early production.

Now that Dan was satisfied with the engine design, he started making tooling and fixtures to facilitate the manufacturing process. He realized that good tooling was necessary both for manufacturing efficiency and to produce consistently good results. By June 1936, a newspaper article states:

> After much experimentation this small model (engine) has been perfected and Mr. Calkin is now preparing to manufacture them in quantity.

Figure 2-7 Model 1 ELF (#113) owned by Carl VanCourt with the first inverted carburetor.

MORE INVOLVED THAN JUST MAKING AN ENGINE

But there was more involved than just making an engine. Dan was breaking fresh ground. There were no other small engines on the market at that time, so there were no small spark plugs, no coils, no battery boxes, no propellers and no airplane designs. Information regarding the performance that a modeler could expect from the engine was also lacking. The Brown Jr. had only been on the market for a short time. Bill Brown had faced similar problems, but the products developed for the Brown Jr. were unsuitable for the ELF because of its small size and unique construction. Dan addressed each of these problems by the end of 1936.

Spark Plug

The spark plug was a major problem. Dan recognized that the existing spark plugs of the day were just not suitable for his small engine. The internal volume of these plugs was so great that it would be impossible to obtain an adequate compression ratio. Fortunately, Dan's father was a dentist and had some knowledge of ceramics. With his help a small $\frac{1}{4}$ inch diameter spark plug was produced having 32 threads per inch.

Referring to an article about spark plugs by Doc Nichol in the November-December 1966, *Engine Collectors Journal*, Doc says:

> Once, when I visited his plant in a downtown Portland building, he (Dan) had a small piece of aluminum holding some very interesting spark plugs. I asked him about these and he related a story of the early ELF spark plugs as he showed these experimental plugs which led to the development of his ELF plug. These early plugs had several types of insulators, varying from glass tubing, mica layers, various resins, to crude home baked porcelain and finally the ELF plug as pictured in Figure 2-8. Dan explained that without equipment, it was not profitable to manufacture his own porcelain insulators, so he contracted to have Champion manufacture an insulator of his design and specification. Then he added a sour note, that when Champion came out with their own line of small plugs, they were copies of his design. The early Champion V plug was of a similar design. The plug base design was the same and the insulator shape was markedly similar. And most important, the method of "sealing" the insulator was also the same. The ELF spark plug was the first of the miniature ($\frac{1}{4}$ x 32 size) spark plugs to be manufactured.

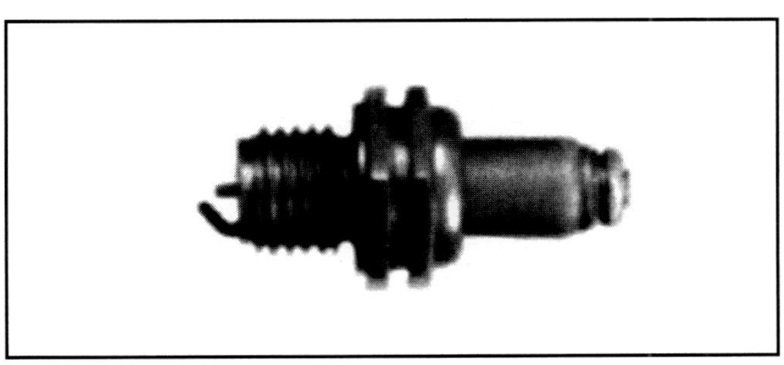

Figure 2-8 ELF 1/4-32 spark plug made prior to the use of Champion made insulators.

Champion Spark Plug Company

In 1934 Dan started a correspondence with the Champion Spark Plug Company of Toledo, Ohio, to inquire about the availability of a small spark plug. Upon hearing that nothing suitable was available, Dan sent his design for a small plug to the Champion Spark Plug Company and asked them to make only the insulator for his plug. At first, they weren't interested in the project. In order to convince the people at Champion that he was a serious customer, Dan sent them a complete, ready to run, ELF engine. Champion was suitably impressed and responded by agreeing to make the insulator portion of the plug according to Dan's drawing. The engine was not returned; Champion stated they were keeping it to show visitors at their Toledo, Ohio, plant. The remaining parts of the plug and its assembly were left for Dan to do.

The ELF Spark Plug

Dan manufactured his own ELF spark plugs, using the insulators bought from Champion, until February, 1938, when the Champion V-2 spark plug became available. The Champion V-2 spark plug was an obvious knock-off of the ELF plug in size, shape and the use of the $\frac{1}{4}$ x 32 thread. From talking with the people involved in the ELF operation, one realizes that there was a lot of hard feeling over Champion's use of Dan's design to make a model engine plug. This dissatisfaction led to Dan's later avoidance of the use of Champion plugs at times when plugs were available from other manufacturers. Refer to Chapter 11 for more information about spark plugs.

Spark Coil

When Dan made his first boat engine, a Model T Ford spark coil and two large dry cell batteries were used for ignition. These added so much weight that the 40 inch boat would only go about 3 or 4 m.p.h. Obviously, that wouldn't do for a model airplane. The smallest coils of the day were still much too heavy and drew too much current for a small engine such as an ELF. Josephine wrote to England and ordered a book about small engine ignition system design. With the help of the book and his own ideas Dan designed a coil that was much smaller and lighter than anything available either here or in England. An uncommon thing about the coil was that it would work on only one pen cell (1.5 Volts) rather than the 3 Volts required by most other coils which followed. Tooling was also made to facilitate the winding of the coils. One thing Dan did was to stretch a length of magnet wire to obtain the smaller diameter he required. Regarding the actual making of the coils a November 1936 article in the Portland newspaper states, in part:

> Mrs. Calkin is officially the 'spark coil expert,' and makes all the coils that go with the engines.

According to Mrs. Calkin, they continued to make coils until the company closed its doors in 1956.

There were at least six known versions of ELF coils produced in addition to some early experimental units. See Chapter 10 for descriptions and photographs of the coils.

Battery Box

For modelers needing a battery box for their ELF, a simple design was included with the instruction sheet. See Figure 2-9.

Engine Mounting System

Details for mounting the engine in an airplane were also included in the instructions. Figure 2-10. Other methods of mounting were shown on the plans for the ELF three foot airplane.

Propeller

The propeller was another challenge. Since no suitable propellers were on the market at the time, it was up to Dan to design, test and manufacture something suitable for the ELF. The results of the propeller study were incorporated in an eight page booklet by Dan. *Notes on Model Airplane and Propeller Design* was listed on early product price lists for 50 cents. One page was devoted to propeller design, enabling a hobbyist to carve his own. The standard propeller for the Model 1 ELF had a twelve inch diameter and a six and one-half inch pitch. By today's standards, this is a very large propeller for such a small engine. Designs for both two and three bladed props were shown. The scimitar shape

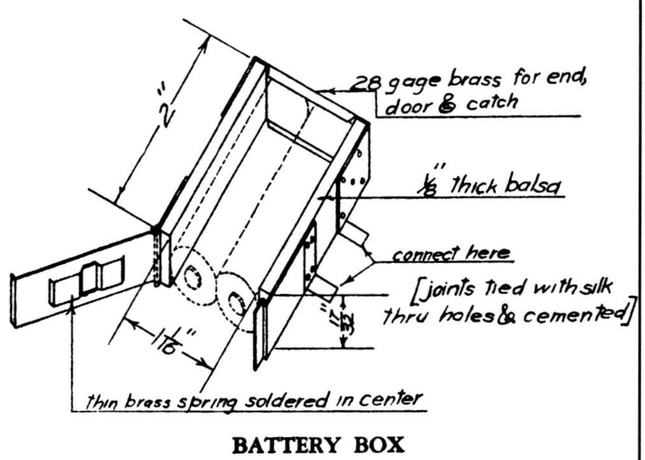

BATTERY BOX

Figure 2-9 A suggested method for making a battery box. (from the 1937 ELF instruction sheet)

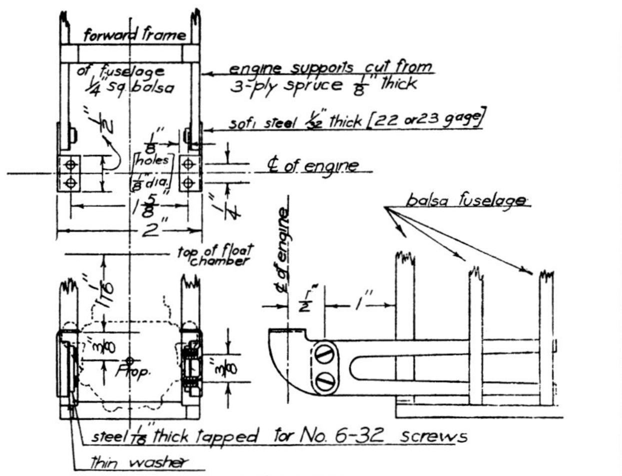

MOUNTING

Figure 2-10 One method for mounting the Model 1 ELF. (from the 1937 ELF instruction sheet)

Figure 2-11 Dan Calkin releasing his 3 foot gas model. (Calkin family photo)

was common in those days, but is rarely seen now. If a model builder required a different pitch, the necessary information for making custom designs was included. The Calkin booklet is reproduced in Appendix C.

The ELF Engine Company produced their own propellers starting in 1936. Dan built a machine to carve the 12 inch x $6\frac{1}{2}$ inch propellers, then each was hand balanced, sanded and finished. ELF owners were advised to use an incandescent light as a strobe to check RPMs. If the engine didn't achieve the desired speed, the tips of the prop were to be trimmed until it did. Chapter 13 contains additional information about ELF propellers.

Airplane Plans & Designs

The booklet *Notes on Model Airplane and Propeller Design* covers more than propeller design. It also includes the design criteria for a stable airplane, giving modelers enough information to design their own airplanes. This small document is of a very high technical quality, while still being easy to read and understand.

If the modeler wanted a proven airplane design, Dan also sold plans for his ELF three foot airplane for fifty cents.

The three foot span airplane mentioned in the Ripley's "BELIEVE IT OR NOT" article is the same one that Dan designed and built for testing his engines. See Figure 2-11 for a picture of Dan launching the airplane for a test flight. The original version of the plane was pictured and described in the June, 1935, *Model Airplane News* in Charles Grant's "Gas Lines" column as follows:

> Though gas model airplanes are complicated and difficult to construct, model fans have made great progress in this line. Here is an unusual one built by Daniel Calkin of 3055 Northeast Everett St., Portland, Maine (sic). It is only thirty-six inches in span, one of the smallest ships which has yet been built. It is powered by an ELF engine and weighs complete thirteen ounces. The engine weighs four ounces.

Dan's little ELF three foot span airplane has at least three known versions. The first two differ mainly in the engine mounting and the location of the fuel tank. In the earliest version (date unknown), the fuel tank was positioned right behind the fire wall bulkhead. The second version, dated July 15, 1935, had a different engine mounting and the fuel tank location was moved to the center wing section. A third version, with a

Figure 2-12 A reproduction of the ELF three foot airplane built by Dan Lutz.

Figure 2-13 Showing the ELF engine test stand for accurately determining engine performance. (Calkin family photo)

January 15, 1938, revision date, restored the tank behind the firewall again and placed the rubber bands, which hold the wing in place, inside the fuselage. A reproduction of this airplane, built in 1997 by master modeler Dan Lutz, is shown in Figure 2-12 and on the dust jacket.

Engine Performance

For additional technical information about the engine, Dan had available, for fifty cents, a chart comparing engine horsepower, torque and fuel consumption over a wide range of speeds. In order to provide this accurate engine performance data, Dan designed and built the engine test stand seen in Figure 2-13. From his test data, Dan plotted the engine's performance as shown in Figure 2-14.

100 Hour Test

The aircraft industry required a 100 hour engine operation test to provide some assurance of reliability. Dan chose to apply this test to his own engines. The test was performed without incident and, after the test, the engine was disassembled and was found to have negligible wear. Then the 100 hour test was repeated; still, no wear was noticeable. In later years other model airplane engine manufacturers used the 100 hour test.

THE FIRST ENGINE SOLD

The mention of Dan's ELF powered airplane in *Model Airplane News* was responsible for the first ELF engine being sold. The customer was Doctor John Paul Young, an intern from Sacramento, California. Doc Young wrote to Dan as a result of seeing the engine mentioned in the June, 1935, *Model Airplane News* (that was on the newsstands weeks earlier). Because the magazine article mistakenly said Dan lived in Portland Maine, rather than Portland, Oregon, the letter was slow getting to Dan. His reply of May 10, 1935, gave details of the engine and listing the price at $21.50. Dan sent a second letter dated

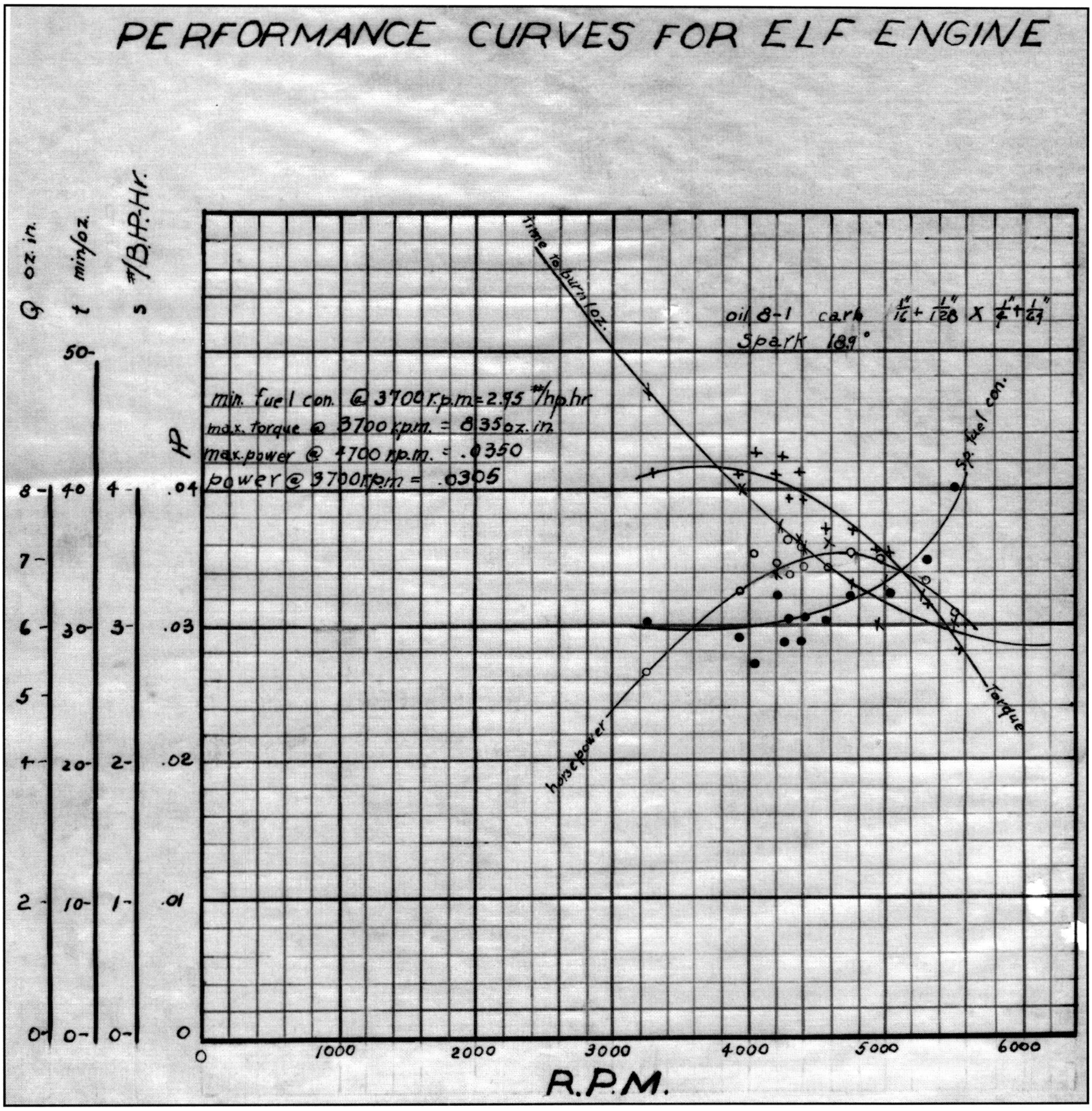

Figure 2-14 A model airplane engine manufacturer that actually measured engine performance was very rare, but Dan Calkin did it from his earliest days. Most companies just made exaggerated claims. (Calkin family scrap-book)

May 27, 1935, acknowledging the receipt of the $21.50 and giving additional details of the engine and promising delivery by early July. Other documentation from the Calkin scrap book lists the sale of this first engine as July 29, 1935.

A third letter from Dan to Doc dated December 16, 1935, states in part:

If you will return me your engine I will put it in first class condition free of charge. Your engine was the first one I sent out and I have since learned a great deal more about engine building.

Manual of the

ELF
TRADE MARK

AIRCRAFT ENGINE

written by

Daniel Calkin, B. S. in M. E., M. S.

TABLE OF CONTENTS

THE

ELF
TRADE MARK

AIRCRAFT ENGINE

One-half Actual Size

Designed Especially for Model Airplanes

A Gasoline Engine Weighing

ONLY 4 OZ.

Built by The

ELF ENGINE CO.

3055 N. E. Everett St. Portland, Ore.

Figure 2-15 The original Model 1 instruction sheet and brochure, both printed in late 1935.

Figure 2-16 Alex Imrie holding his reproduction "Miss San Diego" powered by a reproduction Model 1 ELF built by Mike Patience. (Alex Imrie photo)

Other ELF Engine Users

Elbert Weathers was not only a well known modeler and early user of ELF engines, he was also listed as a distributor of ELFs. To help promote the sale of ELFs, Elbert designed several small planes especially for the ELF and re-covered his famous "Miss San Diego" model and added ELF ENGINES in large letters on the fuselage sides. See Figure 2-16 for a photo of Alex Imrie's reproduction of the "Miss San Diego". Alex is the "Vintage Corner" columnist for the English *Aeromodeller* magazine. He featured this airplane and ELF engine in the March 1989 *Aeromodeller.*

According to the notes in the Calkin family scrap book, Dr. Young won a first and a third place trophy with his ELF during 1935.

The March 1936 *Model Airplane News* contained an extensive write-up of Weathers' ELF powered plane with technical details about the

Figure 2-17 Dr. Calkin holding Cecelia Jane, his first grandchild. (Calkin family photo)

Figure 2-18 Dan Calkin with his baby daughter, Cecelia Jane. (Calkin family photo)

Figure 2-19 Dan Calkin sits atop his boat "Josephine" before its first water trials. In the background is the Sellwood Bridge on the Willamette River about 6 miles upstream from the center of Portland. (Calkin family photo)

ELF engine. The "Miss San Diego" design also appeared in the May 1936 issue of *M.A.N.* and Weathers' famous "Tubby" appeared in the September 1936 *M.A.N.*

Frank Zaic, in his *1938 Model Airplane Year Book,* describes five ELF powered airplanes, including three designs by Frank Ehling.

Another modeler using ELF powered planes was Carl VanCourt. Carl was mentioned in several issues of *M.A.N.* and *Air Trails* with his airplanes along with Ken Willard and other modelers. Carl later became an ELF employee and was an important contributor to this story. Phil Foster, one of the first ELF employees, tells us:

> Also during this time Dan would get visits from Carl VanCourt. Carl was a dental technician—skilled in handwork, crown casting, dentures, lost wax investment casting etc. He was also a model airplane builder. He made many of the ELF 3 foot span models but with neat variations such as lightening holes in the metal engine mounts and silk fillets for the vertical, which technique I have been using for sixty years. Carl was full of fun, lots of smiles and very energetic compared to Dan who was generally very quiet.

Figure 2-20 Phil Foster is enjoying the sun, while Dan is engrossed in some problem. Phil's wife-to-be, Bonnie, took this photo while they were cruising along the Willamette River near the Columbia River (Phil Foster Photo)

DAN CALKIN AT HOME

A New Model Introduced

Near the end of 1937, on Sunday, December 19 to be exact, the Calkins introduced a new model to the world with the birth of their daughter Cecelia Jane. (Jane is also Josephine's middle name). Figure 2-17 shows Dr. Calkin holding Cecelia and looking like a proud grandfather. The baby's father is holding her in Figure 2-18.

Dr. Calkin bought a new Packard 120 in 1937. Dan did not own a car and usually drove his father's car. Most people probably assumed it was owned by Dan.

The Good Ship "Josephine"

Sometime during this period Dan started building a full sized boat. He designed and built a 21 foot

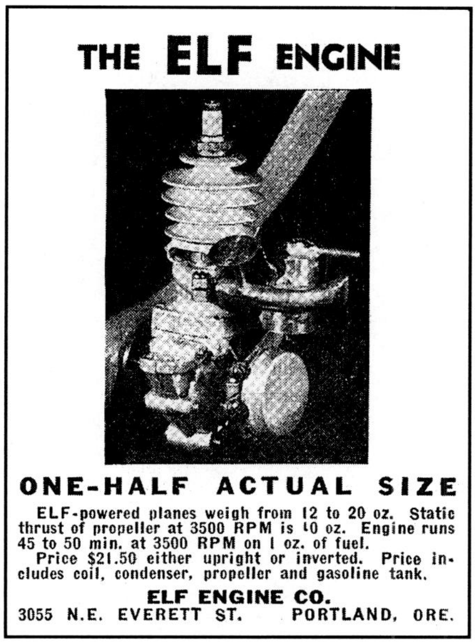

Figure 2-21 (Left) The first ELF advertisement in the January 1938 *Model Airplane News*. It was repeated in February. The other ad (right) appeared in the March 1938 issue.

mahogany cabin cruiser named Josephine powered by a Johnson outboard motor. Several of the "experts" of the day told him he couldn't power a cabin cruiser with an outboard motor, but he did and it was quite successful. Dan did much of the work in the attic and garage of the Everett Street house. The boat's interior was designed to be fitted with bunks and a galley. The boat was taken to the river about 1939 and was used by family and friends to cruise the many rivers of the area. The windows and other details were finished at a later date. Figures 2-19 and 2-20.

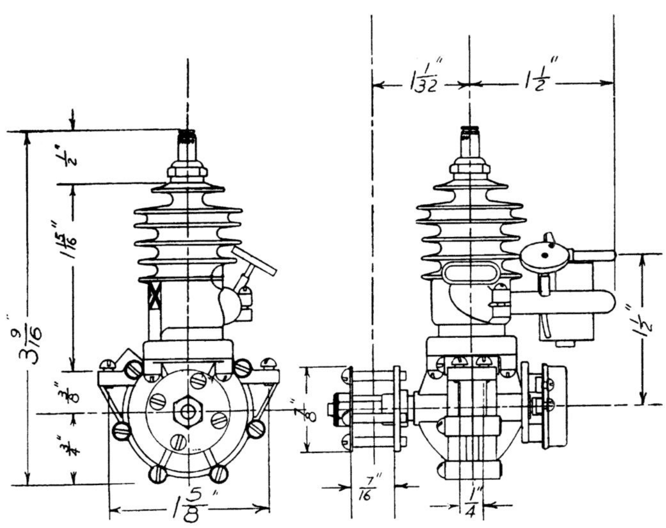

Figure 2-22 The revised Model 1 ELF. Note the change in the needle valve angle.

ADVERTISING

The year 1937 saw ELF engines produced in limited quantities, but advertising was all by word of mouth and free "ads." Dan was a master at writing letters to magazines and newspapers soliciting free publicity. During 1937, articles featuring ELF engines appeared in *Model Airplane News, Air Trails, Flying Aces, Model Craftsman, Popular Aviation*, Ripley's BELIEVE IT OR NOT, and many newspapers.

Many articles and pictures of ELF powered airplanes sent to the magazines by loyal ELF users also helped to spread the word.

The First Advertisement

In January, 1938, the first of three paid advertisements appeared in *Model Airplane News*. See Figure 2-21. For $21.50 you could buy an "ELF for either upright or inverted (operation). Price includes coil, condenser, propeller and gasoline tank." Additional ads appeared in the February and March issues of *Model Airplane News*. Sales of the ELF continued through 1938 and 1939 and were so good that a waiting list was necessary. Customers were notified two weeks before the scheduled shipping date to send their money if they were still interested in buying an engine.

Figure 2-23 The 1938 prototype slant plug ELF single.

Figure 2-24 The right side of the 1938 slant plug prototype.

By early 1938, Dan had made two small changes to the Model 1 ELF. The first was to substitute the new Champion V-2 spark plug for the ELF plug, made necessary because Champion would no longer supply Dan with insulators. Dan still had over 100 unused insulators on hand, but making the spark plugs was so labor intensive that it wasn't desirable to continue making the plugs if it could be avoided. The Champion plug was much more cost effective to use.

The second change was to redesign the carburetor by moving the needle valve from its horizontal position to a 60 degree angle from the horizontal. This change made the needle valve adjustment more accessible. Compare Figure 2-22 to Figure 2-5.

Now that the tooling for the production of the Model 1 ELF Single engines was completed and engines were being produced and sold, Dan turned his talents to the design of a successor engine.

ENGINE DEVELOPMENT
1938 Slant Plug Engine Design

Times were changing however, and, due to changes in contest rules, fuel economy was no longer an important item, so Dan set about designing a completely new engine. The new .097 cu.in. engine was to utilize all new parts — nothing from the Model 1 ELFs would be retained, although the experimental engine did use some Model 1 parts. It is a credit to Dan that he recognized the need for change and was able to modernize the ELF offerings.

The experimental slant plug engine was really a hybrid containing most of the important elements of a new design while using existing Model 1 ELF parts for the external parts. The engine had a sand cast aluminum cylinder, head and two piece crankcase with identical front and rear castings. Like the Model 1, the cylinder had a very thin steel liner with drilled ports for intake, transfer, bypass and exhaust. The piston had two small rings. The four long screws used to secure the head also went through the cylinder fins and screwed into the crankcase. This was a beam mount engine with but a single screw hole in each mount. This is the engine featured in Figures 2-23 and 2-24. A second engine of this kind was nearly finished by Dan Calkin. Dan Lutz finished the second prototype engine. It now resides in his collection.

The crankshaft was of a different construction than the Model 1 engines. The main part of the shaft was machined in one piece from one-

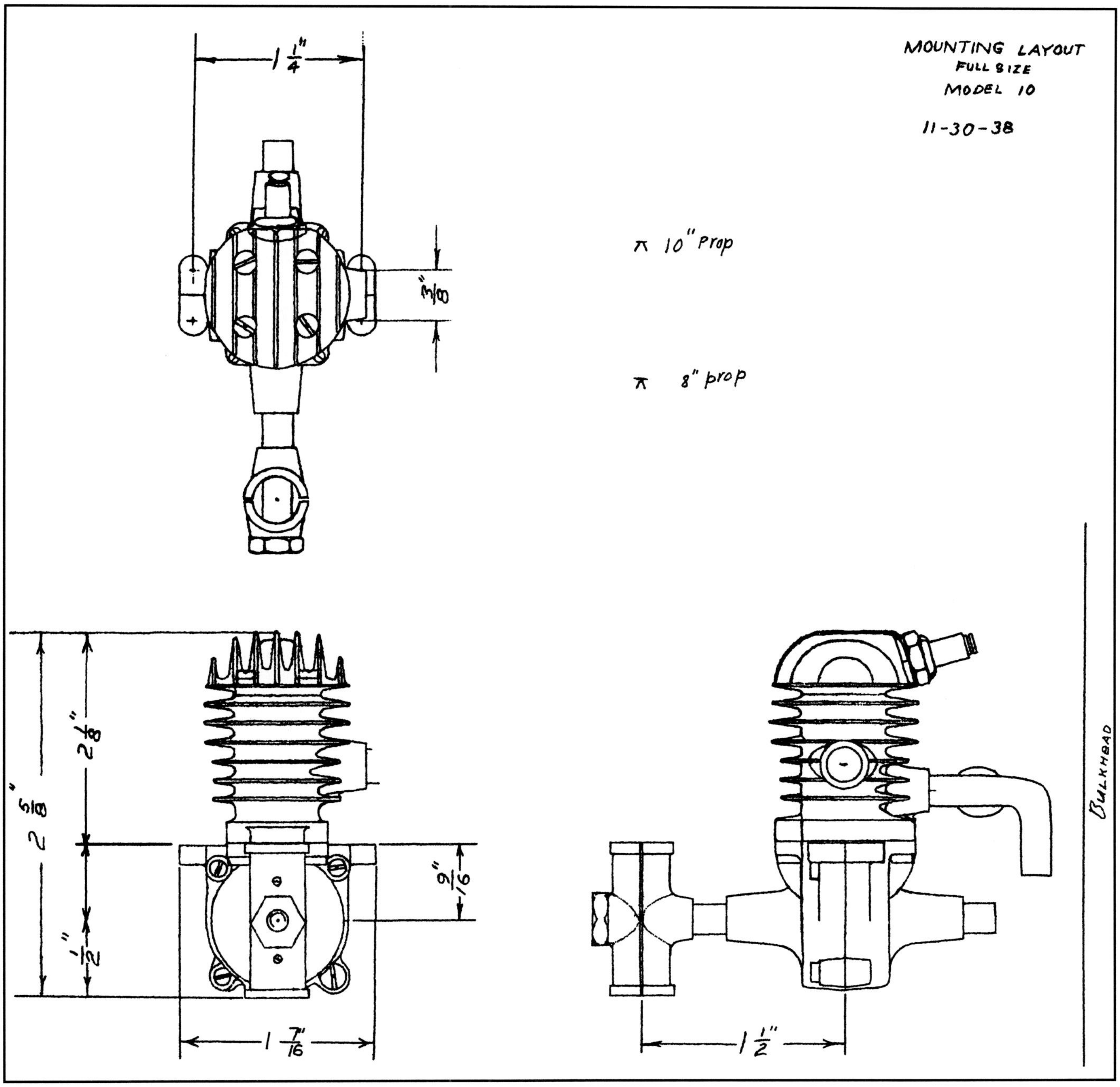

Figure 2-25 The original drawing of the Model 10, Series 1 ELF. (Calkin family scrap-book)

quarter inch steel stock, but it had soldered on counter-weights and shroud. A cam for the breaker points was filed on the rear end of the shaft. The crankshaft had internal threads for a center prop screw rather than external threads for a nut as used by previous Calkin engines. When the crankshaft was assembled with the crankcase, two internal, Model 1 type thrust washers were used, one at each end.

The new engine's most prominent feature was the spark plug that angled almost straight back (actually 15 degrees from the center line of the crankshaft). This distinctive feature was later copied by the manufacturers of the *Kopper King* and the *Perky* engines.

Externally, a Model 1 timer was attached to the rear of the engine. A straight intake tube was used with a Model 1 needle valve.

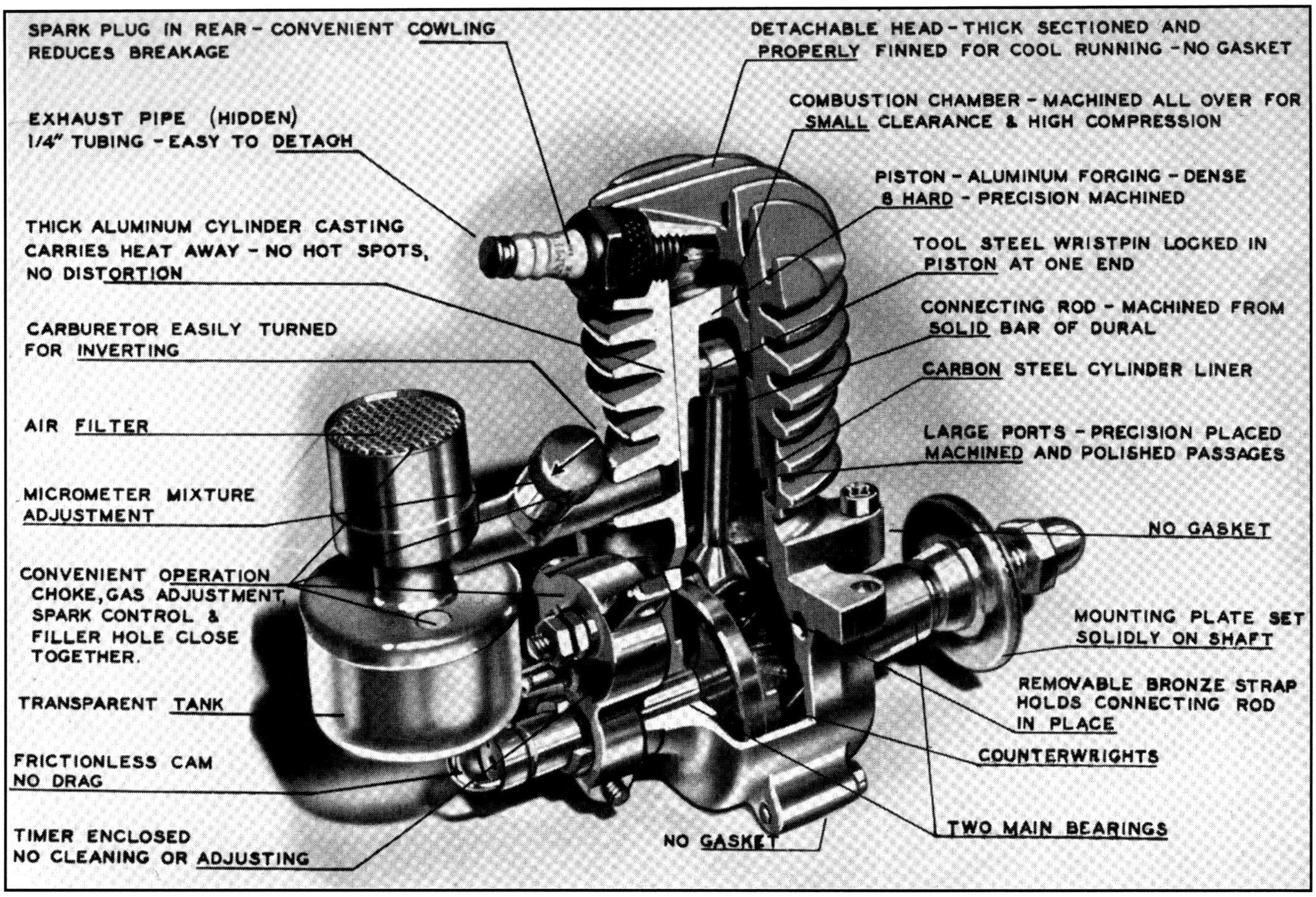

Figure 2-26 The smooth nose ELF Model 10 Cutaway, Series 1. Serial number 52- (third number was cut away). (From the 1940 ELF brochure)

Figure 2-27 The cutaway Model 10, Series 1 used for the above.

Figure 2-28 "Lucky" Lucas showing off his new airplane with the first production Model 10, Series 1 ELF engine. (Phil Foster photo)

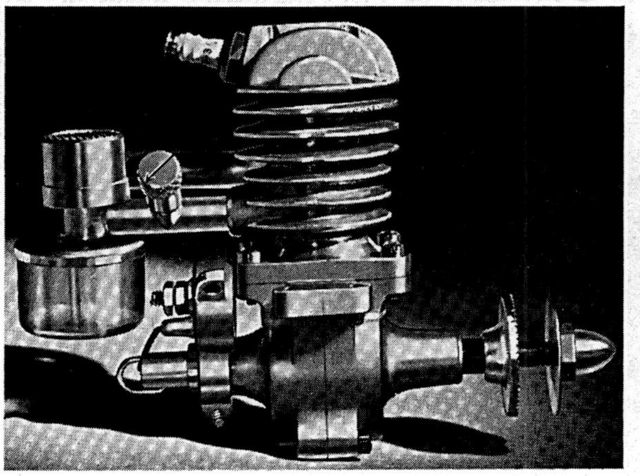

New ELF Model 10 SINGLE

Displacement .097 cu. in.
New Design — Lighter Weight

The new Elf Single is a modern high compression engine designed for planes with wingspread up to 4 feet weighing up to 2 pounds. With more features than you ever thought possible in a class "A" motor, the new Elf weighs but 3 ounces—yet _develops 12 to 17 ounces of thrust!_ Inspect the new Elf carefully. Notice the trim lines of its design and convenient location of all controls. It is so accurately made and runs so cool that only light No. 20 oil is used for lubrication.

SPECIFICATIONS

TYPE: 2 cycle, 4 port.
DISPLACEMENT: .097 cu. in.
GROUND R.P.M.: 7,500.
BORE: $\frac{15}{32}$ inches.
STROKE: $\frac{9}{16}$ inches.
CYLINDER: Cast aluminum lined with carbon steel.
PISTON: Forged aluminum, "micro-machined".
BEARINGS: *Two* main bearings.
TIMER: Elf "No-Drag", fully enclosed.
LUBRICATION: 1 part No. 20 oil to 3 parts gasoline.
AIR FILTER on carburetor intake.
GAS TANK: Transparent, ⅛ oz. cap.
WEIGHT: 3 ounces. Flying wt. 6¼ oz.
PRICE: Engine only.............. $19.00
Complete with Elf coil, condenser and propeller............$21.50

New ELF Model 20 TWIN

Total Displacement .198 cu. in.
Finest, Most Powerful Class "A" Engine You Can Buy

If you are out to scoop all class "A" competition in sight, the new Elf Twin is the engine for you. This little powerhouse is not for beginners, but advanced builders only. It is designed for planes up to 5½-foot wingspread and weighing up to 4 pounds. Displacement is .198 cu. in., weight 5 ounces and it _develops 28 to 34 ounces of thrust!_ The Twin is a combination of two Elf Singles and all but three parts are interchangeable. Its inherent advantages with two cylinders are smoother running—better cooling—faster starting and a smooth, steady output of power. More power to you with the new Elf Twin!

SPECIFICATIONS

TYPE: 2 cycle, 4 port.
CYLINDERS: 2—opposed.
TOTAL DISPLACEMENT: .198 cu. in.
GROUND R.P.M.: 7,500.
BORE: $\frac{15}{32}$ inches.
STROKE: $\frac{9}{16}$ inches.
PISTON: Forged aluminum, "micro-machined".
BEARINGS: 12-ball thrust bearing, *two* main bearings.
TIMER: Elf "No-Drag", fully enclosed at rear of engine.
LUBRICATION: 1 part No. 40 oil to 3 parts gasoline.
AIR FILTER on carburetor intake.
GAS TANK: Transparent, ¼ oz. cap.
WEIGHT: 5 ounces. Flying wt., 8¾ ozs.
PRICE: Engine only........... $28.75
Complete with Elf Twin coil, condenser and propeller.....$31.50

Figure 2-29 The 1940 brochure touting the all new Model 10 and 20 ELF engines. These pictures were glued on the boxes the engines were shipped in. See Chapter 14.

The propeller drive washer, propeller front washer and prop bushing were from the Model 1 ELF, with the prop nut being replaced with a machine screw.

When comparing the Model 1 ELF, the 1938 slant plug prototype engine and the Model 10 slant plug single, one sees the continuous progression of the ELF engine development. Without this engine there seems to be little continuity in engine design.

The Model 10 Design

The 1938 prototype was the basis for the new Model 10 design. By November 30, 1938, Dan produced a drawing (See Figure 2-25) showing the basic engine that would become the next production engine. Not shown in the drawing are the timer, carburetor and propeller hardware. Evidently those designs had not yet been finalized. This prototype engine led directly to the slant plug ELF Single and Twin (Models 10 and 20, Series 1).

THE ILL-FATED SMOOTH-NOSE SINGLE AND TWIN

The Calkin scrapbook shows a photo with the caption "First flights on the Model 10, October 1939." This is the photograph shown on page *xvi* in this book. Dan built a polished, cutaway Single for advertising. This is the engine used in the 1940 brochure as shown in Figure 2-26. The same

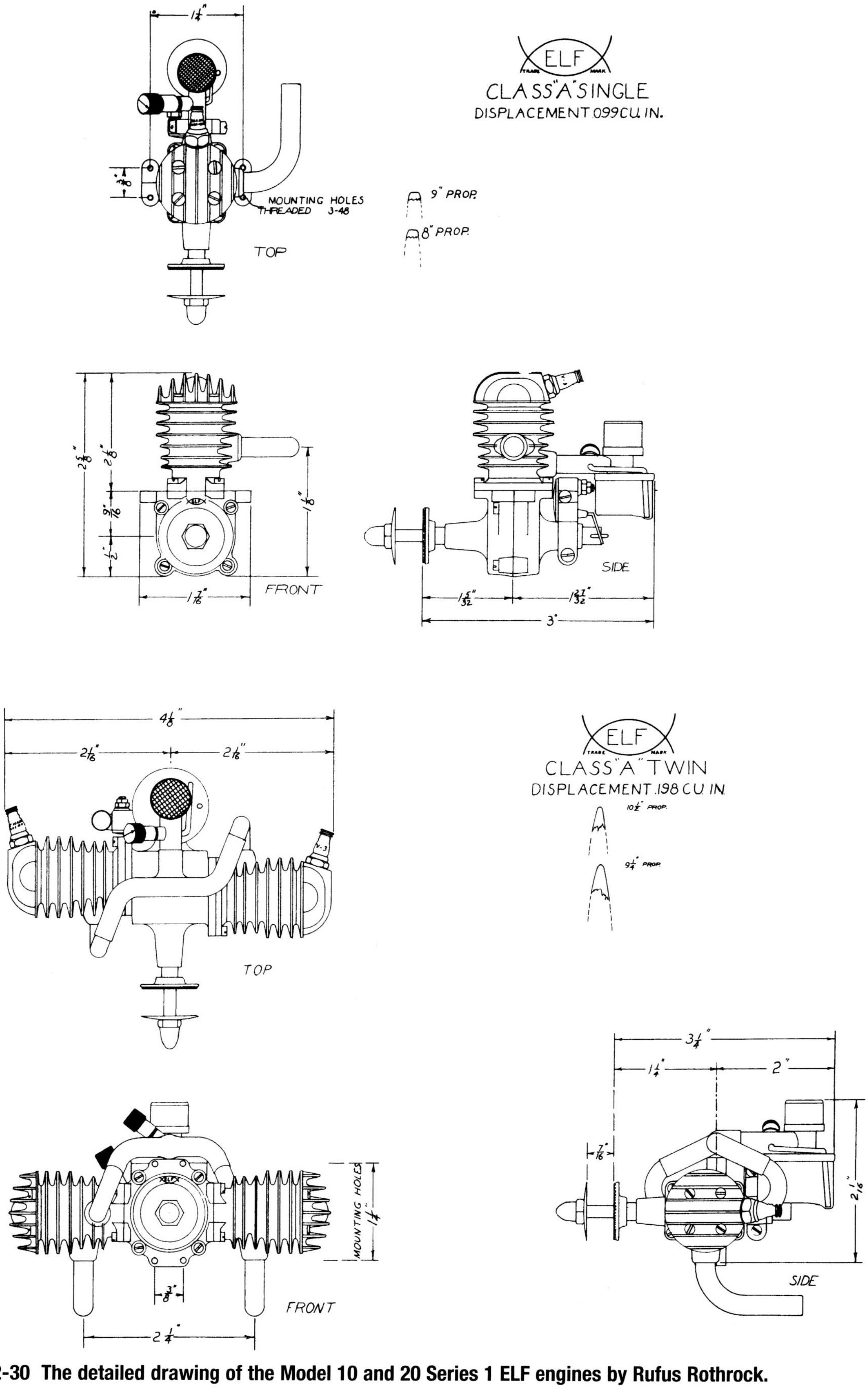

Figure 2-30 The detailed drawing of the Model 10 and 20 Series 1 ELF engines by Rufus Rothrock.

Figure 2-31 The prototype Model 30, Series 1 ELF Twin #TX

engine is also shown in Figure 2-27. By this time Dan already had the new style engines built and ready to ship. The first beam mount, smooth nosed, slant plug Model 10 Single, serial number 502, was delivered on October 19, 1939, to Leonard "Lucky" Lucas. Leonard immediately built a 32 inch span free flight for his new engine and took it to a local contest. See Figure 2-28. He believes it was the smallest free flight built up to that time. None of the other contestants knew what the engine was as Leonard had completely cowled in the engine and painted a large **?** on the side of the cowl to add to the mystery. Later in the contest,

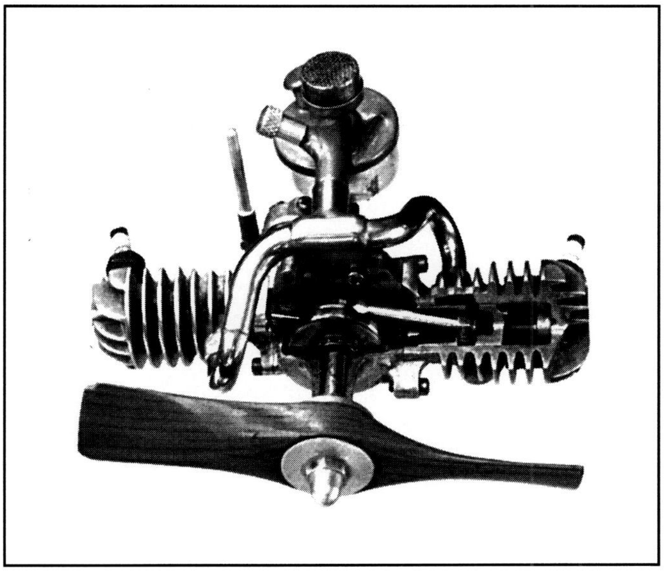

Figure 2-32 This is the cutaway Twin, Model 20, Series 1 #T34

Figure 2-33 The airplane that Phil Foster built to test the new ELF Twin. (Phil Foster photo)

Leonard removed the cowl to show off his new possession: the first production Model 10 ELF.

The ELF Twin Development

The development of the Twin followed that of the Single. The prototype Model 20 was #TX as shown in Figure 2-31. The Model 20 Twin, serial number T3, was delivered about six months later on March 28, 1940, to Carl VanCourt. A cutaway, polished and chrome plated Model 20 was also built and appears in Figure 2-32.

One new development in Dan Calkin's engines was the use of an aluminum piston without rings. Dan had said that no rings could be found which didn't reduce the engine's power. His use of an aluminum piston with a steel liner meant that he had to very carefully choose a casting alloy for the piston to assure that it had a low coefficient of expansion. Most aluminum alloys expand more with heat than steel does, and if the piston were to expand more than the liner the piston would tend to bind when the engine was hot.

Although the crankcase halves were very similar, the new engines ended Dan's practice of using identical front and rear crankcase halves. They differed in that the front half had the ELF logo, while the rear half had a flat surface for the serial

Figure 2-34 A look into the inner workings of the Model 20 ELF as shown in the 1940 ELF brochure. (Calkin family photo)

Phil Foster Explains ELF Manufacturing

Sets-ups for the following were:

1. prop nuts	turret lathe
2. prop washers	turret lathe
3. prop drive-tapered hole	turret lathe
4. con rod	6" lathe, drill press
5. con rod retaining strap	kick press punch
6. thrust bearings (12 ball)	kick press punch
7. crankshafts, single and twin	6" lathe
8. counterweight disc	kick press
9. counterweight assembly	bench set-up
10. crankcase, single and twin	10" South Bend
11. cylinder bore and face	turret and 10" lathe
12. liner bored and finished	turret and 10" lathes
13. head bore and face	drill, tap for spark plug
14. piston wrist pin	6" lathe
15. cylinder liners	special punch from inside
16. carburetor parts-needle valve-filler	turret lathe
17. celluloid tank	Josephine did assembly
18. intake manifold for twin	kick press
19. timer case	6" lathe
20. timer cam (wire & eccentric)	jig & 6" lathe

number corresponding to the logo location on the front half.

Rufus Rothrock was hired to produce the three view drawings shown in Figure 2-30 for which he was paid with a Model 10 ELF, #241.

The new engines were built. Orders for the engines started coming in and Dan started shipping them.

About the time Dan started shipping the new engines, flight testing revealed a weakness in the front crankcase. Phil Foster tells the story:

> I built a shoulder wing model for the new ELF Twin with the smooth nose crankcase. It was first flown at a grass field east of Portland, called Troutdale Airport, the east end of which stopped at the Sandy River. The river also provided a down draft that stopped our models from flying away as they rode thermals down the field. I have photos of this model (Figure 2-33) with the smooth nose ELF Twin installed. Dan and I drove to the field to fly the Twin by ourselves. The field had 18 inch grass. We fired up the engine and set the clock timer for a very short flight. I hand-launched and the plane did a 100 foot loop and came in tangent to the tall grass and stopped, except for the broken prop, undamaged—or so it seemed. We found on preparing for a second flight that the engine would not turn over. We could see nothing wrong but we went home. Shop inspection verified our guess that the forward crankcase had been bent. Remedy—heavy stiffening ribs added to the forward crankcase for Twin and Singles.

Engine Redesign

The bent crankcase and crankshaft must have caused Dan a considerable amount of grief. The engine changes required rework to the front crankcase pattern to add four reinforcing ribs. Dan also decided to modify the cylinder porting requiring changes to the pattern for the cylinder and piston castings. The reason for the cylinder redesign was probably to coax a little more power out of the engine. Indeed, the redesigned engines produced about 500 RPM more than the earlier type. The competition in the small engine field was increasing and the ELF was no longer the only show in town. The transfer porting on the smooth nose engines used a .1 x .3 inch slot through the piston that connected to the transfer port on the bypass side (right side) of the engine. The new piston had no slot or holes for fuel transfer, instead all the porting was built into the cylinder sleeve and the cylinder casting. On the redesigned engine the cylinder ports were uncovered by the top and bottom of the piston when the piston was near the top and the bottom of its stroke. Externally, the modified cylinder had a slight bulge at the bypass resulting in shallower bottom fins in the bypass area.

The rework of the smooth nose engines was fairly extensive. The cylinder casting and liner, piston, crankshaft and the front crankcase were all changed. A reinforced drive washer with a beveled rear face was introduced and the short straight exhaust pipe replaced the curved tube.

Magazine advertising had already been contracted for and new sales literature had been printed before the redesign took place. Thus the advertising continued to show the earlier style engine for many months after the front crankcase had been reinforced and the porting modified. About 34 each of the smooth nose Singles and Twins were made and sold to customers. Most of the engines were sold to local modelers and Dan recalled and replaced nearly all of the these engines with the new, Series 2, reinforced-crankcase engines. It is known that a number of the castings for the smooth nosed engines were melted down to make new castings. Possibly one or more engines had only the crankcase replaced, but this has not been verified.

Lucky Lucas remembers giving back his #502 smooth nose Single and receiving an all new engine (#553) with the strengthened front end and improved porting.

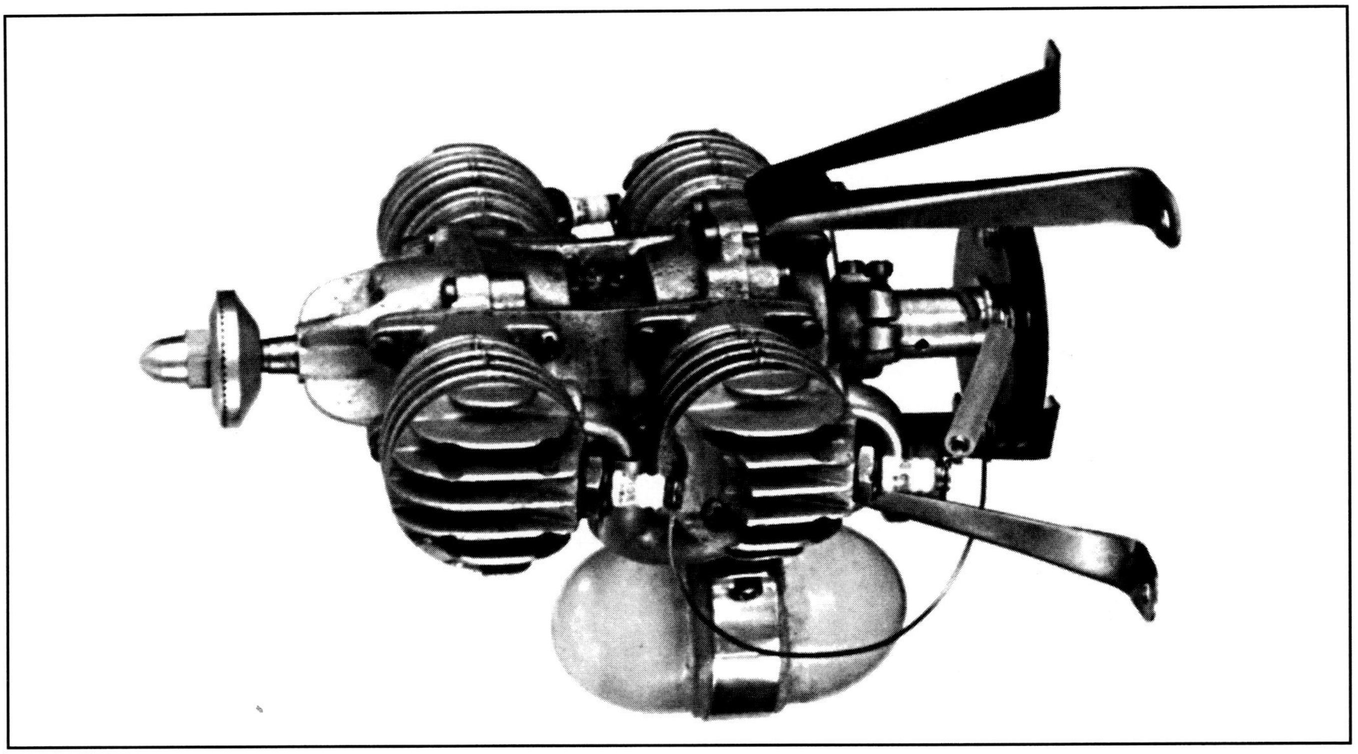

Figure 2-35 The prototype, and much modified, ELF Four (#299).

The engines with un-reinforced (smooth-nosed) crankcases will be referred to as Series 1 engines in the later explanations of the various ELF production engines. (See Chapter 6).

The smooth nosed Singles and Twins are among the rarest of the ELFs. Only four of the Singles and four of the Twins are known to still exist.

Phil Foster worked as a machinist for Dan and talks about methods at Everett Street:

> The Model 1 ELF engine was no longer in manufacture when I worked there although many (Model 1) engines were still being shipped and rebuilt due mostly from getting dirt in the exhaust ports and mangled carburetors.
>
> The first slant heads (engines) were being sent out with the smooth (un-ribbed forward) crankcase.
>
> I did machine work on all of the engine parts but Dan did all of the production tooling set-up, ran through samples and then turned the operation and checking over to me. Dan made production runs of fifty engines as a general rule.
>
> Dan had an old South Bend lathe that he converted to a turret lathe. It was enameled black with gold filigree decorations. It had a lever to open and close the head stock collet(s). It had a turret with four to six tools and wrench release and detents for indexing the tool(s) and a lever for running the tool(s) into the work piece. The cross feed was also lever operated so I could knurl a piece and cut it off. We used kerosene lubricant for machining aluminum and cutting oil emulsion for steel—none for brass.

Phil continues:

> Runs were for fifty engines, Singles or Twins, but I would usually make 200-300 of the minor parts in order to make

the set-up pay off. Dan did all the finishing of the mating parts—piston fit and fit check against a pressure gauge. A special ground file was used to finish the mating cylinder and crankcase. Later he started and ran each engine. Each filter can had two screens with lambs wool inside—dirt was a big problem—engines were flown in all kinds of environments—no regular flying fields were available.

ELF Four Development

About 1939-1940 Dan started to develop the four cylinder ELF. All the machine work and design were done in complete secrecy. Dan Totten and Dan Calkin did it all. I later did some of the production work since the Four used standard cylinder assemblies, front and rear crankcase castings and timers. New, were the distributor, fuel tank, intake manifold and, of course, the split side-crankcase and coupler crankshaft. All needed new tools. There was great excitement developing about this family of engines and interest was growing in the hobby world. Unfortunately Dan's production rate was low and the engine costs were relatively high compared to other engine makers. Dan did not have automatic machines and all castings required his expert skill—no automatic die casting.

Production System

A word about how Dan kept his production system in order. He had some file boxes relating to each engine, containing

Figure 2-36 First ad for reinforced Model 20. January 1941 Model Airplane News.

ELF "Single"

Model 10

Class "A"—Displacement .097 cu. in.

A modern high compression engine designed for planes up to 4-foot wingspread and weighing up to 2 pounds. Notice the trim lines of its design. Location of the spark plug behind the cylinder head simplifies the cowling problem and protects the plug from breakage. All controls are grouped together *behind* the engine for greater safety and convenience.

All interior surfaces and moving parts are "micro-machined" for accuracy and perfect fit. *There are no rings or gaskets in an Elf engine.*

The "Single" can be inverted in 30 seconds without extra parts. Like all Elf engines, it has a reputation for quick, easy starting. An Elf-powered plane can be released in 10 seconds under full power.

SPECIFICATIONS

TYPE: 2 cycle, 3 port.
DISPLACEMENT: .097 cu. in.
OPERATING SPEED: 7,500-8,000 R.P.M.
POWER: 1/12 H.P.
BORE: .468 in.
STROKE: .564 in.
CYLINDER: Cast aluminum lined with carbon steel.
PISTON: Forged aluminum, "micro-machined".
BEARINGS: *Two main bearings.*
TIMER: Elf "No-Drag", fully enclosed.
CARBURETOR: Micrometer adjustment.
AIR FILTER on carburetor intake.
GAS TANK: Transparent, ⅛ oz. capacity.
MOUNTING: Radial.
WEIGHT: 3 oz., Flying wt. 5½ oz.
PRICE: Complete with 1½ oz. Elf Coil and condenser..... **$16.50**

ELF "Twin"

Model 20

Class "A"—Total Displacement .198 cu. in.

The finest, most powerful class "A" motor you can buy. Ideally suited for contest flying and advanced builders, the Elf "Twin" delivers the power of a large motor within the limitations of class "A" displacement. It was designed for planes up to 5½-foot wingspread and weighing up to 4 pounds. Although it is a combination of two Elf "Singles", the "Twin" produces much more power than a single of the same *total* displacement because its greater cooling area enables it to run cooler, take in more fuel on each stroke and use light No. 40 oil for lubrication. It naturally runs more smoothly than a single cylinder engine due to the even distribution of weight and perfect balance of all moving parts.

SPECIFICATIONS

TYPE: 2 cycle, 3 port.
CYLINDERS: Two, opposed.
TOTAL DISPLACEMENT: 198 cu. in.
OPERATING SPEED: 7,500-8,000 R.P.M.
POWER: ⅙ H.P.
BORE: .468 in.
STROKE: .564 in.
PISTONS: Forged aluminum, "micro-machined".
BEARINGS: 12-ball thrust bearing, *two main bearings.*
TIMER: Elf "No-Drag", fully enclosed.
CONTROLS: Grouped together behind the engine.
CARBURETOR: Micrometer adjustment.
AIR FILTER on carburetor intake.
GAS TANK: Transparent, ¼ oz. capacity.
MOUNTING: Radial.
WEIGHT: 5 oz. Flying wt. 8¾ oz.
PRICE: Complete with Elf "Twin" Coil and condenser. **$31.50**

**All three models shown ¾ actual size.*

The Mighty ELF "Four"

Model 40

Class "C"—Total Displacement .396 cu. in.

The Elf "Four" brings super power to class "C". America's finest model motor was developed from the Elf "Twin" and clearly demonstrates the superiority of high efficiency, small bore motors with multiple cylinders. The "Four" was designed for planes up to 10-foot wingspread, radio control planes, large model boats and racing cars. It is the first high speed motor in class "C" and the smoothest-running motor on the market. Its surging power easily swings a 12 or 13 inch prop at 7,500 R.P.M.

SPECIFICATIONS

TYPE: 2 cycle, 3 port.
CYLINDERS: Four, opposed.
TOTAL DISPLACEMENT: .396 cu. in.
OPERATING SPEED: 7,500-8,000 R.P.M.
POWER: ⅓ H.P.
BORE: .468 in.
STROKE: .564 in.
PISTONS: Forged aluminum, "micro-machined."
BEARINGS: Three main bearings 12-ball thrust bearing.
TIMER: Elf "No-Drag", fully enclosed.
DISTRIBUTOR: Elf design, fires cylinders in pairs.
CONTROLS: Grouped behind and below engine.
CARBURETOR: Micrometer adjustment.
AIR FILTER on carburetor intake.
GAS TANK: Streamlined transparent. Capacity ¾ oz.
MOUNTING: Radial.
WEIGHT: 9 oz. Flying wt. 13 oz.
PRICE: Complete with Elf "Four" Coil and condenser **$49.50**

Figure 2-37 The ELF engines featured in the 1941 brochure.

Figure 2-38 The first ad for the Model 40 ELF Four. April 1941 *Model Airplane News* (Note: The distributor was later made smaller on the production engines as well as this prototype, #299. Intake manifold was also changed.

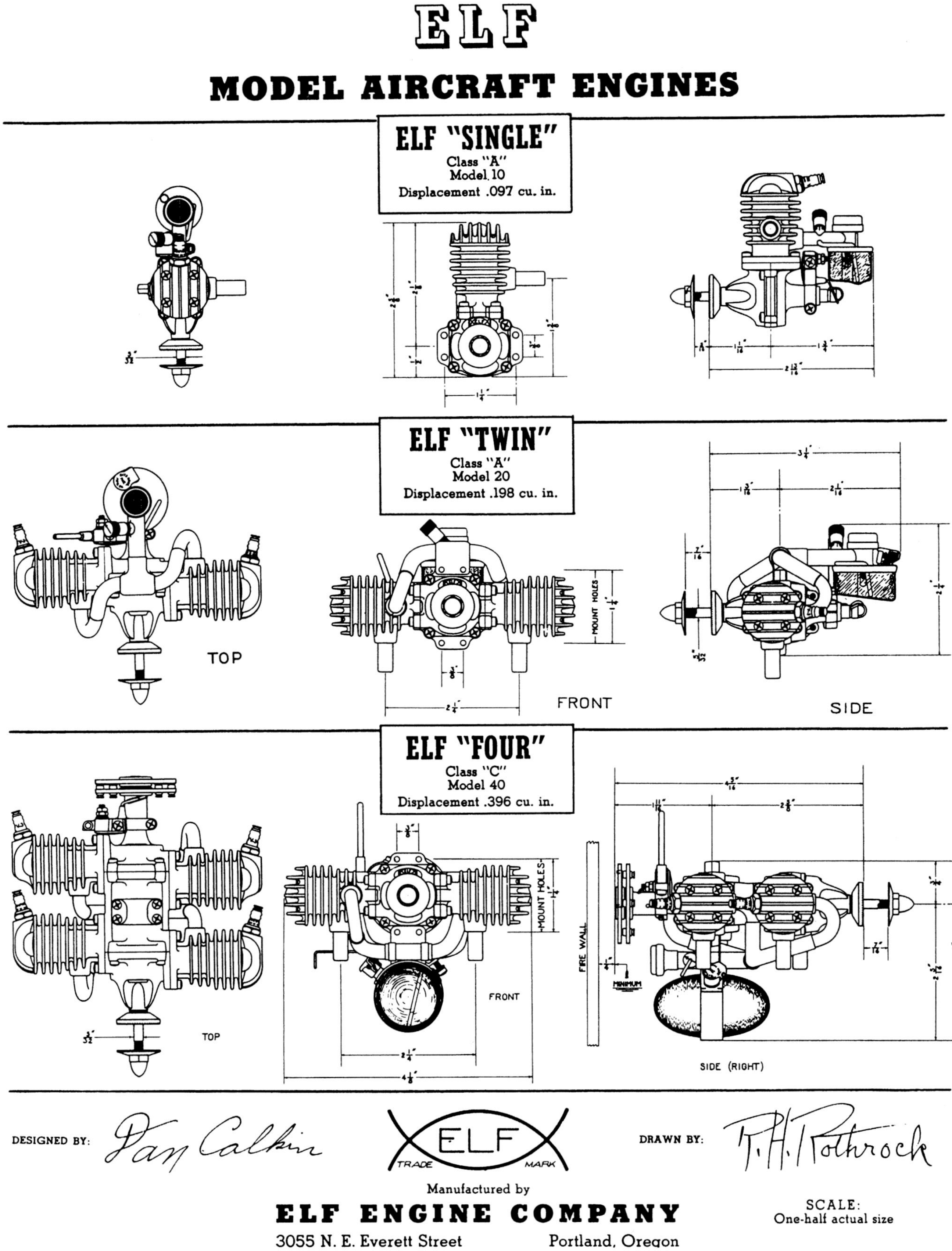

Figure 2-39 Rufus Rothrock's accurate drawing of the Models 10, 20 and 40, Series 2 engines. (Printed here at 85% size)

5x8 cards. Each carried an engine component manufacturing directions, The card had the tooling call out, material, which machine, drawing of the part, step-by-step set-up and all available tolerances. Included were production guides such as rates of feed, lubricant and cutting tool shape. Most important was a record of time to make the part, start and stop time and the number of parts made. This record system was required of parts assemblies, i.e. carburetor, cylinder, crankcase and counter-weights, etc., etc.

During this 1939-1940 period Dan and Josephine let Cecelia (CeCe, we called her) run around the shop. Of course there were many no-no's but we never had any problems with her getting in the way. She was a happy little girl.

Phil also remembers that Dan would occasionally take him and the other employees upstairs to the parlor to listen to Josephine play the piano.

The year 1940 brought a sad note due to the death of Dan's father, Dr. Ralph A. Calkin.

Engine Production Resumes

The improved Twin was introduced before the Single, probably about October 31, 1940. The first ad for the new Twin appeared in the January 1941, *Model Airplane News*, continuing in the February and March issues. See Figure 2-36. Obviously, the Twin was emphasized over the Single. The first improved Single, serial number 03, was shipped on March 29, 1941. The April 1941 M.A.N. introduced the new ELF Four, but also showed both the Single and the Twin. (See Figure 2-38) Dan continued to refer to the new engines as the Model 10 for the Single and Model 20 for the Twin. The engines with the reinforced crankcase and modified porting will be referred to as Series 2 engines in Chapter 6 on the production engines.

ELF Four Introduction

The prototype ELF Four was serial number 299. See Figure 2-35. This prototype engine was photographed and used in a Comet "Zipper" and featured in the 1941 ELF brochure. (Figure 2-40) In April 1941, the Four Cylinder ELF engine was

Figure 2-40 Dan Totten launching a "Zipper" powered by an ELF Four. (From the 1941 ELF brochure).

Figure 2-41 Carl VanCourt with his ELF powered "Zipper". (Carl VanCourt photo)

introduced in advertising as the Model 40. (Figure 2-40) Dan delivered the first production Model 40, serial number 375, to his friend and right hand man, Carl VanCourt on May 30, 1941. This special engine had a chrome plated intake manifold, tank mount and motor mounts.

Rufus Rothrock received an ELF Single (#238) for his detail work on the distributor for the Model 40. Probably Rufus' help was needed because Dan became engrossed in a design project for Jim Walker. About a year later Rufus also made a beautiful three view drawing of the Models 10, 20 and 40 for Dan. (Figure 2-39) For this work Rufus received two ELF Twins (#363 and #536) on March 18, 1942.

Western Modelcraft Supply Company

After the Model 1 production ended, Dan contracted to Western Modelcraft Supply Company to make the propellers for all ELF engines. Western Modelcraft Supply Company consisted of Rufus and Dick Rothrock.

Dick Rothrock, the younger brother of Rufus, in addition to working on propellers, became an ELF employee about 1941. His main job was machining cylinder liners using the Sunnen hone. The bore had a very slight taper to allow the piston to move easily while producing a good seal at top dead center. When he finished a batch of 100 cylinder liners all the inside diameters were within 0.001 inch of each other.

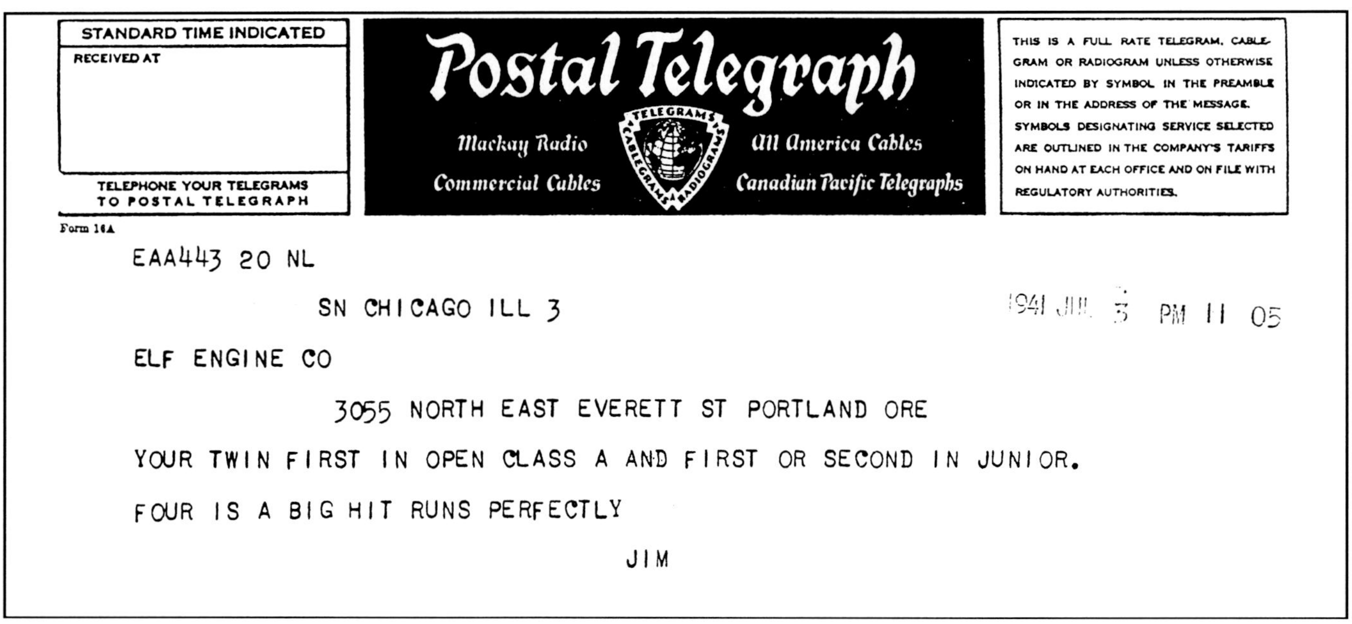

STANDARD TIME INDICATED
RECEIVED AT

TELEPHONE YOUR TELEGRAMS TO POSTAL TELEGRAPH

Postal Telegraph

Mackay Radio — All America Cables
Commercial Cables — Canadian Pacific Telegraphs
TELEGRAMS CABLEGRAMS RADIOGRAMS

THIS IS A FULL RATE TELEGRAM, CABLEGRAM OR RADIOGRAM UNLESS OTHERWISE INDICATED BY SYMBOL IN THE PREAMBLE OR IN THE ADDRESS OF THE MESSAGE. SYMBOLS DESIGNATING SERVICE SELECTED ARE OUTLINED IN THE COMPANY'S TARIFFS ON HAND AT EACH OFFICE AND ON FILE WITH REGULATORY AUTHORITIES.

Form 16A

EAA443 20 NL

SN CHICAGO ILL 3 — 1941 JUL 3 PM 11 05

ELF ENGINE CO

3055 NORTH EAST EVERETT ST PORTLAND ORE

YOUR TWIN FIRST IN OPEN CLASS A AND FIRST OR SECOND IN JUNIOR.

FOUR IS A BIG HIT RUNS PERFECTLY

JIM

Figure 2-42 The telegram from Jim Walker at the 1941 National Model Airplane contest telling about ELF successes. (Calkin family scrap-book)

Figure 2-43 The Calkin family out for a walk near the Everett Street home. (Calkin family photo)

Figure 2-44 Dick Rothrock, a B-24 pilot, built this glider with a pen knife from scrounged materials while stationed in the South Pacific during the early part of WW II. Balsa came from the trees in the background. (Dick Rothrock photo)

Bogus ELF Engines

Dan heard that someone was getting defective ELF castings and parts, assembling a few engines and selling them to hobby shops in the area. Dan sent Leonard Lucas around to the hobby shops to buy up these engines and destroy them. The person assembling the bootleg engines was a former part-time ELF employee. (Note: That person's name does not appear anywhere in this book).

THE JIM WALKER CONNECTION

Soon after moving to Portland, Dan met Jim Walker, owner of the American Junior Company. Jim was an avid modeler, experimenter, expert flyer, radio control pioneer and perhaps the greatest promoter the model airplane hobby has ever had.

During the summer of 1936 a modeler named Oba St. Clair from McMinnville, Oregon, read about Dan in an issue of *Model Airplane News.* On one of his regular visits to Portland, Oba visited Dan at his Everett Street basement shop. This was the first time that Oba had met another modeler. Dan patiently showed Oba the airplanes and engines he had been working on.

Dan suggested that he and Oba go to the golf course where he knew they would find Jim flying his free flight models. Oba met Jim and was thrilled to meet another model builder. Oba had in mind a way to build a model airplane that could be connected by strings to a handle held by the pilot. He could then control the airplane as it circled around him.

Oba built his controllable airplane and named it "Miss Shirley" after his daughter. On another trip to Portland, he drove to Jim Walker's American Junior factory to ask Jim if the airplane had any possibilities. Upon arriving at American Junior, he learned that Jim was at Dan Calkin's place and drove there. Once there, Oba showed his airplane to Dan and Jim, explaining all about it and how it could be flown. Jim asked a lot of questions

Figure 2-45 Carl VanCourt with his Culver Cadet. (Carl VanCourt photo)

about the model. Oba had planned a demonstration flight for Jim but it was raining that day.

Many readers know what happened after that. Jim Walker patented his famous U-Control system in 1942 and went around the country promoting it. Years later, Leroy Cox, owner of L.M. Cox Manufacturing Company, was sued by Walker for not licensing his U-Control system for the Cox ready-to-fly airplanes. Cox had heard about St. Clair's prior work and decided to fight the law suit. With Oba's help, Cox won the suit and the court ruled that the U-Control patent was void and invalid. (Note: Credit for the Oba St.Clair story goes to Dale Kirn and Charles Mackey. (See "Pioneers of Control Line Flying" by Charles Mackey, © 1995).

Dan Calkin and Jim Walker had a long association, but were never really friends. Each recognized the other's talents and abilities and they worked closely on a number of projects. The full extent of Dan Calkin's participation in Walker's projects may never be known, but the ELF ledgers show many entries of payments from American Junior for consulting engineering. In turn, Jim Walker was responsible for getting Dan interested in radio control. Dan retained this interest for the rest of his life. One of the projects Dan did for Walker was to design and build a control stick assembly for Jim to use in flying radio controlled models. Jim used the control stick assembly for several years. A photo of the assembly appears in the next chapter. (Figure 3-4)

Carl VanCourt tells of going to American Junior with Dan. Jim had a machine that vibrated badly and asked Dan to try his magic on the machine. On his next trip to American Junior, Dan fixed the machine and got it running beautifully smooth. Dan told Carl that he could have fixed it easily on the previous trip but didn't think Walker would pay him anything if he fixed it that fast.

WORLD WAR II

The new engines were well received and sales climbed through 1940 and continued climbing through 1941. By the latter part of 1941 sales were brisk and the future looked bright. ELF advertisements prominently featured lists of contest wins by Carl VanCourt and Fred "Red" Burkhart. (Figure 2-36) Carl is shown in Figure

2-41 with his ELF Four powered free flight. He always caused a stir by flying to a contest in his Culver "Cadet." (Figure 2-45)

At the 1941 National Model Airplane Contest held in Chicago, ELF engines were used by first place winners in both Class A Open and Junior class free flight. Jim Walker sent a telegram to Dan telling him of the success of the ELF engines in competition. (See Figure 2-42)

World War II started on December 7, 1941, and with that the business climate rapidly changed. ELF production continued at a brisk rate well into 1942. It appears that some dealers were ordering in anticipation of a shut down of civilian engine production. The few, mostly part-time, ELF employees and associates were soon in military service. Phil Foster went into the Army Corps of Engineers, Carl VanCourt went into the Navy to become a Chief Machinists Mate, Dick Rothrock went into the Army Air Force to become a B-24 pilot. Leonard Lucas had already joined the Flying Tigers as a pilot and was flying P-40s in China. Rufus Rothrock didn't see military service because he was too old, but he did go to work for a company that was doing work for Boeing Aircraft. Dan was not drafted because of his chronic asthma and other health problems. Despite the war, life went on. Figure 2-43 caught Dan, Josephine and Cecelia, "CeCe," out for a stroll near their Everett Street home.

In 1942 Dan's mother, Etta, (the family called her Gram) bought a house on the outskirts of Portland and put the Everett Street home up for sale. The sale of the Everett Street house was completed on August 21, 1942.

Leaving Everett Street also meant finding another location for the shop. In anticipation of the sale of the Everett Street house, Dan rented a 20 foot by 40 foot space in downtown Portland.

Figure 2-46 Dan Calkin in 1936.

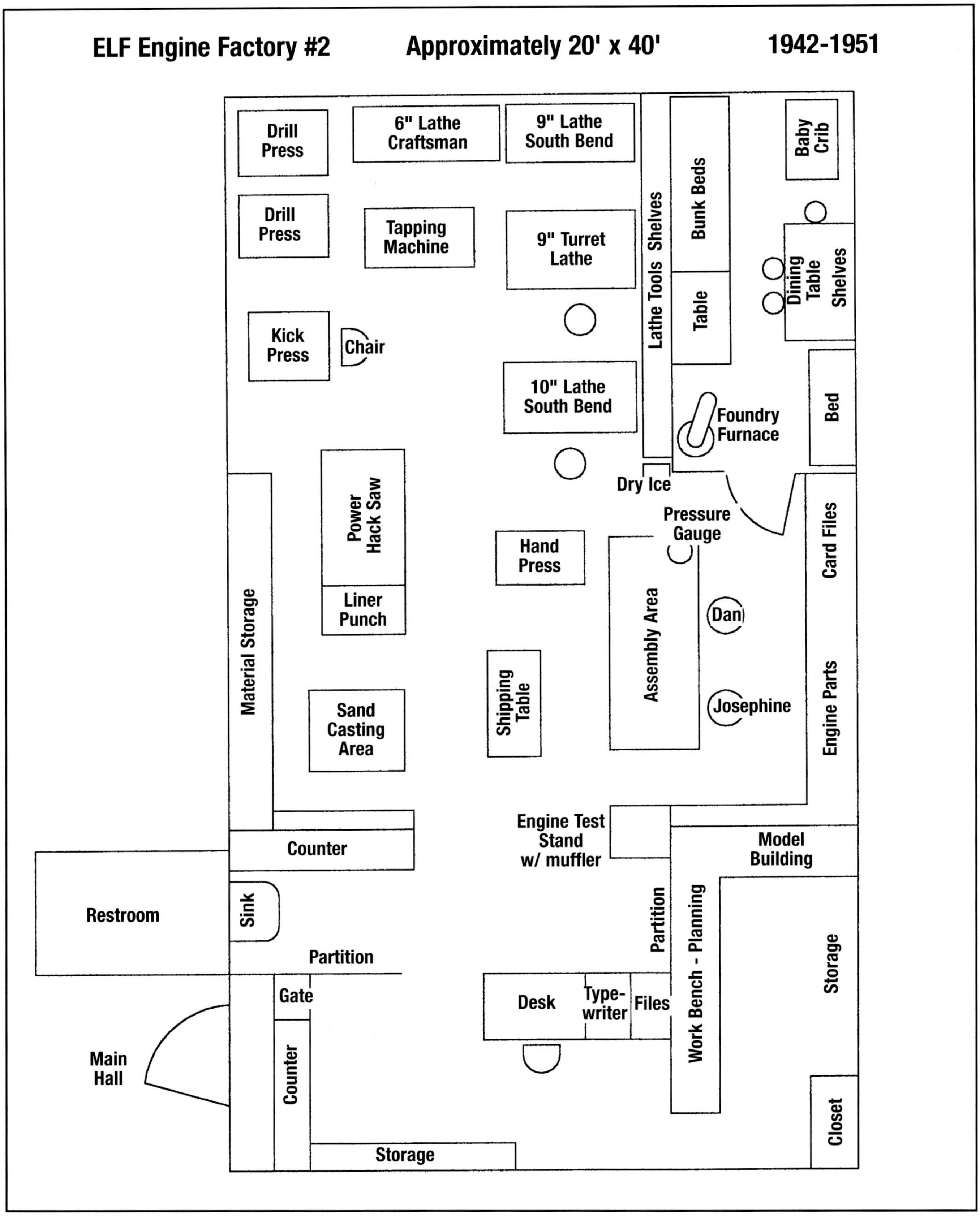

Figure 3-2 The floor plan for the ELF Engine Company at the corner of Washington and 2nd Street in Portland, Oregon. Elf had Room 217 in the rear of building. Sketch by Phil Foster, assisted by Cecelia Calkin Jackson. Computer redrawn by John J. Brown. Jr.

ELF ENGINE COMPANY

PHONE ATWATER 4661
206 S.W.WASHINGTON ST.
PORTLAND, OREGON

Chapter 3 THE MIDDLE ELF YEARS

WASHINGTON STREET

The New Shop

The new location for the ELF Engine Company was in the Kraemer Building in downtown Portland at the corner of Washington and Second Street at 206 SW Washington Street. See Figure 3-1.

This was an historic old building that originally housed the Portland Savings Bank and later the Commercial National Bank. The building itself was an ornate, stately old landmark that had been built in 1888. Dan rented an 800 square foot space for the shop in Room 217 at the rear of the second floor. The ELF Engine Company's shop was only two blocks away from the ocean-going vessels in the Willamette River. The floor plan evolved through the years, but Figure 3-2 (opposite page) shows the final configuration. More on the shop later. The rent on the shop was $35 per month initially, rising to $40 a month in 1946.

Figure 3-1 The Kraemer Building in downtown Portland. (Calkin family photo)

The New Home

After Dan's mother sold the Everett Street home, the family all moved to her new house on the outskirts of Portland. The grand piano was moved to the new Calkin family home, address Rt. 1, Box 820 River Road, (now 17409 River Road) near Oregon City. This was about 12 miles outside Portland. The River Road home overlooked the Willamette River, and Dan and his family could just walk down the hill to their waiting cabin cruiser to cruise the rivers of the area. Dan and Josephine commuted from their River Road home to their downtown business via the interurban trol-

Figure 3-3 The Calkin family home on River Road near Oregon City. (Calkin family photo)

ley car running between Portland and Oregon City. Figure 3-3 shows their home in a winter scene.

WARTIME SHUTDOWN

This move to the new shop occurred during the spring of 1942, but soon Dan could not get materials for his engines as all available stocks of materials were taken for the war effort. Records show that the last ELF engine sold before the wartime ban on civilian production was an ELF Four #464 to a Cpl. Charles H. Gilligan. The date of the sale was November 12, 1942.

The war years were difficult for the ELF Engine Company. No engines could be produced because of wartime restrictions and the lack of materials. Dan did get some small government contracts. A few girls were hired for assembly and ELF produced electronic connectors and cases for radio crystals. Dan also subcontracted to make control boxes for drone aircraft through his association with Jim Walker. Overall, the ELF Engine Company profited very little from the wartime work.

TRIP TO WASHINGTON, DC

Dan took the control stick that he had designed and built for Jim Walker to Washington, DC, hoping to get a contract to build them for the government, but he was unsuccessful. (The control stick is shown in Figure 3-4) He hired a patent attorney to research the control stick for a possible patent application. A patent search was made regarding the control stick, and Dan's patent attorney, Mr. McDougall, submitted a patent application on May 10, 1943. It is unknown whether or not a patent was actually granted.

It is known that Dan never patented anything related to his engines.

ENGINE PRODUCTION RESUMES

In October 1944, an article in *Air Trails* magazine announced that some manufacturers of engines had been allowed to resume limited production of engines for aeronautical studies and said that ELF was one of the companies. The first ELF Twins were shipped on June 3, 1944. Singles were being shipped by the end of the month. Shipments of four cylinder engines were delayed, awaiting redesign, until 1947.

Dan had used the shutdown period to improve tooling and manufacturing processes to allow him to reduce the price of his engines. Unfortunately, sales were hampered because the War Production Board told Dan that he had to maintain the prewar prices for engines, although other engine manufacturers did not have this same requirement. Thus, the ELF engines were at a competitive disadvantage compared to other engines as they could not raise prices. When production resumed in 1944 there was a great demand for engines for awhile, and ELF was not able to capitalize on it.

The first engines produced were, with one exception, exact copies of the engines previously made, probably using existing pre-war parts. The exception was the replacement of the pre-war cast aluminum piston with a new two-piece machined piston made of bar stock and tubing.

Figure 3-4 The control stick assembly designed and built by Dan Calkin for Jim Walker. Dan applied for a patent on the device. (Calkin family photo)

New Two-Piece Piston

The new piston consisted of a piston skirt and top which could be screwed together. The skirt was machined from aluminum tube, the top from bar stock. To assemble the new piston, the piston top was connected to the connecting rod using the wrist pin. Then the piston skirt and top were screwed together and the top machined. The piston was then ready for the final machining of its outside diameter to match the cylinder.

The new piston was credited with increasing the power output, probably by raising the compression ratio slightly. It also covered the ends of the wrist pin, preventing scuffing of the cylinder, and production of the new piston was less labor intensive than the cored, cast, and machined piston used previously. If the piston exhibited excessive wear, the skirt could be replaced easily and cheaply. This two-piece piston can be identified by removing a cylinder and looking to see if the ends of the wrist pin can be seen. If the wrist pin ends can be seen, that identifies an engine made before November 1942. If not, the engine was made after that time. If you look closely at the top of one of these pistons you should see a serrated circle about $\frac{1}{8}$ inch in from the edge around the circumference. This is the edge where the two parts were screwed together.

Engine No. Type
Owner's Name
Address

ENGINE PARTS		
Spark Plug, Champion V-III	$ .50	
Propeller Mounting Plate	.25	
Propeller Washer	.10	
Propeller Nut	.20	
Piston (fitted to cylinder)	.90	
Con Rod Strap	.10	
Con Rod	.80	
Cylinder Casting	2.30	
Cylinder Liner	.90	
Wristpin	.30	
Cylinder Head	1.25	
Single Crankcase	2.15	
Single Crankshaft	2.80	
Shaft Collars	.25	
Twelve Ball Trust Bearing	.80	
Bearing Race	.25	
Twin Crankcase	2.60	
Twin Crankshaft	3.45	
Screws	.10	
CARBURETOR PARTS		
Single Gas Tank	.45	
Twin Gas Tank	.45	
Adjustment Needle	.25	
Air Filter	.20	

TIMER PARTS		
Extension	.30	
Casting	.70	
Fixed Point	.75	
Spring Point	.50	
Trigger	.10	
Stop	.10	
Cam	.10	
Lid	.10	
Washers	.10	
REPAIRS		
Take Apart, Clean, Reassemble, and Test Run		
Single	.75	
Twin	1.25	
Repolish Bearing Surfaces	.25	
Rescrape Crankcase and Cylinder Joints	.25	
Straighten Crankshaft Quill	.25	
Clean and Adjust Timer	.25	
Clean and Adjust Carburetor	.25	
Assemble Timer Points	.30	
Repair Twin Manifold		
Repair Carburetor		
Repair Timer		
Postage		
C.O.D. Fee or Insurance		

Remarks:

ELF ENGINE CO., 206 S. W. WASHINGTON ST., PORTLAND 4, ORE.

Figure 3-5 The ELF repair sheet (1000 printed, October 10, 1945)

Flared Exhaust Pipe

A more noticeable change was to the exhaust pipe. The pre-war, short, straight exhaust pipe was replaced with a straight, flared exhaust pipe. This was usually an extra cost accessory item, although there were periods when all engines were supplied with them.

Workers Rehired

With the war coming to a close, Dan started preparing for an expected big jump in engine sales. To help with engine production Phil Foster was hired in April, 1945, right after his discharge from the Navy. Rufus Rothrock was rehired on September 5, 1945, and his brother Dick on November 14, 1945. Government regulations required that the war veterans had to be given their old jobs back. Rufus did production engine work as well as experimental radio control work making very small relays and motors of Dan's design. The others all worked on producing parts for engines.

In anticipation of a surge of orders, Dan ordered larger quantities of raw materials than normal and a new six-panel instruction sheet was printed. Dick Rothrock did the inked drawings for the inside pages of the instruction sheet for which he was paid $25.00. See Figure 3-6.

New Lathe

To facilitate production, a new Model A, 9 inch South Bend lathe was purchased on September 30, 1946, at a cost of $428.94.

Machined Cylinder Fins

After engine production had resumed, Dan shipped 97 engines before he changed the tooling to make a new style cylinder and head. The mod-

ELF

Model Aircraft Engines

"TWIN" ELF "SINGLE"

"FOUR"

INDEX

Figure 3-6 The front of the six page 1946 Instruction sheet. The inside pages had inked drawings by Dick Rothrock. (500 printed November 21, 1946)

ified castings were used starting with a Single #377 on November 9, 1945. The new cylinders were slightly larger for better cooling and in order to accommodate the machining of the top three fins. The fins were initially turned on a lathe. A few years later they would be milled. Twins with machined fins were first shipped on December 18, 1945. In Chapter 6 on the ELF production engines, the side-port engines with machined cylinder fins are called Series 3 engines. Cylinders are discussed in detail in Chapter 7.

The ELF engines had been criticized by some who said they looked crude compared to the mass produced die-cast engines of the period. Dan had purposely chosen a rough sand cast surface for his engines because he felt that the uneven surface provided better cooling than a smooth surface.

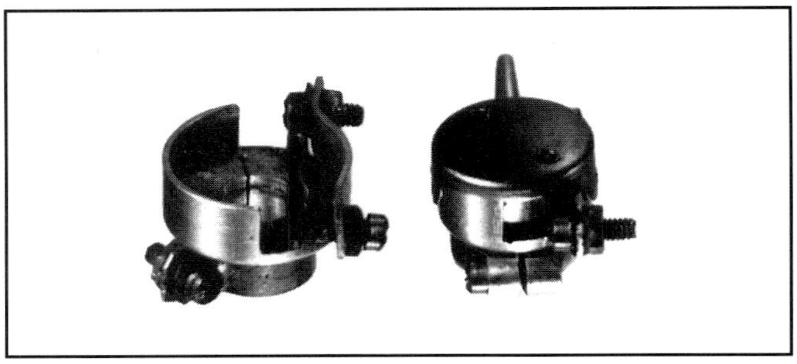

Figure 3-7 Two experimental timers used in the development of the round timer. Timer on the right was on ELF Single #314 sent to Carl VanCourt for testing.

In order to machine the fins, Dan had to enlarge the diameter of the cylinder castings slightly. He couldn't machine something off without first adding something extra on. This required reworking the cylinder patterns and tooling and additional machining steps on every production engine. All ELFs now had machined fins. The machined fin look was retained throughout all subsequent production.

Ted Enticknap tells of buying a new ELF for a very expensive $21.50 compared to $12.95 for a die cast *Super Cyclone*. When Ted's father saw the sand cast ELF he "hit the ceiling" and criticized Ted for wasting his money on such a crude (looking) engine. Obviously he was only looking at the exterior and he didn't know of the precision of the interior. Ted relayed his Dad's concerns (anger) to Seattle's ELF distributor, Harold Weeks, at the Bon Marche Department store. Mr. Weeks probably helped to convince Dan that he could sell more ELF engines if the exterior appearance could be enhanced. Dan reluctantly changed to machined fins in order to improve the marketability of his products.

5/8 Inch Prop Washers

When production resumed in 1944 the first engines shipped were equipped with the pre-war type ¾ inch diameter propeller washers but by the summer of 1946, the diameter was changed to ⅝ inch. At the same time the size of the hole was increased to ¼ inch. A new longer prop nut with a ¼ inch diameter threaded sleeve, extending into the prop, was added at the same time. See Figure 8-21.

By now several companies were manufacturing propellers for model airplane engines, but most props had a ¼ inch hole in the center. The new prop nut allowed the ELF engines to use the commercially available propellers without an adapter and also allowed Dan to shorten the crankshaft. The propeller hardware is discussed in detail in Chapter 7.

The production of the Singles and Twins continued otherwise unchanged until near the end of 1946.

The Round Timer

The next change to the engines occurred in late August, 1946, for the Singles and October, 1946, for the Twins when the round timer was introduced, replacing the cast offset timer with the wire cam. The new timer had a round, machined aluminum case with a cast aluminum arm. It was used in conjunction with a new split cam that was pressed on the aft end of the crankshaft. The single cylinder engines and the Twins used the same timer, having one set of breaker points. Figure 3-7 shows two prototype round timers. Coincidental with the introduction of the new timer on the Model 10 engines was the change to a six-ball thrust bearing. The prototype Single for the new timer and ball bearing was #314 that Dan sent to Carl VanCourt on March 21, 1945, for testing and evaluation. The thrust bearing actually used in this one engine had holes for 12 balls but only 6 bearings were installed in alternate holes. The Twins continued to use the standard twelve-ball thrust bearing.

A MOVE TO CALIFORNIA?

At this time Dan and Carl VanCourt were exchanging correspondence about the possibility of moving the company to Southern California, a move that Josephine definitely favored. Carl looked for locations in both the San Diego and Los Angeles areas. Though Carl was still in the Navy, stationed in San Diego, California, there was serious discussion about making Carl a senior member of the company. According to the plan, Carl would pretty much run the company, leaving Dan more time to develop products and increase

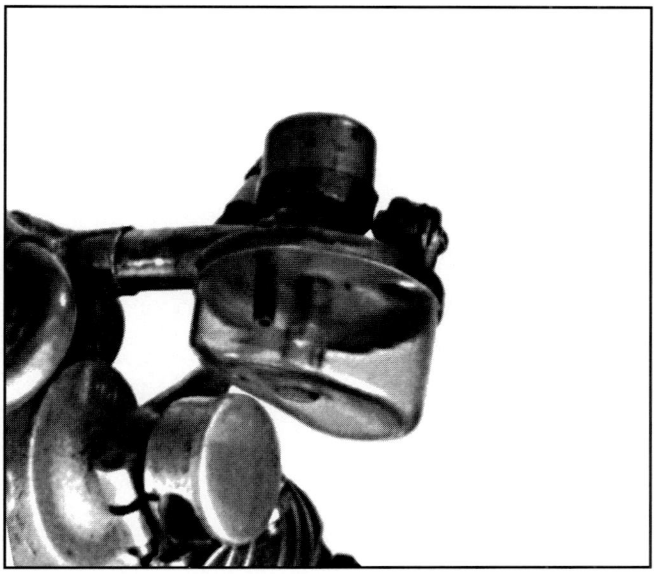

Figure 3-8 The new double beveled celluloid tank for the Twin with the round timer. This tank provided better clearance with the new timer.

Figure 3-9. A proposed larger tank for the ELF Twin. Made by Dan Calkin, tested by Carl VanCourt.

production, but nothing came of these discussions. Both Dan and Josephine, but especially Josephine, were anxious to relocate to Southern California. Dan thought that the drier weather and warmer climate would improve his health, while Josephine longed for the sunshine there.

PROBLEMS!

Soon however, there were problems: some vital materials were very difficult to obtain, such as the seamless Shelby tubing used for the cylinder liners and the tungsten points used in the timer assemblies.

Many new engine companies had sprung up and competition was very keen. The expected sales increases did not materialize and all ELF employees were laid off on December 8, 1946. For this reason, the planned move to California never materialized. This was a really low period for ELF Engine Company. Dan again had to do all the work with only Josephine to help. To get by, they were forced to take out a loan to pay the salaries of the workers and for the purchases of materials and tooling.

Health problems had always plagued Dan, but the winter of 1946/47 was especially bad, and Dan had to almost close his shop for about six months because of respiratory problems. The spring of 1947 saw Dan's health improve to the point that he went for several months without missing a day of work. In this shop, as in all his shops, there was a cot where he could lie down and rest when he didn't feel well.

THE NEW FUEL TANK

The new round timer was larger in diameter than the earlier cast timer and did not fit the engines particularly well due to interference between the timer and the celluloid fuel tank. To solve this problem, on the Model 20 engines the bottom of the tank was changed from a single bevel to a double bevel. Figure 3-8. At that time most of the engines were being used for control-line flying instead of free flight and the modelers wanted larger fuel tanks. So, in mid-February 1947, the new double-beveled tank from the Twin was adapted to the Model 10 engines, becoming its standard tank. This gave the Single a much greater fuel capacity while also solving its interference problem. Various ways to enlarge the fuel tank on the Twins were tried but nothing was implemented. One of the larger tanks tried for the Twin is shown in Figure 3-9. Carl VanCourt suggested this style tank in 1946. Dan built it for him and Carl tested it. No others were made.

Special Features

SAME CONSTRUCTION AS LARGE AIRCRAFT ENGINES. Aluminum cylinders with thin hardened steel liners. Aluminum pistons, "micro-machined," with special leak-proof construction with no external wristpin hole and no rings or grooves. Piston shaped for 100% bearing surface at operating temperature.

The cool running of the ELF and the effective oiling of its bearings make practical continuous operation with high-power racing fuels or glow plug fuels. The lower operating temperature is due to the aluminum-cylinder-steel-liner construction copied from large aircraft engine design.

CASTINGS: Copper aluminum alloy cast in sand molds. No gaskets.
CONTROLS: Grouped behind and above the engine for effective cowling.

LUBRICATION: No. 40 oil. Gives higher power output because of less drag. Its use is possible because the ELF runs cool and has very small bearing clearances.

UNRETOUCHED PHOTOGRAPH - ¾ ACTUAL SIZE

TYPE: 2 cycle, 3 port.
CYLINDERS: 2 banks of 2 cylinders opposed.
TOTAL DISPLACEMENT: .396 cu. in.

OPERATING SPEED: 8,000-15,000 R.P.M.
POWER: ½ H.P.
BORE: .468 in.
STROKE: .564 in.

BEARINGS: 3 main bearings, 12-ball thrust bearing.
WEIGHT: Bare, 9 oz. Flying wt. ignition with two twin coils, 15 oz.
MOUNTING: Radial.
CARBURETOR: Micrometer adjustment. Air filter on carburetor intake.

GAS TANK: Transparent. Capacity 4 minutes. Metal tanks available for alcohol.
EXHAUST PIPES: Flared. Exhaust manifold easily attached.
TIMER: ELF "No-Drag." Self-cleaning points.

THE WORLD'S FINEST MODEL ENGINE

Dan Calkin's ELF FOUR

One-cylinder engines are no longer used in automobiles or airplanes. This is because multi-cylinder engines are vastly superior in smoothness, power, reliability, and responsiveness. These advantages are now available for your model airplane.

If you are interested in radio control, you will like the FOUR. With its perfectly balanced reciprocating parts and double power impulse, the FOUR produces vibrationless power that does not disturb delicate control mechanisms. The high power and low weight of the FOUR leave you plenty of reserve for carrying equipment. If you enjoy racing and stunting, consider these advantages of the FOUR. Its compactness makes it easy to cowl and permits perfect streamlining of your airplane. It throttles smoothly and responds instantly with terrific power so that you have absolute control of your plane at all times.

Its beautiful appearance and resemblance to a full size power plant offer inspiration for authentic miniature reproductions of all types of aircraft.

Whatever type of airplane you model, the FOUR is the modern way to power it.

ELF ENGINE COMPANY, 206 S. W. Washington Street • Portland 4, Oregon

ELF
Model Aircraft Engines

"FOUR"

"TWIN" ELF TRADE MARK "SINGLE"

CATALOGUE No. 449

Figure 3-10 The brochure touting the ELF Four, printed November 1948. Engine in photograph is # 911.

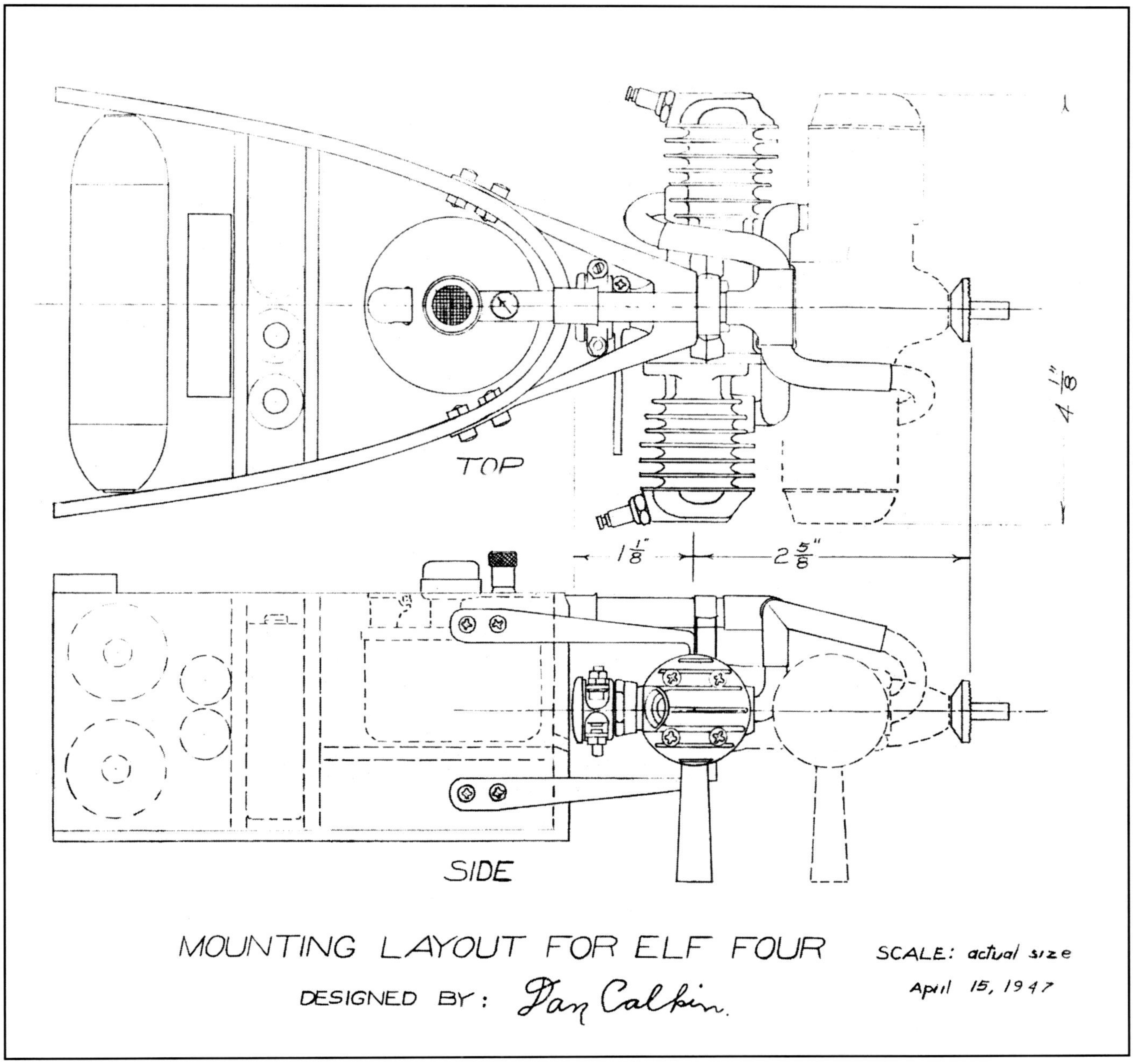

Figure 3-11 A suggested method for mounting the new 1947 ELF Four in an airplane. Notice that it could also be used by a Twin equipped with the Model 40s manifold extension and carburetor. (First printed September 26, 1947)

THE SLOTTED EXHAUST PORT

Starting on April 27, 1947, the exhaust porting in the cylinder liner was changed from three square holes to a slot. This was done by repositioning the punch, after making the three holes, to punch out the two bridges between the holes. The punch used for this purpose is shown in Chapter 15 Figures 15-41, 15-42 and 15-43.

THE NEW ELF FOUR

The four cylinder ELF had been out of production for nearly five years. The distributor system used in the earlier four cylinder ELFs could no longer be used because the sparking between the distributor plates caused radio frequency interference that would make it impossible to use the engine for radio controlled airplanes. A second version of the round timer appeared in May of 1947 on the new four cylinder engines. This timer had two

ELF ENGINE COMPANY - - -

ELF "SINGLE", ELF "TWIN", ELF "FOUR" MODEL AIRCRAFT ENGINES

206 S. W. WASHINGTON ST.
PORTLAND 4, OREGON

Figure 3-12 This is the new letterhead Calkin had printed on December 9, 1948.

sets of points. The new timer assembly, lacking a distributor, solved the problem. However, it did require the use of either four single coils or two twin coils plus two condensers. The earlier Four with the distributor had required only a single coil and condenser.

The re-introduction of the four-cylinder ELF involved a large amount of refinement to find just the right combination of variables to get the most out of the new engine, but the new ignition system and carburetor presented the biggest problems. The first ELF Four of the new series, #555, was sold on May 9, 1947. A brochure was mailed out in November, 1948, announcing the new engine. Figure 3-10. The brochure also mentioned that a metal tank was available for use with alcohol fuels.

A novel mounting system was devised based on the one that Dan was using on his own model at the time. The engine was somewhat awkward to mount in an airplane because of its long intake tube and large fuel tank. The new method devised for mounting the engine required a rounded nose on the airplane to allow the long intake manifold to position the tank inside the airplane. Notice that this carburetor and mounting system could also be used with the ELF Twin. Figure 3-11.

TIN PLATING

In a letter dated August 19, 1948, to Carl VanCourt, Dan encouraged Carl to set himself up an electro-plating outfit immediately. He said, "We are doing all our own plating now. Our plating is tin which doesn't require polishing and makes a pleasing appearance with the aluminum." Up to this time Dan had always sent the carburetors and manifolds to an outside plating shop to be plated with cadmium. The records show that Dan had purchased the materials for the tin plating solution on May 19, 1947. Tin plating probably started soon after that as the supply of cadmium plated carburetors and other plated parts was exhausted.

1947 CHANGES IN THE INDUSTRY

Two significant changes occurred in the model airplane industry near the end of 1947. The first was the big drop in engine prices led by Ohlsson and Rice. The other was the introduction of the glow plug by Ray Arden.

ENGINE PRICE CUTS

A shake-up of the model airplane engine business began in 1947. The December 1947 *Model Airplane News* advertisement by the Ohlsson and Rice Company announced a big price reduction on the very popular *Ohlsson* engines. For example, the price of the *Ohlsson .23* was dropped from $16.50 to $9.95. Similar price cuts were announced for their other engines as well. Ohlsson's price cut was made possible by their wartime acquisition of high volume machinery for war contracts. When the war ended, Ohlsson kept this machinery and used it to mass produce engines and soon was producing as many as 7,000 engines a day. Many engine manufacturers were soon forced out of business, unable to compete at that price. The ELF Engine Company was similarly impacted, but survived.

THE GLOW PLUG

In late 1947 big changes were sweeping the model engine world. In the same December, 1947 issue of *Model Airplane News* that announced the drop in the *Ohlsson* engine prices, Ray Arden introduced the "revolutionary new" glow plug, taking the model airplane industry, especially the engine manufacturers, by storm. The glow plugs were used successfully by the west coast team in the big East-West Competition on September 6 and 7 and suddenly almost any engine could be operated without the bulky, heavy, problem-prone spark ignition system.

Like other manufacturers, Dan soon offered engines especially made for glow plug operation. The first response to the glow plug craze was to make available a stamped aluminum tank, replacing the celluloid tank, to customers who wanted to use methanol based (glow) fuels. Many modelers found out, to their dismay, that the methanol based glow fuels would quickly destroy the celluloid fuel tanks. The metal tanks became available in late 1947.

During this period, the Calkin family continued to live at the River Road home. Dan did take time out from building ELFs to build CeCe a tree house that she loved.

MILLED FINS

In early 1948, Dan Calkin came up with a method to reduce the time required to machine the cylinder fins. By milling the fins, instead of turning them, he was able to machine all three fins in one operation. See Chapter 15 for photos of the tooling needed for this operation. If you examine the tooling marks very carefully on an engine with milled fins you will see the circular tool marks around the cylinder. All subsequent production engines used cylinders with milled fins.

ELF DEVELOPMENT CONTINUES

In June of 1948, the diameter of the exhaust port in the cylinder casting was enlarged from 0.250 inches to 0.275 inches.

On August 19, 1948, Dan wrote to Carl VanCourt to announce that "a new HIGH COMPRESSION ELF Single, Model 10-10" had been sent to him under separate cover (#1568). Dan explains further:

> I have spent 6 months going over the engine, fixing up details, and bringing its power up to competition, just keeping a few engines in stock to take care of any stray orders. The unit you have at hand is a finished product and I believe it is competitive in every respect.
>
> The engine you have will turn close to 9,000 R.P.M. with an 8" diameter 6" pitch prop and 11,000 or better with an 8" x 4" using gasoline and outboard motor oil mixed 3-1. It runs on a mixture of half gasoline, half nitro-propane mixed 3 to 1 with No. 40 oil with the ignition shut off or with a glow plug at slightly increased R.P.M. Now this is as far as I got. On straight alcohol it really starts to roll for a few seconds and then slows way down from pre-ignition. I have ordered some cold plugs to try out, maybe you can find some other solution. I think the engine is good for another 1,000 R.P.M. on some hot fuel if we can find it.

From the above it appears that Dan had finally been able to bring the Single up to the point where it was really competitive again, but the engine was never advertised and the modeling community didn't hear about it. The perception among most modelers at the time was that the ELF engines were nice but the power output did not compare to other better known engines of the day. Little did they know!

Only eight of these Singles were shipped after the one sent to VanCourt. The improved single cylinder engines all bore serial numbers in the 1500s (#1562, 1564, 1565, 1566, 1569, 1573, 1577 and 1578) Naturally, all the performance improvements made to the Singles were soon introduced into the Twins and Fours.

Almost immediately after introducing the high compression heads, Dan realized that the high compression could cause problems. For one thing, they required a much hotter spark to the plug to make it work. Dan stopped using these

Figure 3-13 The Model 10 #1562 won by "Lucky" Lucas with the ELF vane-type fuel pump installed.

heads after shipping only eight engines and returned to the standard heads.

Pressed-Together Piston

The first ten engines sold with the large exhaust continued the use of the two-piece piston that was screwed together. The subsequent ELF engines had a new two piece piston that was pressed together rather than screwed, thus reducing machining time. Chapter 7 has photos of all pistons.

Certificate of Guarantee.

This Certificate of Guarantee becomes effective upon receipt of the attached Guarantee Registration Card at the ELF Engine Co.

It certifies that the ELF Engine which you have purchased was thoroughly tested before leaving the factory and is guaranteed to give satisfactory performance in every way. This guarantee applies to the original purchaser.

ELF ENGINE COMPANY
206 S.W. Washington St. Portland 4, Oregon

TEAR OFF AND MAIL THIS CARD TODAY

ELF.................................. No..............................
NAME..
ADDRESS...
Name of Dealer..
Date of Sale...

Figure 3-14 Guarantee card. (1000 printed December 1, 1949)

The Fuel Pump

Dan built a few engines with an engine-driven vane-type fuel pump. The pump replaced the timer cover on the rear of the engine. To drive the pump, a piano wire shaft with a 90 degree bend

ELF ENGINES, October,1949

		SINGLE	TWIN	FOUR
GLOW PLUG		$15.50	$25.50	$52.50
IGNITION less coil & condenser		18.50	28.50	56.50

ACCESSORIES

		SINGLE	TWIN	FOUR
FUEL PUMP		$4.50	$4.50	$4.50
FLARED EXHAUST PIPE	ea.	.15	.15	.15
MOUNTING BRACKETS	pr.	.80	.80	.80
DOUBLE TIMER		2.80	2.80	3.80

Figure 3-15 The first price list for ELF glow plug engines and for the new fuel pump.

Figure 3-16 Model 10 ELF with stamped metal tank.

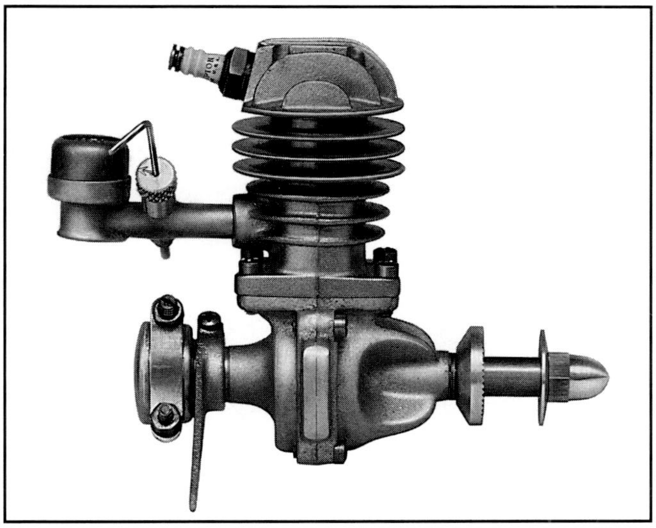

Figure 3-17 Model 10 ELF with tankless carburetor and air filter. (Calkin family photo)

Figure 3-18 The Model 10 ELF with the bell-mouthed intake.

on the front of the pump engaged a $1/32$ inch x $3/64$ inch slot cut into the aft end of the crankshaft. The pump allowed the fuel supply to be as much as fourteen inches from the engine. In a letter dated November 16, 1949, Dan stated that he was planning to offer a fuel pump as an accessory that could be used on any of the slant plug ELF engines. The price for the pump would be $4.50. There is no evidence that any accessory pumps were ever sold.

Dan mailed an announcement of the pump-equipped ELFs (Figure 3-19) to all ELF customers and dealers but it generated little enthusiasm. Probably the ELF customers really did not understand the advantages of the fuel pump. This was a product many years ahead of its time. Only five pump-equipped ELFs were ever shipped; a Single (#1562) which was a gift to the local "Stardusters" club for use as a contest prize, two Twins (#897 and #906) and two four-cylinder engines (#934 and #953). Figure 3-13 shows the #1562 ELF Single with the pump attached and with the tankless carburetor. The fuel pump was listed at $4.50 on the October 1949 price sheet. (Figure 3-15) With the fuel pump the tankless carburetors were introduced, as the pump could not be used with the conventional intake tube with fuel tank attached.

The pump-equipped Single was won by Leonard "Lucky" Lucas, who built a large control-line airplane for it. He set a control-line endurance record with a flight of 7 hours and 15 minutes that ended only when he had to quit due to fatigue. During the flight, he pumped fuel to the airplane through a 50 foot long piece of fuel line to a large fuel tank. The fuel flowed by gravity feed to a float carburetor from a First Model ELF. The float controlled the flow of fuel from the airplane's internal tank to the engine.

Rounded Fins

About October, 1949, another small change was made to the cylinder cooling fins. Dan devised a way to round the edges of the fins to replace the square edges previously used. All subsequent engines have the rounded fins.

◄ **FLASH ANNOUNCEMENT** ►

ELF PRESSURIZED JET CARBURETION!!

- Manifold Fuel Injection With Pump -

Sensational!! *Revolutionary!!*

Unbelievably smooth running, no matter how you stunt your airplane - never a miss in dives, loops, upside down flying!

Perfect throttling - from barely turning over to roaring, dynamic power.

Simplified engine mounting - use any gas tank placed anywhere on airplane, engine easily mounted on any fire wall.

ELF pressurized jet Four	$64.50
ELF pressurized jet Twin	34.50
ELF pressurized jet Single	21.50

ELF ENGINE CO., 206 S.W. WASHINGTON ST., PORTLAND 4, ORE.
ATWATER 4661

Figure 3-19 This announcement of the new ELF pump drew little response. (500 printed June 2, 1949)

Steel Timer Case

Also changed in October 1949, was the material used for the timer case. After this date the case material was steel, rather than aluminum.

FIRST ELF GLOW PLUG ENGINES

For small engines like the ELF, the change to glow plug ignition was especially important, because the weight and bulk of the timer, condenser, battery, coil, switch and wiring represented a much greater percentage of the airplane's weight than it did with the larger engines. Additionally, the more powerful fuels used with the glow plug equipped engines meant that those engines produced more power than an engine with spark ignition. The glow plug thus greatly improved the power-to-weight ratio. From then on until production ceased, Dan Calkin offered both spark ignition and glow plug versions of his engines. The only exception to this being the ELF Six that was offered only as a glow plug engine. Several options were offered that could be used with both spark or glow plug ignition.

Keep in mind that the use of the engines was rapidly changing from free-flight to control-line airplanes. Control-line models needed much larger fuel tanks than did free-flight models.

First, Dan offered an engine with the standard carburetor, but with a stamped aluminum fuel tank to replace the celluloid tank. Figure 3-16.

SELECT THE NEW HIGH PERFORMANCE ELF TO SUIT YOUR AIRPLANE

	SINGLE	TWIN	FOUR
STANDARD EQUIPMENT--includes timer, carburetor & spark plug(s)	$21.50	$33.50	$59.50
GLOW PLUG EQUIPMENT--includes tankless carburetor & glow plug(s)	17.50	29.35	54.25
ECONOMY EQUIPMENT--includes tankless carburetor, no plugs	16.75	27.85	51.25

PARTS PRICE LIST

	SINGLE	TWIN	FOUR
TIMER complete	$2.95	$2.95	$3.95
CASE	.70	.70	1.10
LID	.05	.05	.05
HANDLE	.25	.25	.25
FIXED POINT	.30	.30	.30
SPRING POINT	.95	.95	.95
WASHERS	.05	.05	.05
CAM	.60	.60	.60
TANKLESS CARBURETOR with AIR FILTER	1.65	1.65	1.65
TANKLESS CARBURETOR with BELL MOUTH	1.65	1.65	1.65
CARBURETOR with PLASTIC TANK	2.95	3.35	3.95
CARBURETOR with METAL TANK	2.95	3.35	3.95
ADJUSTMENT NEEDLE	.30	.30	.30
AIR FILTER	.25	.25	.25
GASKETS	.15	.15	.15
TANK, PLASTIC	.45	.65	.95
TANK, METAL	.45	.65	.95
MANIFOLD	--	1.95	4.95
MANIFOLD EXTENSION	--	.35	--
CRANKCASE	2.50	2.75	6.70
CRANKSHAFT	3.80	4.70	6.10
BALL RACE	.10	.10	.10
BALL RETAINER	.75	.75	.75
PROP NUT	.20	.20	.20
PROP WASHER	.05	.05	.05
PROP MOUNT	.30	.30	.30
COLLARS	.15	.15	.15
PISTON ASSEMBLY (includes CON ROD, STRAP and WRISTPIN)	2.50	2.50	2.50
CYLINDER & LINER	4.30	4.30	4.30
CYLINDER HEAD	1.40	1.40	1.40
MOUNTING BRACKETS	pr. .80	.80	.80
FLARED EXHAUST PIPE	.15	.15	.15

Figure 3-20 The December 1949 price sheet lists metal tanks and tankless carburetors with either an air filter or bell-mouth. Noticeably absent are the fuel pump and the double (two speed) timer. (1000 printed December 1, 1949)

Most control-line users had no need for the attached tank, celluloid or metal, so the next offering was a tankless engine. This consisted of a standard carburetor with air filter, but with the fuel tank and tank top deleted. (Figure 3-17). The next offering was an optional bell-mouthed carburetor. (Figure 3-18). All these versions are listed on the December 1949 price list, Figure 3-20. All three options were of interest to glow plug users, but these same options were sold on some spark ignition engines as well. See Chapter 9 for more information about carburetors and tanks.

Because the stamped metal tanks and tankless carburetors were in production for only a short time, they are rather scarce.

It is interesting to note that Dan Calkin states, in his log book on April 11, 1950, that the ELF spark ignition Twin #965 was the "best engine ever." Then, a month later he wrote of another spark ignition Twin #938. "Even better than 965."

There have been persistent rumors that Dan did not like glow plugs but, in his personal letters to his friend Carl VanCourt, he praised them and advised Carl to use them also.

HALF A ENGINES

Dan Calkin gave some thought to making a half size ELF to compete with the new half-A engines that were appearing on the market. In a June 29, 1950, letter to Carl VanCourt he says:

> I am much interested in the small engines. Jimmy (Walker) gave me a *Baby Spitfire* but I haven't run it yet. Have you got any ideas on them? I considered making a Single 1/2 size. Maybe the main appeal of these engines is the price.

There is no evidence that any half size ELF engines were ever made.

A NEW CHILD, A NEW HOME

The Calkin family celebrated the birth of their son on May 16, 1949, and named him Edwin Daniel Calkin, Jr. A few months later Dan decided to stop commuting by moving his wife, daughter and new-born son into his shop in the Kraemer Building. The desire to eliminate the commute each day was also a factor, but the most important reason was that Dan's health seemed to be much better at the shop than at the River Road home. Perhaps this arrangement also made it easier for them to care for their new son. Another factor was his desire to be on his own, away from his mother, as they sometimes got on each other's nerves. By this time Dan and Josephine had been living with Dan's mother for 15 years. The new "home" did not have a place to put the piano so it stayed at Dan's mother's house.

Figure 3-21 shows Josephine on a shopping trip in downtown Portland, pushing the stroller with their son, "Win."

Figure 3-21 A street photographer caught Josephine out shopping with her baby, "Win". (Calkin family photo).

CECELIA DESCRIBES LIVING IN SHOP

Living in the Kraemer building was described this way by Dan's daughter, Cecelia, in an August 1, 1994, letter to the author:

> The Kraemer building was an unusual and interesting place to live. The room he (Dan) rented was 40 ft. long and 20 ft. wide with windows looking out over the roof of a one-story building next door. Dan partitioned off a 12 ft. by 6 ft. corner of the room for living quarters. He made built-in bunk beds with army cots and built-in folding tables with space above for a kitchen-dining area. Cooking was done on an electric hot plate and a Bunsen burner. There was a dry-ice box for cold storing food. The furnace for melting aluminum to pour castings was also in that room. There was a public rest room in the hall outside the main room. A small sink inside the main room supplied water. The rest of the room housed his business. The whole building was steam heated.

A little further explanation, perhaps, is needed to explain the living arrangements at the Washington Street shop. Dan's health problems were always with him and he was continually exploring ways to minimize them. Many of his problems were caused by dust. Dan systematically set about trying to build a "clean room," or dust-free area, by doing everything he could to eliminate materials which would hold dust. He had no upholstered furniture and no carpeting. The floor covering was linoleum. Army cots were used for beds in order to be rid of mattresses which retained dust.

It is interesting to note that the foundry area was also located in the "living quarters."

The rest of the ELF operation was described by Cecelia this way:

> He had three or four small lathes, two drill presses, a power hack saw, a kick press, and other machinery for manufacturing the ELF engines.
>
> During this time his wife and daughter helped with the business. Though Cecelia was only about 12 or 13 years old, Dan taught her to work with the equipment. First, she learned to clean up and oil the lathes and drill presses. He had begun automating some of the machining of the ELF and taught her how to run the lathes.
>
> Dan was a master craftsman as well as an outstanding designer of things. Every part of each engine was machined to tolerances of thousandths of an inch so carefully that no gaskets were needed. He started by sand casting aluminum cylinders, crankcases and heads, then cutting some of the fins into the cylinders, and machining all fitted surfaces precisely. He drilled out the cylinders and lined them with hardened steel liners for the pistons to travel in. To make liners, steel tubing was cut to the correct length, port holes were die cut into the walls, then they were heated to a red hot temperature and plunged into cold water. When they were cold, the liners were pressed into the cylinders.
>
> The crankshafts were made from .250 inch steel flat stock, rough cut on the power hacksaw to the correct length and machined to precise measurements so the connecting rods could be securely connected to them. Thrust bearings were made of brass disks with steel balls tapped gently into tiny holes. Each engine was test run on a stand before being sold.

Cecelia remembers watching her mother spin the thrust bearings in her hand to see if they turned freely enough.

Through the years the family had enjoyed occasional outings on their boat, "Josephine," but finally the time came to sell it and let someone else

Major Change in Elf Engine Eliminates Intake Manifolds

Automatic Induction Valve Designed by Dan Calkin Has Many Advantages

We believe this new induction valve is the ideal construction for model airplane engines. It combines the best features of the side port and rotary valve engines with none of their disadvantages and has additional advantages of its own!

Unique Characteristics

The very simple construction of these new engines makes them rugged and reliable. This simplicity of design is especially noticeable on the multi-cylinder engines which have no intake manifolds. They are consequently lighter in weight and have greatly reduced frontal area.

Further advantages the automatic induction valve has over other types of engines are:

1. Lubrication of bearing and cylinder is far more effective because there is never any vacuum in the crankcase. This reduces wear and also contributes to the easier starting.

2. Power is equal to or better than the power of other types of engines because the valve automatically provides a full crankcase charge at any speed.

3. This providing of a full crankcase charge without blow-back through the carburetor improves running and idling speeds and is the main factor in improved starting.

4. The maximum available time for drawing in air in each stroke is utilized, consequently higher air velocities in the carburetor can be used, which in turn make the engine less affected by tank position or fuel level. This is especially important in control line flying.

5. The fuel mixture gets richer as the engine speed increases over the full speed range. This unique characteristic makes the engine very easy to start and also allows a very simple method of throttling control. To slow the engine down (either glow plug or ignition) it is only necessary to richen the mixture or choke the engine and the engine runs smoothly at the desired reduced speed without fourcycling.

Perfect Glow Plug Operation

All ELF Engines run equally well on glow plug or spark coil ignition. Glow Plug and Ignition Models are identical except for plugs and timer.

Same Construction As Large Aircraft Engines

Cylinder construction is the same as large aircraft engines: aluminum cylinders with thin hardened steel liners. Aluminum pistons, "micro-machined", have a special leak-proof construction with no external wristpin hole and no rings or grooves. Piston is shaped for 100% bearing surface at operating temperature.

No. 40 oil is used for lubrication. It gives higher power output because of less drag. Its use is possible because the ELF runs cool and has very small bearing clearances.

New ELF Six

To supply the demand of modelers who want more power than the Four offers, we are buildin*g* a limited number of Six cylinder engines. Write to us for details.

Vibrationless Power For Radio Control

The ELF Single is carefully designed for the smoothest running possible in a one cylinder engine. For the ultimate in smooth power, we recommend the ELF Multi-Cylinder Engines. They are an incomparable power plant for radio controlled airplanes. With their perfectly balanced reciprocating parts and multiple impulses, the Class A Twin, the Four, and the Six produce vibrationless power that does not disturb delicate control mechanisms. Their high power and low weight leave you plenty of reserve for carrying equipment.

If your specialty is racing and stunting control line planes, consider these advantages. The stronger suction on the jet of these engines makes them less sensitive to tank position and maneuvering of the airplane. Consequently they will stunt without missing and run their tanks dry with scarcely any variation in speed. Conventional balloon and wedge tanks give excellent results.

The beautiful appearance of ELF Engines and their resemblance to full size power plants offer inspiration for authentic miniature reproductions of all types of aircraft. Their compact and simple design makes them easy to cowl. Easily attached exhaust pipes throw oil clear of the airplane. They also increase the power of the engines and keep out dirt.

Whatever type of airplane you model, ELF is the modern way to power it.

TRADEMARK ELF TRADE MARK

PRICES:

	Single *Class A*	*Twin* *Class A*	*Four* *Class C*	*Six* *Class C*
GLOW PLUG: Tank, mounting brackets and plugs included.	$12.50	$21.50	$45.50	Custom Built.
IGNITION: Tank, mounting brackets, plugs and timer included.	$15.50	$24.50	$49.50	Write us for details.

ELF ENGINE COMPANY, 206 S. W. Washington Street • Portland 4, Oregon

Figure 3-22 The flyer announcing the new ELF reed valve engines. Notice the introduction of an ELF SIx on a custom order basis. (Printed November 1950).

PARTS PRICE LIST

ELF ACTION VALVE ENGINES

	SINGLE	TWIN	FOUR
TIMER complete	$2.95	$2.95	$5.95
CASE	.70	.70	1.10
LID	.05	.05	.05
HANDLE	.25	.25	.25
FIXED POINT	.30	.30	.30
SPRING POINT	.95	.95	.95
WASHERS	.05	.05	.05
CAM	.60	.60	.60
TANKLESS CARBURETOR ~~[illegible]~~	1.65	1.65	1.65
~~[illegible]~~	~~[illegible]~~	~~[illegible]~~	~~[illegible]~~
~~[illegible]~~	~~[illegible]~~	~~[illegible]~~	~~[illegible]~~
~~[illegible]~~	~~[illegible]~~	~~[illegible]~~	~~[illegible]~~
ADJUSTMENT NEEDLE	.30	.30	.30
~~AIR FILTER~~	~~[illegible]~~	~~[illegible]~~	~~[illegible]~~
~~GASKETS~~	~~[illegible]~~	~~[illegible]~~	~~[illegible]~~
~~TANK, PLASTIC~~	~~[illegible]~~	~~[illegible]~~	~~[illegible]~~
TANK, METAL	1.00 ~~[illegible]~~	1.00 ~~[illegible]~~	1.00 ~~[illegible]~~
~~MANIFOLD~~	--	~~[illegible]~~	~~[illegible]~~
~~MANIFOLD EXTENSION~~	--	~~[illegible]~~	--
CRANKCASE	2.50	2.75	6.70
CRANKSHAFT	3.80	4.70	6.10
BALL RACE	.10	.10	.10
BALL RETAINER	.75	.75	.75
PROP NUT	.20	.20	.20
PROP WASHER	.05	.05	.05
PROP MOUNT	.30	.30	.30
COLLARS	.15	.15	.15
PISTON ASSEMBLY (includes CON ROD, STRAP and WRISTPIN)	2.50	2.50	2.50
CYLINDER & LINER	4.30	4.30	4.30
CYLINDER HEAD	1.40	1.40	1.40
MOUNTING BRACKETS	pr. .80	.80	.80
LONG ~~FLARED~~ EXHAUST PIPE	.25	.25	.25
THROTTLE FOR CARBURETOR	1.00	1.00	1.00

Figure 3-23 The August 1950 price list with many changes from the previous list.

enjoy it. On October 13, 1949, the boat was sold for $625.00 to Henry Grill of Portland, Oregon.

THE ACTION VALVE ENGINES

In August of 1950, a major change was made in the ELF line. A system Dan called an Automatic Induction System was introduced, replacing the previous side-port style engines. This new induction system, that Dan later called the Action Valve System, is more commonly referred to as reed valve induction. (Figure 3-22) The reed valve engines constitute the last major change in the ELF line and are referred to as Series 4 engines in Chapter 6. This change necessitated a major redesign of the engine. The pattern used for casting the cylinder was changed to eliminate the external intake port from the castings. The pattern for casting the rear crankcase section was changed on the Singles and Twins to accommo-

Figure 3-24 The ELF pencil. Two gross of these were ordered July 13, 1951. (Thanks to CeCe for this one.)

date the reed valve induction system. This also eliminated the external induction manifold as used on the Twins and Fours, with a resulting reduction in materials and labor. The intake tube for the Fours and Sixes was moved to the center crankcase extension section.

Dan continued to offer both spark ignition and glow plug ignition engines for sale. The engines were basically identical, and any glow plug engine except the Six, became a spark ignition engine simply by adding the timer assembly (breaker points) and cam and installing a spark plug in place of the glow plug. All the glow plug engines had the rear crankcase machined to accept a timer, in the event that was desirable.

In order to use up about 100 remaining machined cylinders for side port engines in which Dan had invested a lot of casting and machining time, the early production of reed valve engines all used these old style cylinders with the intake port plugged. Some old cylinders had not yet been drilled for the intake port so those were left undrilled. (Figure 7-10)

The reed valve engines made with the side port cylinders are sometimes referred to as "composite" engines. There were 6 Singles, 13 Twins, 16 Fours and one Six of the "composite" engines made and sold.

Reed Valve ELF Single

The single-cylinder versions of the Action Valve series of engine are rather rare because, at the time of their introduction, there were other Class A engines that dominated the field. The *Arden* was certainly a prime example. Who has ever heard of an ELF Single developing the same power to weight ratio as a .099 ball bearing *Arden* glow model? This was the reason that Dan Calkin favored multi-cylinder designs. Here was a field that he dominated.

The ELF Six

For many months Dan had been experimenting with a new ELF engine having six cylinders. This engine was a real engineering triumph for him. The toughest part was making the crankshafts; they required a lot of work. To assemble the engine so that it turned freely, and to keep everything in proper alignment over the required distances required very exacting tolerances.

The ELF brochure (Figure 3-22) sent out in November, 1950, announced, "To supply the demand of modelers who want more power than the Four offers, we are building a limited number of six cylinder engines." These engines were to be custom built to order. The first ELF Six (Model 60) #1054 was shipped on March 11, 1951. The second engine shipped was really the first one made. It went to Dan's friend Harry Fosbury and was #986. The #986 engine was the only Six made with the old side-port cylinders. The other two Sixes made at Washington Street are #1018 and #1045. The first batch of four Sixes was built in late 1950 and early 1951. Dan wanted to keep the fourth one for himself, but Earl Vivell, a well known model airplane engine distributor from the Bay Area near San Francisco, California, came along and wanted it, so it was sold. Later, Earl Vivell ordered a three bladed prop and exhaust pipes for his Six. (Figure 3-25)

The 1950 Price List

The ELF Six used many standard ELF parts such as the piston, cylinder, cylinder liner, crankcase and center crankcase extension. The new parts were a crankshaft, of course, along with new prop hardware: the prop nut with a 10-32 thread

Figure 3-25 Earl Vivell, (center) explains his ELF Six powered fiber glass R/C plane to a couple of interested spectators at the 1953 National Model Airplane Contest. (U.S. Navy photo) *West Coast Model News*, September 1953.

to match the new crankshaft, a new prop washer with a larger hole and a new drive washer, also with a larger hole.

Dan later made one more batch of eight six cylinder engines.

The metal tanks were deleted being replaced by a firewall-mount, half-round tank (which was really only useful with the four cylinder reed valve ELF). Though the new tank was listed for the Single and Twin, it held far too much fuel for the Twin and couldn't be mounted with the Single.

The flared exhaust pipe was deleted and replaced by the new $5\frac{1}{2}$ inch long exhaust pipe. This pipe was intended to be bent by the user to best fit his airplane. Gone, also, were the bell-mouthed intake tubes.

The only carburetor used on the reed valve engines was the straight tube with needle valve.

ANOTHER MOVE

The year 1951 brought yet another move for the Calkin family and the ELF Engine Co. An article in the August 3, 1951, Portland newspaper the *Oregonian*, documents the fact that the Kraemer building was sold and "...will be wrecked and the lot used for parking." Later, Dan was given a 30 day eviction notice.

The last engine shipped from the Washington Street location was on January 10, 1952. The building was razed during the early part of 1952.

ELF Engine Factory #3 Approximately 26' x 40' 1952-1956

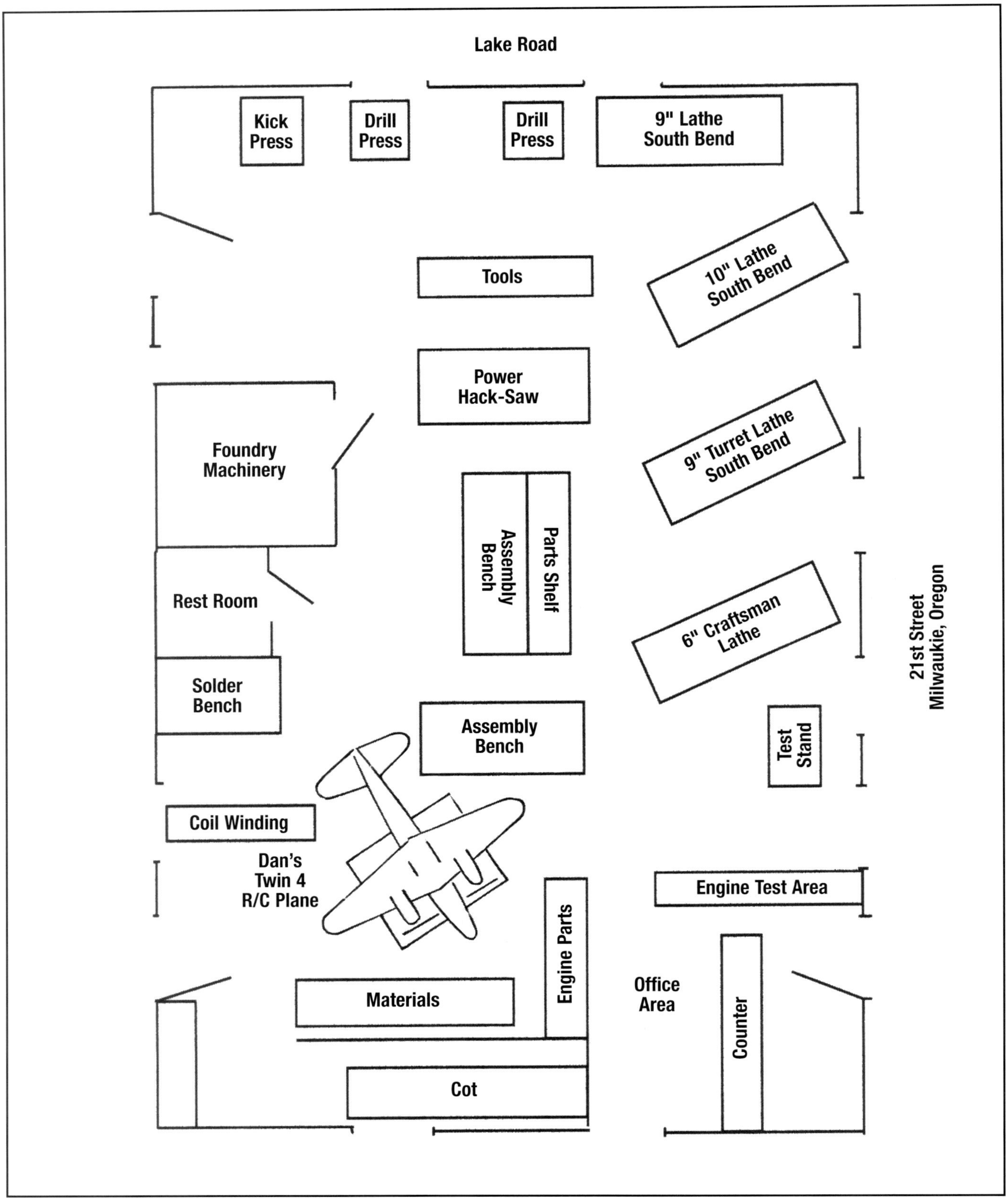

Figure 4-2 The floor plan for the ELF factory on 21st Street in Milwaukie, Oregon. From a sketch by Phil Foster and Cecelia Calkin Jackson, redrawn on a computer by John J. Brown, Jr. Airplane was suspended from ceiling by pulleys .

ELF ENGINE COMPANY

1526 TWENTY-FIRST STREET
MILWAUKIE 22, OREGON

Chapter 4 THE LAST ELF YEARS

TWENTY-FIRST STREET

THE NEW SHOP

For a new shop, Dan's mother bought a small piece of property on the outskirts of Portland and financed the construction of a concrete block building for the ELF Engine Company. Dan paid her $40.00 a month rent. See the floor plan on the opposite page. (Figure 4-2)

The September 28, 1951, *Milwaukie Review* newspaper announces:

Construction began Wednesday on a 1000 square foot cement block building at 21st and Lake Road and will house the ELF Engine Company.

According to Dan Calkin, who owns and operates the building and shop, the company will manufacture and wholesale gasoline engines to power model airplanes. He expects to be operating in the new building in about a month.

Figure 4-1 A 1996 photo of the building Dan Calkin had built in 1952 to house the ELF Engine Company. The entrance to the ELF Engine Company was the door on the left. (Don Edmunds photo)

Figure 4-3 The Calkin family home in Milwaukie, Oregon. (Calkin family photo)

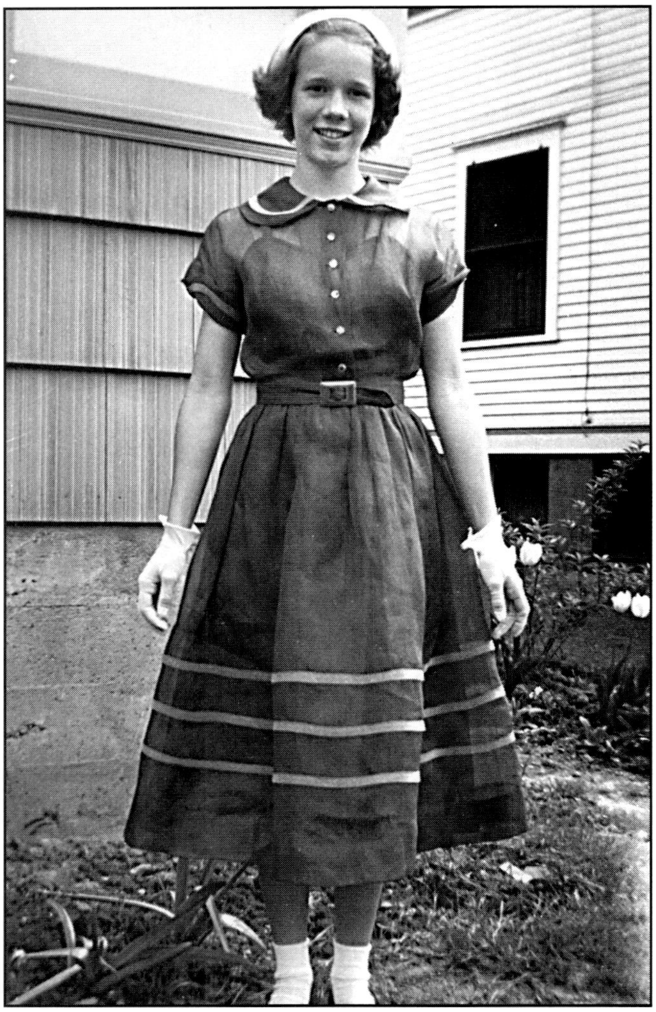

Figure 4-4 Through the ELF years, Cecelia continues to grow. Here she is in 1952. (Calkin family photo)

The building was completed in early December, 1951. It was located at 1526 21st Street (now 11358 21st Street), only a few blocks away from their home. See Figure 4-1. The shop was also only a block away from the Milwaukie Union High School that Cecelia attended.

The first engine was shipped from the 21st Street shop on February 7, 1952. Throughout 1952 the ELF Engine Company produced only Twins and Fours. No single or six cylinder ELFs were made initially. In fact, no more ELF single cylinder engines would ever again be manufactured or sold. The last Single had been sold at the Kraemer building location by August, 1951.

THE NEW HOME

Faced with the necessity of quickly moving both his shop and living quarters from Washington Street, Dan moved his family into a duplex house that his mother had fortunately just finished having built. This was at 706 (now 10404) W. 23rd Street in Milwaukie, Oregon, on the outskirts of Portland. See Figure 4-3. The house was a two story affair with one bedroom upstairs and two downstairs. Dan and Josephine used the upstairs bedroom, while CeCe had one downstairs bedroom and Dan's mother, "Gram," used the other. Both floors had kitchens. Because of his success in making a dust free area at the Washington Street shop, Dan started doing the same at the 23rd Street house. Phil Foster once visited Dan at this home and remarked about how strange the house looked because of the absence of carpeting or any upholstered furniture.

Through the ELF years the children of Dan and Josephine continued to grow. Figure 4-4 shows Cecelia in a new dress she made in her Home Economics class in the Spring of 1952. She was also old enough to be of real assistance in helping Dan make the ELF engines. The Milwaukie High School she attended was only a short block from the ELF factory and she regularly went to the shop every day after school to help.

THE OPEN CAST TIMER

In order to further reduce costs, Dan developed a new timer assembly that required very little machining. The timer case and arm were a single casting. The breaker point assembly used the same parts as the previous round timer. The open cast timer was first produced with a single set of breaker points for the Model 20 engines and first sold on a Twin in May 1952. It was never used on the Singles, although it would certainly fit. The last ELF Single (#5008) had already been sold on July 24, 1951.

A short time later, Dan started production on a different timer casting holding two sets of breaker points for the Model 40 engines. The first sale of a Model 40, (#1163) with the cast open timer, was in November 1952. By this time, most modelers were using glow plug ignition, and sales of spark ignition engines were dwindling. Most spark ignition engines sold were probably operated on glow plugs. See Chapter 10 for more information about timers.

ELF SIX PRODUCTION RESUMES

Dan had a number of inquiries about the six cylinder engine, so in 1953/54 he manufactured a batch of eight engines. With the four engines produced in 1950/51, a total of twelve six cylinder ELFs were made. Ten were sold and two were retained by Dan for his own use. He was using ELF Fours in his radio planes at that time, but thought that a time might come when he would want more power than the Fours provided.

Figure 4-5 Six ELF Sixes in various stages of manufacture. (Calkin family photo)

The eight Model 60 engines manufactured at the 21st Street plant were numbered 1355, 1358, 1363, 1367, 1369, 1373, 1374, 1378. Numbers 1369 and 1373 were retained by Dan for his own use.

Figure 4-5 shows six of the six cylinder engines when they were almost finished. The two engines that Dan kept for his own use were finished at a later time. More photos in Chapter 6.

PLANT AUTOMATION

Dan seems to have reached a point where his costs of operation were very low. All he bought was raw materials and spark/glow plugs. Everything else was made in-house. By this time, Dan had been able to automate enough machines and processes that he was able to produce engines with the help of only Josephine and Cecelia. In a 1952 letter to Carl VanCourt, Dan stated, "I'm working furiously on making the plant automatic which project will not be finished for several weeks yet." Two lathes were set up with servo motors, cams, limit switches, and an automatic tool changer, so the only thing the operator needed to do was to insert the material and press a start button.

The ELF factory was featured in a Milwaukie *Enterprise-Courier* newspaper article of August 19, 1952, and its pictures were used in Figure 4-6, 4-7 and 4-8. Figure 4-6 shows Dan's daughter CeCe holding one of the little connecting rods that she had machined on the automatic lathe. Figure 4-7 catches Josephine at her solder bench. Figure 4-8 shows Dan and CeCe in the shop discussing some aspect of the business.

The Sand Casting Operation

The sand casting operation was similarly automated. Here is what Cecelia said about the sand casting in her August 1, 1994, letter:

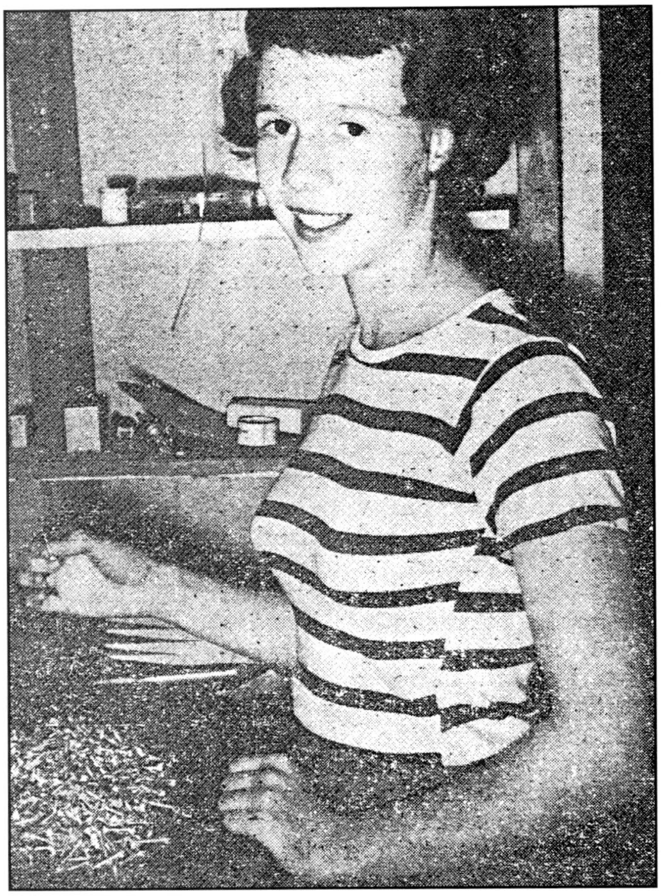

Figure 4-6 Cecelia holding one of the connecting rods that she had machined. (Calkin family scrap-book)

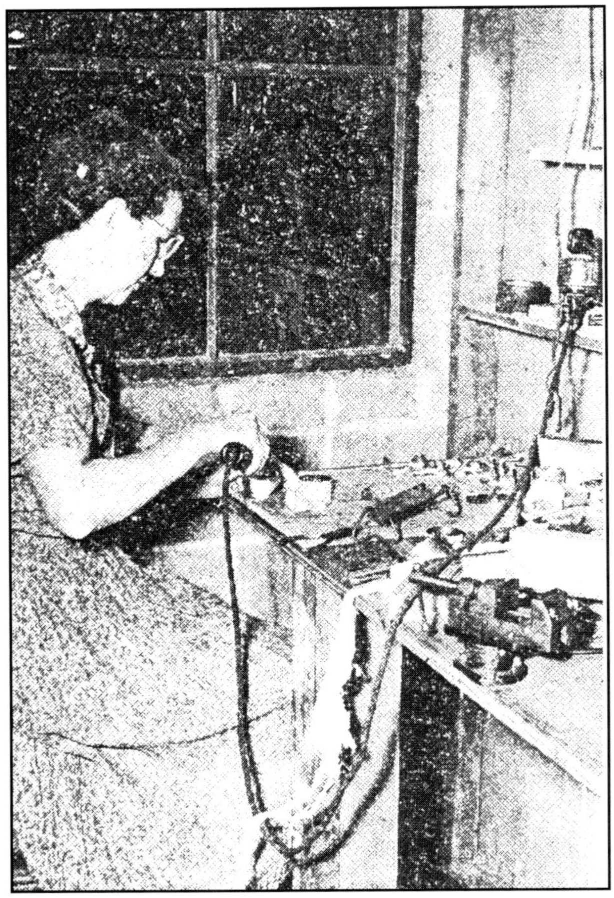

Figure 4-7 Josephine working at the solder bench. (Calkin family scrap-book)

During her high school years Cecelia worked after school making the sand cast molds. Dan automated the sand casting process so all that was needed was placing the moist sand in the mold then pressing the button for the mechanism to compact the sand and gently place it in the tray in the right order for pouring the molten aluminum into the mold. After the metal was cold, the pieces were carefully broken off the "tree." They were inspected and then machined.

Some people have suggested that Dan used the lost wax process for castings since the methods must have been known to him through his father, a dentist. Others have suggested a permanent mold was used. Both are incorrect. Only sand castings were made. Dan did have a "casting machine" to automate the process and assure consistently good results. All the match plate patterns in use at the end of production have been located and all are beautifully made. See Chapter 15 for pictures of the match plates. Only bits and pieces of the casting machine have been located.

The records show that Dan purchased 150 pounds of a #12 aluminum casting alloy on July 14, 1950, and another 200 pounds on May 22, 1951. This alloy contained 8% copper with no other appreciable amounts of other elements.

It is known that Dan worked with very small flasks made of $\frac{1}{4}$ inch aluminum plate, which Cecelia referred to as trays. The match plates were also made of $\frac{1}{4}$ inch aluminum plate with aluminum or brass patterns attached to only one side. This is an uncommon system called symmetrical patterns. For example, both halves of the cylinder pattern were on the top side of the plate. Cecelia would dust ordinary white parting compound on the patterns, then cover the patterns

Figure 4-8 Dan and Cecelia working together in the shop. (Calkin family scrap-book)

with very fine sand, packing the sand around the patterns with her fingers. She would ram up the sand with a small wooden rammer. The patterns were vibrated by an automatic vibrator to provide a little extra clearance to facilitate the removal of the match plate from the sand. The match plate was automatically removed from the sand with a motor driven apparatus to assure that the sand was not damaged by the removal of the pattern. Next she would repeat the above steps, ramming up another flask using the same pattern.

When the two flasks were rammed up, a motor-driven machine would align the two sand molds perfectly. The sprue built into the match plate was half of a $\frac{1}{2}$ inch diameter bar. In order to get the complete sprue, she would ram up two more flasks which fitted against the first two. The sprue halves would now line up exactly, giving a full $\frac{1}{2}$ inch diameter hole down through the sand in which to pour the molten aluminum. Next, another four flasks were rammed up and machine aligned directly above the first set of four. Several additional sets of flasks could be stacked on top of these, producing a stack of castings four or five or more castings high.

Dan always poured the molten metal himself, after carefully checking the color with his practiced eye. The color of the molten metal changes slightly as the metal gets hotter. After skimming off the impurities, Dan would pour the aluminum using a ladle to transfer the molten metal from the 7 inch crucible to the rammed up flasks.

The casting operation was automated to the point that after the flasks were rammed up by hand, the remaining operations, until pouring, were all automatic.

GASKETS

Dan had always prided himself on having a perfect metal-to-metal fit on his crankcases and cylinders, so that no gaskets would be required. The method used to get this very fine mirror-like finish on the mating surface is called lapping. He would rub the machined face of the casting against a flat surface, using a very fine abrasive, in a figure eight pattern until the surface had a mirror finish. In spite of this, it seems that a very few engines did seem to require gaskets. Reed valve engines have less suction than the side port engines and, if there was any leakage at all, the engine would not draw fuel properly. This problem was first noticed in October, 1953, on Model 40 engine #1259. The next occurrence of the problem was in early January, 1954, on another

Model 40 engine, this time on #1255. To solve this problem, Dan put thin (0.004 inch) paper gaskets between the crankcase and each cylinder. A punch and a die set was made to punch out the paper gaskets and, by the end of January, 1954, all ELF engines had gaskets to assure a proper seal between each cylinder and crankcase.

By this time, the old 1937 Packard finally had to be replaced. Dan was without a car for several months, then he bought a used Plymouth. Fortunately, Josephine received a small inheritance that enabled the purchase of a 1953 Packard "Patrician". The first time Cecelia saw the Packard was when her father used it to pick up her and brother Win from summer camp on August 10, 1955.

THE END OF THE ELF YEARS

In early 1955 Dan purchased 25,000 of the 3-48 x 3⁄8 fillister-head Phillips screws used on the ELF engines to attach the cylinder heads to the cylinders. Also purchased were 2750 of the boxes he used to ship engines. These purchases indicate that Dan expected the ELF Engine Company to be in business for many years to come. Who would have expected that the ELF Engine Company would be out of business a year later?

Because Dan was continually plagued with respiratory problems, he couldn't work steadily, and the first few months of 1956 were another bad period. This inability to work may have prompted him to consider working for another company, and probably he would have had fewer sick days away from the damp climate of Portland. There was a serious shortage of engineers in the middle '50s and Dan responded to an ad from Douglas Aircraft Company seeking engineers just to find out what they might offer him.

Dan typically wore a pull-over knit sweater and trousers. The clothes, as well as the shop, were always immaculate. Very often he wore a stocking cap over his hair. When he dressed up in a business suit to go to Douglas Aircraft for a job interview, his son, "Win," was heard to remark, "Look, Daddy's dressed up like a man."

Dan and Josephine both drove to Santa Monica for the job interview. They liked the offer to work at a high paying job, and Dan formally accepted the Douglas Aircraft offer of employment. Josephine was especially pleased with the prospect of moving to Southern California.

Finally, in 1956, after devoting 25 years to the ELF engines, Dan closed the ELF Engine Company. The last engines were shipped on April 11, 1956. At that time, the ELF Engine Company had been in business longer than any other model airplane engine company.

The house on 23rd Street and the shop building on 21st Street were put up for sale in the spring of 1956, and were sold. Dan's mother stayed behind in the Portland area to handle the sale of the properties. She rejoined the family later in Santa Monica, California.

Chapter 5 THE YEARS AFTER ELF

California

A NEW BEGINNING

The offer of employment from Douglas seemed so generous that Dan couldn't refuse it. So, in March 1956 Dan accepted the offer of a position as a controls engineer with the Missile Division of Douglas Aircraft in Santa Monica, California.

Dan traveled to Santa Monica alone while his family stayed behind so CeCe could graduate from high school and Win could finish the first grade. When school ended in June, the family traveled by bus to Santa Monica to rejoin Dan.

Douglas paid all the shipping costs to move Dan and his family to Santa Monica in Southern

Figure 5-1 Win, Josephine and CeCe are shown at the first Calkin home on 24th Street in Santa Monica. (Calkin family photo)

California. All the family belongings, including the grand piano, as well as everything in the ELF shop, were shipped to Santa Monica.

Douglas also paid the bus fare for Josephine, CeCe and Win to travel to Santa Monica.

THE SANTA MONICA HOME

Dan and Josephine looked all around the area for a suitable house to buy, finally settling on a property at 1141 and 1141½ 24th Street in Santa Monica. See Figure 5-1. Notice the Calkin's 1953 Packard "Patrician" sitting in front of the house. They took possession of the property on July 1, 1956.

The property on 24th Street had two houses on it. The main house was a one-story, pink stucco structure and on the back of the lot was another smaller house where Dan's mother would live. All the equipment from the ELF shop was kept in a rented storage unit. One month after moving in, Dan had a garage built on the property for his workshop and storage of the shop equipment.

Figure 5-2 The garage Dan had built behind his home shortly after moving to Santa Monica. The small house next to it was where his mother Gram lived.

A New House

Eight years after moving to Santa Monica, it was obvious that the main house was so badly damaged by termites that it had to be demolished. While their new house was under construction, the family moved into the Lincoln Place Apartments at 1224 Harvard Street in Santa Monica, beginning May 26, 1964. On July 9, 1964, they were able to rent the house at 1127 24th Street, next door to the new house they were having built. They rented there until the construction on their own house was completed. Dan,

Figure 5-3 A recent picture of the house Dan had built to replace the termite-ridden earlier house.

Josephine and Win moved into the new house on January 11, 1965.

Josephine had enjoyed playing the piano for many years, but, after a few years in Santa Monica, she realized that she had lost interest in continuing to do so. As a result, the piano was sold.

CeCe

After moving to California, CeCe enrolled in Pomona College in the fall of 1956. She left home, moving to the college, and attending there full-time for two years with no declared major. While attending Pomona College she met her future husband, Martin Allan Jackson.

Dan felt that since she didn't really have a plan for what she wanted to pursue, she should go to a vocational school and learn some salable skills. CeCe complied with her father's wishes and attended Willis Business School for six months learning typing and other office management skills. After completion of this training, she went to work as a secretary at Douglas Aircraft in Santa Monica, the same plant where her father worked, although they did not work together.

CeCe and Martin were married on November 25, 1960, in Santa Monica, California. He worked for the Arroyo Grande Valley newspapers in California as reporter-photographer until he was drafted into the Army in August 1961. The Army trained him in electronics and he taught missile maintenance and repair at Redstone Arsenal, Alabama, until his discharge in August, 1963.

After working short term for several newspapers in California and Arizona, Martin and CeCe settled down in Hemet, California, where Martin worked for the *Hemet News*. After twelve years there he quit the newspaper business and moved his family to be near his parents. He landed a job with a school district driving a bus at first, then doing electrical and electronic work when the school district learned that he had these skills. He and CeCe also have a small publishing business.

Figure 5-4 Josephine comforting Win after learning of his leukemia. (Calkin family photo)

CeCe and Martin have three children, and these are the only grandchildren of Dan and Josephine Calkin: Allanna Lea was born on May 12, 1963; Ian Allan on June 15, 1966; and Joy Janene on June 9, 1981.

Win

Shortly after moving to Santa Monica, Dan's son, Win, was diagnosed as having leukemia. Figure 5-4 shows Josephine trying to comfort her son.

Win was unable to go to school and missed the second and third grades, but he had learned to read and read everything he could lay his hands on. The following year the school district was able to provide a home teacher, and his schooling became more organized. He was tested and found to be brilliant, like his father. When he was of junior high school age he resumed regular schooling.

Figure 5-5 Photo of Win taken a short time before his untimely death due to the complications of leukemia. (Calkin family photo)

Win was much interested in physics and mathematics as well as history and photography. During his Santa Monica High School years he looked forward to the annual written science and math competition, and each year was one of the top two students in these areas.

In March of 1961, both Junior and Senior E. D. Calkin assembled a Heathkit Stereo Tuner model AJ-30. When finished, the tuner worked, but the dial didn't track properly. Dan and his son correctly diagnosed the problem to be a defective I.F. output transformer. When the replacement transformer arrived and was installed, the tuner worked perfectly.

Win graduated with honors from Santa Monica High School in June 1966, then enrolled at UCLA. He frequently had to miss class for his chemo and radiation treatments, but kept up on his studies and was an honor student. Very few people even knew he had a medical problem. He was often seen riding his bicycle around the campus and the Westwood area and appeared to be healthy. Several of Dan's co-workers donated blood for Win's treatments.

Win was a sophomore and an honor student at UCLA when he died on June 12, 1968, shortly after his 19th birthday. Figure 5-5 shows Win wearing his favorite sport jacket a few months before his death.

The loss of their only son was a devastating blow to both Dan and Josephine. Win had survived for 12 years after the diagnosis of his illness. At the time of his death it was believed that Win had been able to survive leukemia for a longer period than anyone else had previously.

DAN, THE ROCKET SCIENTIST

At Douglas, Dan worked on fuel slosh problems on the Thor missile, as well as the control problems of other missiles requiring higher mathematics in their solution. In a 1976 letter to his granddaughter Allanna, Dan wrote:

> My job as a dynamics specialist in the vehicle controls division consisted of writing equations describing the motion of space vehicles under the action of their control systems. This included the vibration of the vehicle, the sloshing of fuel in the tanks, the effects of the astronauts moving about, wind gusts, air pressure, etc.
>
> The branch of mathematics used is called differential equations. Usually a set of these equations are required for each problem, and they are written in the form (called a matrix) that is readily handled by the big computers. The answers given by the computers are in the form of tables of numbers or graphs which show the motion of the vehicle.

A master's degree is the minimum schooling requirement, most staff members have PhD's.

The company encourages on-the-job training, in fact, requires it for vehicle dynamics specialists.

Doctoral and post-doctoral courses are given each semester in the plant. The company also helps engineers go back to school to obtain advanced degrees.

Dan was highly respected in his field and occasionally conducted classes for this in-plant training. He also tutored the young men in the department. He further commented in the 1976 letter:

The main engineering tool today is the giant electronic computer. The machine itself is a large collection of transistors, memory cells, card readers, tape recorders, printers, etc. This is called the 'hardware'. The machine must be told exactly what to do to solve each problem. This information is placed on cards in the form of punched holes representing numbers. Decks of the cards are called 'software' and they are really more important than the machine itself. Desk calculators and slide rules are used for checking. The slide rule has been almost replaced by the hand held calculator.

ABE GALLAS

Abe is a long time model builder, and was a co-worker of Dan Calkin. Although Abe was familiar with the ELF engines and worked with Dan at Douglas for a long time, he never realized that the two were connected while they worked together.

Abe tells a story of how he first met Dan Calkin. Abe had just graduated as a mechanical engineer from MIT, and was hired by the Douglas missile division where Dan worked. On his first day on the job, Abe's boss showed him a 16 mm movie of the large truck and trailer used to transport a Nike-Hercules missile. The transporter was

Figure 5-6 (Left to Right) Gram, Dan and Josephine. Painting above is of Josephine when she was 24. (Calkin family photo)

Figure 5-7 Josephine and Dan enjoying an outing at Disneyland. (Calkin family photo)

Figure 5-8 The last home of Dan and Josephine Calkin at 16342 Angler Lane in Huntington Beach, California.

undergoing tests at the Aberdeen Proving grounds in Maryland. A serious problem developed when the truck carrying a missile drove across "speed bumps." The trailer would bend up in a "V" that also broke the missile. The U.S. Army naturally was anxious for Douglas to resolve the problem. Abe was asked if he'd ever had any college courses in dynamics, and he admitted he had taken one class on the subject. His boss asked him to look into the problem. Abe freely admitted to me that, at the time, he had no idea at all regarding what could be causing the problem, or what could be done to fix it.

The next day Abe's boss suggested to him that he should go out to the parking lot to examine the missile transporter that was parked there. Abe went out to look, and when he got there he saw a man already looking over the transporter, measuring and taking notes. Since Abe had no clue as to how to proceed, he introduced himself to the note taker and found it was Dan Calkin. Abe told Dan of his assignment, and admitted he had no idea at all of what to do. He asked Dan if he could accompany him to find out what he was doing. Dan explained all about the problem's possible causes and cures.

Abe told me that if he had been pressed for a solution to the transporter problem he would have recommended adding hundreds of pounds of weight to strengthen the trailer.

Dan realized that the problem was caused by a low frequency vibration, and recommended that hundreds of pounds of weight be removed from the trailer in order to raise its resonant frequency and increase the spring rate, allowing the trailer to bend instead of breaking. The company followed Dan's advice, and the problem was solved.

RON SPORING

Another model builder and Douglas colleague of Dan's was Ron Sporing. Ron was in charge of the group that designed the control system for the Zeus missile, a very successful missile. Ron credits Dan Calkin's mathematical model that was used to create the control system design for much of the success of the Zeus.

Ron tells the story that whenever there was a serious problem with the performance of any of the missiles, the design group would all gather around to discuss possible causes and fixes. Very often, Dan had little to say, but his opinions were greatly respected. When someone would say, "What do you think, Dan?" he would quietly outline a series of steps, and they usually followed

Dan's recommendations. When Dan talked, people listened.

Dan would sometimes become a little irritated whenever a project he was working on required some machining. The problem was, that to run a machine, you had to belong to the union. Dan would try to finagle a way to do the machining himself, but was usually not successful. It rankled him that he had to wait for someone else to do the machining when he felt that he could probably do it better and faster.

On May 14, 1970, Dan presented a paper he authored entitled, "Linear Estimation of Structural Parameters From Dynamic Test" to the Industry Coordinating Committee at Las Vegas. He was also a contributor to several other scientific papers.

DAN'S HEALTH PROBLEMS

Dan continued to have health problems until it was finally determined that he had a fungus infection on one of his lungs. An operation was performed to remove half of the affected lung and his health steadily improved after that.

TRANSFER TO HUNTINGTON BEACH

Douglas transferred Dan to the Huntington Beach Missile Facility and the Calkins bought a house near the plant at 16342 Angler Lane in Huntington Beach. They moved in on December 5, 1969. Figure 5-8. Dan's mother, "Gram," continued to live in the little house on the 24th Street property in Santa Monica until her death in June, 1973.

The property in Santa Monica was kept as a rental for several years, then sold.

Dan worked at the Huntington Beach Douglas plant until he retired on May 31, 1972.

A retirement luncheon was held for Dan at the Meadowlark Country Club near the Douglas Plant. Then, a few days later, a retirement party was held at Sam's Sea Food Restaurant in Seal Beach. Several hundred people attended the party including two vice-presidents, two directors and four chief engineers.

Josephine wrote a letter to her daughter CeCe to tell her about the retirement dinner. She said:

> It was WONDERFUL, the affection and esteem in which they held Dan! Person after person came up to Dan saying, 'I don't see how the company is going to get along without you' and 'How am I going to work problems without your help?' and they MEANT it....
>
> I was so PROUD of Dan. He looked so distinguished and received the applause and praise so exactly right. I of course have always known that he was TOPS and it was nice to have others know it too.

Figure 5-9 The last photo of Dan taken 2 weeks before his death. He was at a party preparing to toss a game piece. (Calkin family photo)

Figure 5-10 Josephine Calkin sitting in her kitchen in 1995.

Figure 5-11 Dan Calkin's daughter Cecelia "CeCe" Calkin Jackson. A 1998 photo. (Calkin family photo)

After his retirement, Dan continued doing work for Douglas as an outside consultant.

As time permitted, he worked on designing and building an auto-pilot for a six foot wing span model. The model was powered by two wing-mounted ELF Fours. Dan built an elaborate three section catapult to launch the craft. Strip rubber of the type used to power rubber-powered models was used to provide the power to launch the airplane. He wanted the model to fly as smoothly as full size airplanes, but although it flew beautifully, he was never fully satisfied with its performance. The paint job was never finished, but Dan found that the sanding sealer alone was fuel proof.

Dan died of a massive heart attack on March 2, 1978. He was buried four days later at the Woodlawn Cemetery in Santa Monica, California, next to his mother and his son. The photo in Figure 5-9 was taken just two weeks before his death.

THE SHOP EQUIPMENT

Since moving to Southern California, Dan had kept the equipment from the ELF shop in his garage, but when he found out that he would be transferred to Huntington Beach, he placed an ad in the local newspaper and he sold most of the machinery. He didn't want to move all the machinery to his new home. These sales all occurred between September 11, and October 13, 1968.

After his death, the last lathe, drill press, band saw, meters, drill bits, vice, etc. were sold. The sale occurred on September 9, 1978. His big twin engine radio controlled plane was given away soon afterward, after Ron Sporing had removed the engines and radio from it.

JOSEPHINE

While living in Huntington Beach, Josephine kept very busy as a volunteer at the local public library, and was involved with P.E.O., a national sorority. But in early 1997, she had a series of health problems and it became necessary for her to be in a nursing home to get the care she required. In June 1997, she and her daughter, CeCe, decided that the Calkin home on Angler Lane should be closed up and sold. The house was sold the very first day it was listed for sale. On July eighteenth, CeCe, with her husband Martin and sixteen year old daughter Joy, started emptying out the house. I was invited to come to the house to look for possible material for this book. Many interesting items were found including photographs and ELF company records which were borrowed and copied.

On Christmas Eve, 1998, I last visited Josephine at the nursing home and I brought her a poinsettia plant. I also brought along a copy of the manuscript for this book, finished except for final proofing and some photographic work. I was pleased to be able to show her the (almost) finished book. She seemed pleased also.

After Christmas of 1998, Josephine's health rapidly deteriorated and she passed away on January 23, 1999. She was buried at Woodlawn Cemetery in Santa Monica, with her sister, Pheobe Peck, her son, Edwin "Win" Daniel, Jr, Dan's mother, Etta, and her husband, Edwin Daniel Calkin.

Chapter 6 ELF PRODUCTION ENGINES

ALL THE MODELS AND PRODUCTION CHANGES

Introduction

There is a need to approach this section cautiously and realize that there is an enormous amount of information contained here that may take some time to fully assimilate. The caution is: Don't jump to conclusions. Read the descriptions carefully and refer to items referenced for a more complete understanding of the models, types, production changes and options. A study of Chapters 7-13, where the changes are detailed, would be helpful.

This section may cause some concern among ELF owners. There is a good chance that some owners may have ELF engines considered prime examples that may not conform to the following descriptions. If an engine differs from the descriptions, it doesn't necessarily mean that that engine is not a prime example of an original ELF. Many things can cause an engine to differ.

First of all, re-check the serial number, in good light, with some magnification. Some numbers are very, very difficult to tell apart. Either you or Dan may have mis-read the number. Refer to Appendix A: Serial Numbers.

Your engine may have been returned to the factory for repairs and new, probably later, parts were factory installed to replace the originals. From 1944 to 1966 about 200 engines were returned to ELF for repairs. Unfortunately, only three serial numbers of the 200 engines that were listed in the repairs log book were recorded.

An engine may have been sold to a dealer, then returned to ELF by the dealer and resold again, perhaps at a much later date. A few engines may have sat around the shop for some time before Dan got around to fixing some problem and selling them. Dan probably saw nothing wrong with using old style parts after a new part had been introduced. Most manufacturers did some of that.

The parts on your engine may have been swapped with parts from another engine or another owner may have put on non-original parts, or reproduction parts. For example, any needle valve from any slant plug engine would fit any other slant plug engine. Any of the three kinds of timers would fit any other engine.

Some things are certain, however. Some changes made by the factory required making permanent changes in the tooling. For example, when Dan Calkin started machining the cylinder fins, he modified the patterns and tooling and it was no longer possible for him to make the unmachined cylinders. So any slant plug engine with machined fins must have been made after November 9, 1945; if not the whole engine, at least the cylinder. The earlier unmachined cylinders could not be machined without making them undersized. An engine with unmachined fins could have been sold after this date, however.

A number of similar changes were made; use them as a guide.

The descriptions of the engines start with the original configuration of each engine by Model and Series. Known production changes are listed. Accessories for the engines are described separately in Chapter 12. Additional details of engine construction are described in Chapters 7, 8, 9, 10 and 11. Other related items are described in Chapters 13, 14 and 15.

The ELF tooling and descriptions of some manufacturing procedures are described in Chapter 15. Experimental work on a four stroke engine and radio control experiments are covered in Chapters 16 and 17.

The title of this book is "DAN CALKIN and his ELFS (he never called them ELVES)." Neither did he call his Model 1 ELF a "Corncob" or his pre-war sideport ELF Four a "Goose Egg" Four. The side port Twins and Fours were never referred to as "Pipe Models" and he certainly did not call his new round timer of 1947 a "Banjo" timer. The author considers these slang terms to be inappropriate and they will not be seen elsewhere in the book. While we don't always know what terms Dan Calkin did use, we can at least use the words that are the most descriptive without being slang.

Figure 6-1 A basket of ELFs. These are some of the ELF engines used to produce this book.

Model 1 ELFs									
Description page	—	—	—	—	—	—	—	89	
Engines produced	—	—	—	—	—	—	—	400	400
Start shipping	—	—	—	—	—	—	—	7/29/35	
End Production	—	—	—	—	—	—	—	1939	
Slant Plug ELFs									
Porting	Side Port	Side Port	Side Port	Side Port	Side Port	Reed Valve	Reed Valve	Reed Valve	
Reinforced Front Case	No	Yes	Yes	Yes	Yes	Yes	Yes	Yes	
Cylinder fins machined	No	No	Yes	Yes	Yes	Yes	Yes	Yes	
TImer Assembly	1st Cast	1st Cast	1st Cast	Round	No, Glow	Round	2nd Cast	No, Glow	
Model 10 ELFs	Series 1	Series 2	Series 3A	Series 3B	Series 3G	Series 4B	Series 4C	Series 4G	
Description page	91	93	95	97	99	101	—	103	
Engines produced	34	348	125	111	2	6	0	4	630
Start shipping	10/19/39	3/29/41	11/9/45	8/26/46	3/14/50	11/2/50	—	11/18/50	
End Production	12/18/40	10/12/45	8/12/46	4/20/50	3/21/50	8/6/51	—	7/24/51	
Model 20 ELFs	Series 1	Series 2	Series 3A	Series 3B	Series 3G	Series 4B	Series 4C	Series 4G	
Description page	105	107	109	111	113	115	117	119	
Engines produced	34	254	156	122	3	21	60	184	834
Start shipping	3/28/40	10/31/40	12/18/45	10/11/46	2/15/50	10/16/50	5/19/52	10/6/50	
End Production	10/5/40	10/30/45	9/12/46	4/24/50	3/4/50	4/7/52	11/14/55	3/6/56	
Model 40 ELFs	Series 1	Series 2	Series 3A	Series 3B	Series 3G	Series 4B	Series 4C	Series 4G	
Description page	—	121	—	123	125	127	129	131	
Engines produced	0	55	0	50	8	46	66	55	280
Start shipping	—	5/30/41	—	5/9/47	12/2/49	8/23/50	11/7/52	8/23/50	
End Production	—	11/12/42	—	6/29/50	5/2/50	4/7/52	12/20/55	4/11/56	
Model 60 ELFs	Series 1	Series 2	Series 3A	Series 3B	Series 3G	Series 4B	Series 4C	Series 4G	
Description page	—	—	—	—	—	—	—	133	
Engines produced	0	0	0	0	0	0	0	12	12
Start shipping	—	—	—	—	—	—	—	3/11/51	
End Production	—	—	—	—	—	—	—	9/20/54	
Total Produced	—	—	—	—	—	—	—	—	2158

Figure 6-2 ELF engine models, series, production dates and production figures.

Note Wartime Shutdown: No Model 10s shipped from August 10, 1942 to July 17, 1944. No Model 20s from April 5, 1942 to June 3 1944. No Model 40s shipped from November 2, 1942 to May 9, 1947.

Start Here

Refer to the chart above for the correct page number, if you are looking for information about a particular ELF. If the engine is any of the slant plug variety, **answer these four questions first**: Is it a side port or reed valve engine? Is there a reinforced crankcase? Does it have machined cylinder fins? What kind of timer does it have, if any? With that you can look straight down to the page number for that engine and all its production changes, options and accessories. Go for it.

Figure 6-3 Model 1, Top; #6, Middle; #339, Bottom; #213 as shipped, ready to run.

ELF SINGLE, MODEL 1 ENGINES

Specifications

Bore	0.545 inches
Stroke	19/32 inches
Displacement	0.138 cubic inches (2 1/4 cc)
Type	One cylinder, two-stroke, four-port
Overall height	3 1/2 inches
Overall length	2 7/8 inches
RPM	3200-6000, 3500 using an ELF 12 x 6-1/2 propeller
Horsepower	1/30
Weight	4 ounces
Production period	July 29, 1935 to 1939
Quantity produced	About 400 (Known serial numbers range from 6 to 432)
Spark plug	Fig. 11-4 ELF manufactured using Champion insulator, fixed ground electrode, 1/4-32 thread
Cylinder	Fig. 7-1 Sand cast with cored-in ports, unmachined fins, ELF logo on right side
Cylinder liner	Fig. 7-2 Thin steel liner, drilled ports
Piston	Fig. 7-3 Sand cast aluminum with three cast iron rings, two upper, one lower
Wrist pin	Hollow drill rod, pinned to piston
Connecting rod	Fig. 7-4 Cast aluminum, attached to shaft using a riveted brass clip
Crankcase	Fig. 8-1 Two identical aluminum castings, vertically split, engine mounts tapped 6-32
Crankshaft	Fig. 8-2 Tool steel with pressed-on steel counter weights.
Bearings	Fig. 8-2 Two main plain bearings, 2 bronze thrust washers
Prop drive washer	Fig. 8-3 Steel, concave rear face, 5 degree taper to crankshaft
Prop front washer	Fig. 8-3 0.020" thick aluminum
Prop nut	Fig. 8-3 8/32 hex steel, unplated
Timer	Fig. 10-1 Adjustable timer, two-piece stamped round case, cast aluminum arm
Timer cam	Fig. 8-2 Hardened steel, shrunk-on rear end of crankshaft
Carburetor	Fig. 9-1 Gravity feed, brass float chamber, wood float for upright operation
Needle valve	Fig. 9-22 Horizontal needle valve, large brass ratchet wheel, brass ratchet spring
Fuel tank	Fig. 9-3 External cylindrical 0.005" brass, 1" diameter, 1" length with wood filler plug

Production Changes

Carburetor	Fig. 9-1,2 About February 1938, changed to 60 degree angle (about serial number 100)
Spark plug	Fig. 11-7 February 1938, changed to Champion V-2 (about serial number 100)

Options/Accessories

Coil	Fig. 10-6, 10-7
Condenser	Fig. 10-13, #1 and #2
Flywheel	See Chapter 12
Carburetor	Fig. 9-2 Inverted carburetor became available soon after start of production.
Carburetor	Fig. 9-4 Feb. 1938 Optional inverted carburetor with celluloid tank, no float chamber.
Propeller	Fig. 13-1 12" x 6-1/2 prop

Note

Model 1 engines were sold bolted to a plywood mount complete with battery holder, coil, condenser, tank and prop. The engines were all wired, ready to run. See opposite page.
Some modelers removed the float and used the float chamber as a gas tank.

Figure 6-4 Model 10, Series 1 #549

ELF SINGLE, MODEL 10, Series 1 ENGINES

Side Port, Unmachined Fins, SMOOTH NOSE, Offset Cast Timer, BEAM MOUNTING

Specifications

Bore	0.468 inches
Stroke	0.564 inches
Displacement	0.097 cubic inches
Type	One cylinder, two-stroke, four-port
Overall height	2 5/8 inches
Overall length	3.975 inches
Overall width	1 7/16 inches
RPM	7500 with 8 x 5 3/4 propeller
Horsepower	1/12
Weight	3 ounces
Production period	October 19, 1939 to December 18, 1940
Quantity produced	34. Most were recalled and replaced by factory with Model 10, Series 2 engines
Known s/n range	502-563
Spark plug	Champion V-3
Cylinder head	Fig. 7-5,6 Cast aluminum, seven fin, spark plug slanted at 15 degrees to horizontal
Cylinder	Fig. 7-7,8 #1 Sand cast aluminum, unmachined fins, 0.25 inch exhaust port, small bypass, machined flange
Cylinder liner	Fig. 7-9 #1 Thin steel liner, three square (caged) exhaust ports
Piston	Fig. 7-11 #1 Cast aluminum, transfer slot in bypass side, no piston rings
Wrist pin	Fig. 7-13 Solid 9/32 inch diameter drill rod, ends exposed
Connecting rod	Fig. 7-13 Machined aluminum, attached to shaft with clipped-on phosphor bronze strap
Crankcase, front	Fig. 8-4,5 Cast aluminum, no reinforcing webs, ELF logo on front, beam mounts
Crankcase, rear	Fig. 8-6,7 Serial number above crankshaft, beam motor mounts tapped for 3-48 screws
Crankshaft	Fig. 8-13 Tool steel with two soldered-on lead counter weights/shroud, 8-32 thread for prop nut
Bearings	Two main plain bearings with plain thrust bearings
Prop drive washer	Fig. 8-19 #1 3/4" dia. aluminum, concave rear face, 5 degree taper to crankshaft, knurled face
Prop front washer	Fig. 8-20 #1 3/4" diameter, tapered thin aluminum
Prop nut	Fig. 8-21 #1 5/16 hex aluminum, machined to spinner point
Timer	Fig. 10-2 #1 Offset cast aluminum case, enclosed points, adjustable, no-drag timer
Timer arm	Fig. 10-2 #1 Some had knurled timer arm on cast boss, most had no timer arm
Timer cam	Formed piano wire
Carburetor	Fig. 9-5 Straight intake tube, square-top air filter, 1/8 inch filler hole, cadmium plated
Needle valve	Fig. 9-23 First style, without wire extender
Fuel tank	Fig. 9-5 Celluloid, 1/8 oz., 0.8 inch diameter, no center post, crimped to top with gasket

Production Changes

Fuel tank	Date unknown — Glued-in center post, screw under air filter
Timer	Fig. 10-2 #3 Date unknown — Boss for timer arm removed from casting.
Timer arm	Fig. 10-2 #3 Date unknown — Straight, stepped, un-knurled, attaches to clamp screw, but most engines had no timer arm.

Options/Accessories

Coil	Fig. 10-8
Condenser	Fig. 10-13 #2
Exhaust pipe	Fig. 12-3 #1
Propeller	Fig. 13-4

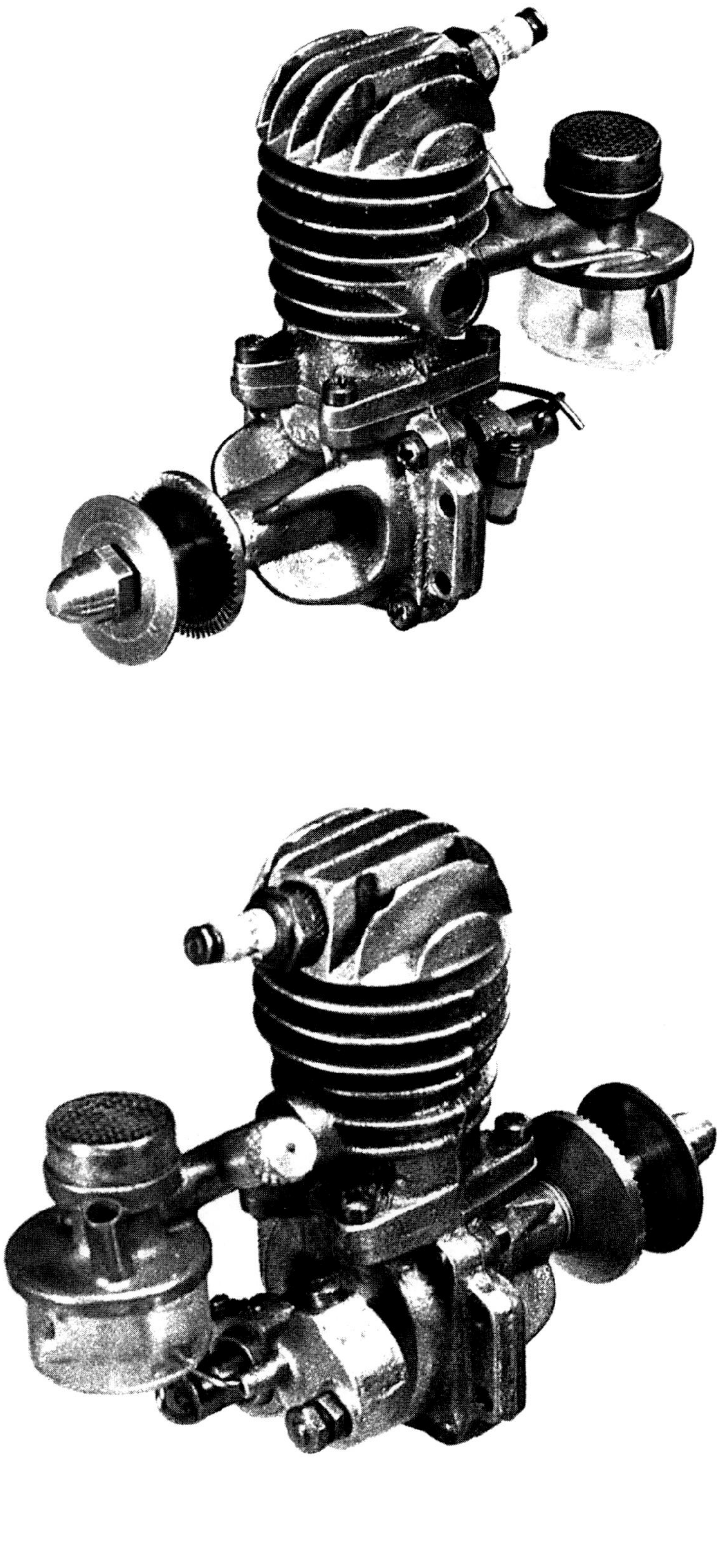

Figure 6-5 Model 10, Series 2 #153

ELF SINGLE, MODEL 10, Series 2 ENGINES

Side Port, Unmachined Fins, WEBBED NOSE, Offset Cast Timer, RADIAL MOUNTING

Specifications

Bore	0.468 inches
Stroke	0.564 inches
Displacement	0.097 cubic inches
Type	One cylinder, two-stroke, Changed from four-port to three-port
Overall height	2 5/8 inches
Overall length	3.975 inches
Overall width	1 7/16 inches
RPM	7500-8000 with 8x6 propeller
Horsepower	1/12
Weight	3 ounces
Production period	March 29, 1941 to August 10, 1942 and June 27, 1944 to October 12, 1945
Quantity produced	348
Spark plug	Champion V-3
Cylinder head	Fig. 7-5,6 Same as Model 10-1
Cylinder	Fig. 7-7,8 #2 New — Bypass enlarged, bypass area fins shallower, 3 internal holes in flange
Cylinder liner	Fig. 7-9 #2 New — Transfer port eliminated
Piston	Fig. 7-11 #2 New — Same as Model 10-1, but transfer slot eliminated
Wrist pin	Fig. 7-13 Same as Model 10-1
Connecting rod	Fig. 7-13 Same as Model 10-1
Crankcase, front	Fig. 8-4,5 New — Four reinforcing webs added, beam mounts eliminated, radial mounts added
Crankcase, rear	Fig. 8-6,7 New — Beam mounts eliminated, new radial mount
Crankshaft	Fig. 8-13 New — Thinner (lighter weight) counter-weights
Bearings	Same as Model 10-1
Prop drive washer	Fig. 8-19 #2 New — Beveled rear face
Prop front washer	Fig. 8-20 #1 Same as Model 10-1
Prop nut	Fig. 8-21 #1 Same as Model 10-1
Timer	Fig. 10-2 #3 Same as Model 10-1 prod. change — Boss removed, arm attached to clamp screw
Timer arm	Fig. 10-2 #3 Same as Model 10-1 production change — but most shipped without arm
Timer cam	Same as Model 10-1
Carburetor assembly	Fig. 9-6 New — Short brass filler tube, square-top air filter
Needle valve	Fig. 9-23 Same as Model 10-1
Fuel tank	Fig. 9-6 Continue Model 10-1 production change — Center post glued-in, screw under air filter

Production Changes

Piston	Fig. 7-11, #3 June 3, 1944 — Two-piece, screwed together
Wrist Pin	Fig. 7-13 June 3, 1944 — Shortened for new piston
Carburetor	Fig. 9-25 1941 — Rounded top on air filter, both types used until production stopped in 1942.
Carburetor	Fig. 9-25 June 3, 1944 — Only round-top filters used

Options/Accessories

Coil	Fig. 10-8 until 1942, Fig. 10-10 from 1944
Condenser	Fig. 10-13 #2, #3, #4
Motor mounts	Fig. 12-7
Exhaust pipe	Fig. 12-3 #2
Exhaust pipe	Fig. 12-3 #4 New — June 27, 1944 — Flared pipe for 0.275" exhaust
Propeller	Fig. 13-4 until 1942, then Fig. 13-5

Figure 6-6 Model 10, Series 3A #314

ELF SINGLE, MODEL 10, Series 3A ENGINES

Side Port, MACHINED FINS, Webbed Nose, OFFSET CAST TIMER

Specifications

Bore	0.468 inches
Stroke	0.564 inches
Displacement	0.097 cubic inches
Type	One cylinder, two-stroke, three-port
Overall height	2 5/8 inches
Overall length	3 3/4 inches
Overall width	1 7/16 inches
RPM	7500 with 8 x 6 propeller
Horsepower	1/12
Weight	3 ounces
Production period	November 9, 1945 to August 12, 1946
Quantity produced	114
Spark plug	Champion V-3
Cylinder head	Fig. 7-5,6 New — Slightly enlarged
Cylinder	Fig. 7-7, 8 #3 New — Slightly enlarged, machined (turned) top three fins and exhaust port
Cylinder liner	Fig. 7-9 #2 Same as Model 10-2
Piston	Fig. 7-11 #3 Same as Model 10-2 production change — 2-piece piston, screwed-together
Wrist pin	Fig. 7-13 Same as Model 10-2 production change — Shortened for new 2-piece piston
Connecting rod	Fig. 7-13 Same as Model 10-1
Crankcase, front	Fig. 8-4,5 Same as Model 10-2
Crankcase, rear	Fig. 8-6,7 Same as Model 10-2
Crankshaft	Fig. 8-13 Same as Model 10-2
Bearings	Same as Model 10-1
Prop drive washer	Fig. 8-19 #2 Same as Model 10-2
Prop front washer	Fig. 8-20 #1 Same as Model 10-1
Prop nut	Fig. 8-21 #1 Same as Model 10-1
Timer	Fig. 10-2 #3 Same as Model 10-2 prod. change — Offset cast timer with boss removed
Timer arm	Fig. 10-2 #3 Same as Model 10-2 production change — Attached to clamp screw, all engines
Timer cam	Same as Model 10-1
Carburetor	Fig. 9-6 Same as Model 10-2 production change — Rounded top air filter
Needle valve	Fig. 9-23 Same as Model 10-1
Fuel tank	Fig. 9-6 Same as Model 10-1 production change — Glued-in center post, screw under air filter

Production Changes

Connecting rod	Fig. 7-13 March 1946 — Thickness increased, centered exactly
Cylinder liner	Fig. 7-9 #3 April 27, 1946 — Slotted exhaust, two bars removed
Crankshaft	Fig. 8-13 May 26, 1946 — Crankshaft shortened for longer prop nut
Prop drive washer	Fig. 8-19 #3 May 26, 1946 — 5/8 inch diameter
Prop front washer	Fig. 8-20 #2 May 26, 1946 — 5/8 inch diameter
Prop nut	Fig. 8-21 #2 May 26, 1946 — Extended with threaded sleeve

Options/Accessories

Coil	Fig. 10-10
Condenser	Fig. 10-13
Motor mounts	Fig. 12-7
Exhaust pipe	Fig. 12-3 #3
Propeller	Fig. 13-5

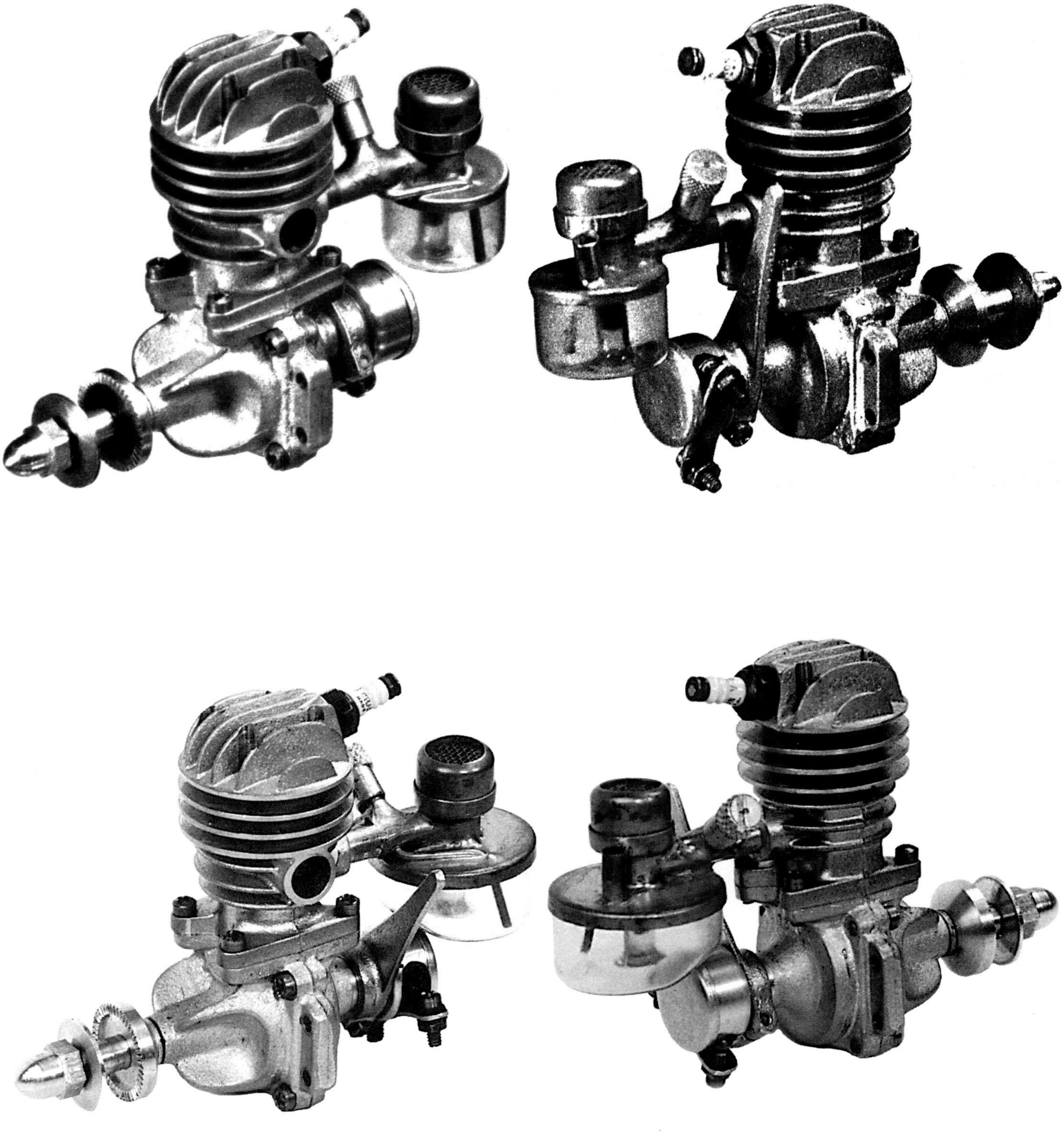

Figure 6-7 Model 10, Series 3B Top #1562 with early small tank. Bottom #1562 with larger double beveled tank.

ELF SINGLE, MODEL 10, Series 3B ENGINES

Side Port, MACHINED FINS, Webbed Nose, ROUND TIMER

Specifications

Bore	0.468 inches
Stroke	0.564 inches
Displacement	0.097 cubic inches
Type	One cylinder, two-stroke, three-port
Overall height	2 5/8 inches
Overall length	3 3/4 inches
Overall width	1 7/16 inches
RPM	7500 with 8 x 6 propeller, 8000 after August 1948 with enlarged exhaust
Horsepower	1/12
Weight	3 ounces
Production period	August 26, 1946 to April 20, 1950
Quantity produced	101
Spark plug	Champion V-3
Cylinder head	Fig. 7-5,6 Same as Model 10-3A
Cylinder	Fig. 7-7,8 #3 Same as Model 10-3A production change
Cylinder liner	Fig. 7-9 #3 Same as Model 10-3A production change — Slotted exhaust
Piston	Fig. 7-11 #3 Same as Model 10-2 production change — 2 piece screwed-together
Wrist pin	Fig. 7-13 Same as Model 10-2 production change — Shortened for new piston
Connecting rod	Fig. 7-13 Same as Model 10-3A production change — Thickness increased, centered exactly
Crankcase, front	Fig. 8-4,5 New — Front case interior modified for use with new 6 ball thrust bearing
Crankcase, rear	Fig. 8-6,7 Same as Model 10-2
Crankshaft	Fig. 8-13 Same as Model 10-3A production change — Shortened for longer prop nut
Bearings	Fig. 8-17 New — 6 ball thrust bearing
Prop drive washer	Fig. 8-19 #3 Same as Model 10-3A production change — 5/8 inch diameter
Front washer	Fig. 8-20 #2 Same as Model 10-3A production change — 5/8 inch diameter
Prop nut	Fig. 8-21 #2 Same as Model 10-3A production change — Extended with threaded sleeve
Timer	Fig. 10-4 New — Round machined aluminum timer, snap on cover
Timer arm	Fig. 10-4 New — Cast aluminum arm
Timer cam	New — Machined steel cam, split, pressed-on shaft
Carburetor	Fig. 9-6 Same as Model 10-2 production change — Rounded top air filter
Needle valve	Fig. 9-13,23 Same as Model 10-1
Fuel tank	Fig. 9-6 Same as Model 10-1 production change — glued-in center post

Production Changes

Cylinder liner	Fig. 7-9 #3 September 5, 1946 — Hardened cylinder liners
Crankshaft	Fig. 8-13 Date Unknown — Holes for wire cam eliminated
Carburetor	Fig. 9-7 February 18, 1947 — Large tank top for large double beveled celluloid tank (like Model 20)
Carburetor	April 1947 — Change from cadmium plating to tin plating
Cylinder	Fig. 7-7,8 #3 January 1, 1948 — Cylinder fins milled instead of turned
Cylinder	Fig. 7-7,8 #3 August 19, 1948 — Larger 0.275 inch diameter exhaust port
Piston	Fig. 7-11 #4, 7-11 October 14, 1948 — Two-piece pressed-together pistons
Cylinder	Fig. 7-7,8 #3 October 5, 1949 — Rounded fins

Options/Accessories

Coil	Fig. 10-10
Condenser	Fig. 10-13 #3,4
Motor mount	Fig. 12-7
Exhaust pipe	Fig. 12-3 #3 Flared to fit 0.250 exhaust
Exhaust pipe	Fig. 12-3 #4 August 19, 1948 — Flared pipe to fit 0.275 inch exhaust
Propeller	Fig. 13-5
Crankshaft extension	Fig. 12-8 New
Metal tank	Fig. 9-8 New — Stamped aluminum replacement for acetate tank
Carburetor	Fig. 9-9 New — Tankless with air cleaner
Carburetor	Fig. 9-10 New — Tankless with bell-mouthed intake
Double timer	No photo available
Fuel pump	Fig. 12-10 New

Figure 6-8 Model 10, Series 3G #1578 with air filter, tankless carburetor and bell-mouthed carburetor.

ELF SINGLE, MODEL 10, Series 3G ENGINES

Side Port, Machined Fins, Webbed Nose, NO TIMER (GLOW)

Specifications

Bore	0.468 inches
Stroke	0.564 inches
Displacement	0.097 cubic inches
Type	One cylinder, two-stroke, three-port
Overall height	2 5/8 inches
Overall length	3 3/4 inches
Overall width	1 7/16 inches
RPM	9000 with 8x6 propeller
Horsepower	1/12
Weight	3 ounces
Production period	March 14, 1950 to March 21, 1950
Quantity produced	2 (#1566, 1578)
Glow plug	Champion VG-3
Cylinder head	Fig. 7-5,6 Same as Model 10-3A
Cylinder	Fig. 7-7,8 #3 Same as Model 10-3B prod. change — Rounded, milled fins, 0.275" exhaust port
Cylinder liner	Fig. 7-9 #3 Same as Model 10-3A production change — Slotted exhaust, hardened
Piston	Fig. 7-11 #4 Same as Model 10-3B production change — Two-piece pressed-together
Wrist pin	Fig. 7-13 Same as Model 10-2 production change — Shortened for 2-piece piston
Connecting rod	Fig. 7-13 Same as Model 10-3A production change — Thickness increased, centered exactly
Crankcase, front	Fig. 8-4,5 Same as Model 10-3B — Machined for 6-ball thrust bearing
Crankcase, rear	Fig. 8-6,7 Same as Model 10-2
Crankshaft	Fig. 8-13 Same as Model 10-3A prod. change — Shortened for longer prop nut, no wire cam holes
Bearings	Fig. 8-18 Same as Model 10-3B — 6-ball thrust bearing
Prop drive washer	Fig. 8-19 #3 Same as Model 10-3A production change — 5/8 inch diameter
Prop front washer	Fig. 8-20 #2 Same as Model 10-3A production change— 5/8 inch diameter
Prop nut	Fig. 8-21 #2 Same as Model 10-3A production change — Extended with threaded sleeve
Timer	None
Timer arm	None
Timer cam	None
Carburetor	Fig. 9-8, 9-9 or 9-10. See Options
Needle valve	Fig. 9-23 Same as Model 10-1

Production Changes

None

Options/Accessories

Motor mounts	Fig. 12-7
Exhaust pipe	Fig. 12-3 #4
Carburetor	Fig. 9-8 Carburetor with metal tank
Carburetor	Fig. 9-9 Tankless with air filter
Carburetor	Fig. 9-10 Tankless with bell-mouthed intake
Propeller	Fig. 13-5
Crankshaft extension	Fig. 12-8
Fuel pump	Fig. 12-10 (Round timer case needed for mounting)

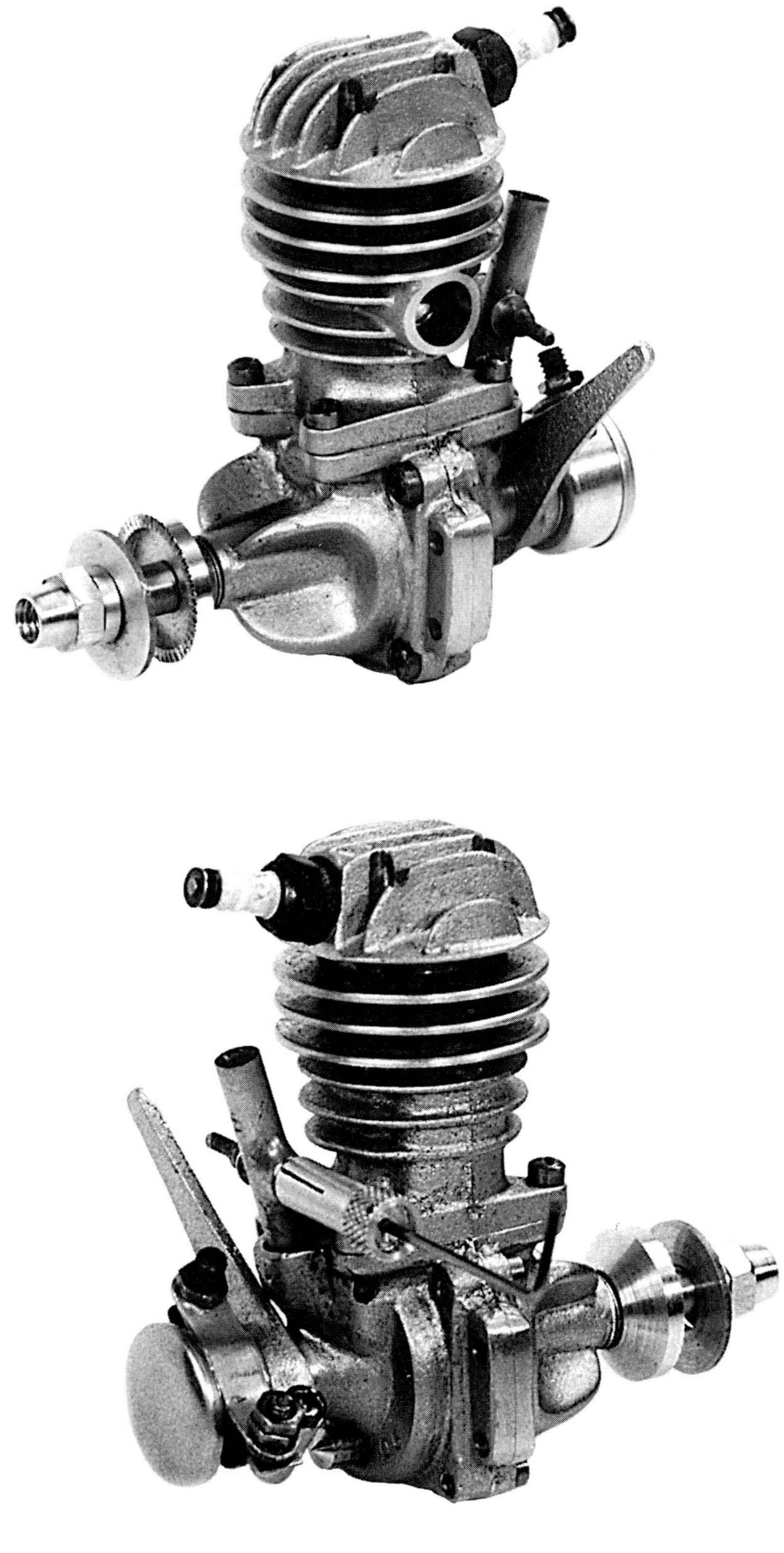

Figure 6-9 Model 10, Series 4B #5006

ELF SINGLE, MODEL 10, Series 4B ENGINES (No Series 4A Singles made)

REED VALVE, Machined Fins, Webbed Nose, ROUND TIMER

Specifications

Bore	0.468 inches
Stroke	0.564 inches
Displacement	0.097 cubic inches
Type	One cylinder, two-stroke, change from three-port to two-port
Overall height	3 3/4 inches
Overall length	3 inches
Overall width	1 7/16 inches
RPM	9000 with 8x6 propeller
Horsepower	Unknown
Weight	3.3 ounces
Production period	November 2, 1950 to August 6, 1951
Quantity produced	6 (Engines # 5000, 5002, 5004, 5005, 5006, 5009)
Spark plug	Champion V-3 or AC 4H
Cylinder head	Fig. 7-5,6 Same as Model 10-3A
Cylinder	Fig. 7-7,8 #3 Same as Model 10-3B prod. change except intake port plugged #5005
Cylinder liner	Fig. 7-9 #4 New — Two-port liner
Piston	Fig. 7-11 #4 Same as Model 10-3B production change — 2-piece, pressed-together
Wrist pin	Fig. 7-13 Same as Model 10-2 production change — Shortened for 2-piece piston
Connecting rod	Fig. 7-13 Same as Model 10-3A production change — Thickness increased, centered exactly
Crankcase, front	Fig. 8-4,5 Same as Model 10-3B production change — Machined for 6-ball thrust bearing
Crankcase, rear	Fig. 8-6,7 New — Houses near vertical intake tube and reed valve
Crankshaft	Fig. 8-13 Same as Model 10-3B production change — Shortened for longer prop nut
Bearings	Fig. 8-17 Same as Model 10-3B — 6-ball thrust bearing
Prop drive washer	Fig. 8-19 #3 Same as Model 10-3A production change
Prop front washer	Fig. 8-20 #2 and 3 Same as Model 10-3A production change but with 0.100 thick washer added
Prop nut	Fig. 8-21 #3 New — Drilled and tapped through
Timer	Fig. 10-4 Same as Model 10-3B — Round machined aluminum timer
Timer arm	Fig. 10-4 Same as Model 10-3B — Cast aluminum arm
Timer cam	Same as Model 10-3B — Machined steel cam, split, pressed-on shaft
Carburetor	Fig. 9-11 New — Straight tube with needle valve
Needle valve	Fig. 9-24 New — Second type with wire extender

Production Changes

Cylinder	Fig. 7-7,8 #4 March 26, 1951 — New 2-port, no intake port (#5000, 5002, 5004, 5006, 5009)

Options/Accessories

Coil	Fig. 10-10
Condenser	Fig. 10-13
Motor mounts	Fig. 12-7
Exhaust pipe	Fig. 12-3 #5
Propeller	Fig. 13-5
Crankshaft extension	Fig. 12-8
Double timer (2 speed)	No photo available

Figure 6-10 Model 10, Series 4G #5006

ELF SINGLE, MODEL 10, Series 4G ENGINES (No Series 4C Singles made)

REED VALVE, Machined Fins, Webbed Nose, NO TIMER (GLOW)

Specifications

Bore	0.468 inches
Stroke	0.564 inches
Displacement	0.097 cubic inches
Type	One cylinder, two-stroke, change from three-port to two-port
Overall height	2 5/8 inches
Overall length	2.8 inches
Overall width	1 7/16 inches
RPM	9000 with 8x6 propeller
Horsepower	Unknown
Weight	3 ounces
Production period	November 18, 1950 to July 24, 1951
Quantity produced	4 (Engines # 5001, 5003, 5007, 5008)
Glow plug	Champion V-G3, AC 4G or Auto-lite M4G
Cylinder head	Fig. 7-5,6 Same as Model 10-3A
Cylinder	Fig. 7-7,8 #3 Same as Model 10-3B prod. change except intake port plugged #5001, 5003, 5007
Cylinder liner	Fig. 7-9 #4 Same as Model 10-4B — Two-port liner
Piston	Fig. 7-11 #4 Same as Model 10-3B production change — 2-piece, pressed-together
Wrist pin	Fig. 7-13 Same as Model 10-2 production change — Shortened for 2-piece piston
Connecting rod	Fig. 7-13 Same as Model 10-3A production change — Thickness increased, centered exactly
Crankcase, front	Fig. 8-4,5 Same as Model 10-3B — Machined for 6-ball thrust bearing
Crankcase, rear	Fig. 8-6,7 Same as Model 10-4B — Houses near vertical intake tube, reed valve
Crankshaft	Fig. 8-13 Same as Model 10-3A production change — Shortened for longer prop nut
Bearings	Fig. 8-17 Same as Model 10-3B — 6-ball thrust bearing
Prop drive washer	Fig. 8-19 #3 Same as Model 10-3A production change — 5/8 inch diameter
Prop front washer	Fig. 8-20 #2 and #3 Same as Model 10-4B — 0.100 thick washer added
Prop nut	Fig. 8-21 #3 Same as Model 10-4B — but drilled and tapped through
Timer	None
Timer arm	None
Timer cam	None
Carburetor	Fig. 9-11 Same as Model 10-4B — Straight tube with needle valve
Needle valve	Fig. 9-24 Same as Model 10-4B — Second type with wire extender

Production Changes

Cylinder	Fig. 7-7,8 #4 March 26, 1951 — 2-port cylinder without intake port (used on #5008 only)

Options/Accessories

Motor mounts	Fig. 12-7
Exhaust pipe	Fig. 12-3 #5
Propeller	Fig. 13-5
Crankshaft extension	Fig. 12-8

Figure 6-11 Model 20, Series 1 #T-7

ELF TWIN, MODEL 20, Series 1 ENGINES

Side Port, Unmachined Fins, SMOOTH NOSE, Offset Cast Timer

Specifications

Bore	0.468 inch diameter
Stroke	0.564 inches
Displacement	0.198 cubic inches
Type	Two cylinder, two-stroke, four-port
Overall height	2 1/16 inches
Overall length	4 3/32 inches
Overall width	4 1/8 inches
RPM	7500 using a 10-1/2 x 6 propeller
Horsepower	1/6
Weight	5 ounces
Production period	March 28, 1940 to October 5, 1940
Quantity produced	32, probably most recalled and replaced by factory with Model 20, Series 2 engines
Spark plugs	Champion V-3
Cylinder heads	Fig. 7-5,6 Same as Model 10-1
Cylinders	Fig. 7-7,8 #1 Same as Model 10-1
Cylinder liners	Fig. 7-11 #1 Same as Model 10-1
Pistons	Fig. 7-10 #1 Same as Model 10-1
Wrist pins	Fig. 7-13 Same as Model 10-1
Connecting rods	Fig. 7-13 Same as Model 10-1
Crankcase, front	Fig. 8-8 New — Cast aluminum, no reinforcing webs, ELF logo on front, radial mount
Crankcase, rear	Fig. 8-9 New — Serial number on rear opposite logo, 2 flanges for radial mounting
Crankshaft	Fig. 8-14 New — Tool steel with two soldered-on lead counter weights/shroud 8-32 thread
Bearings	Fig. 8-17 New — Two plain main bearings, 12-ball thrust bearing
Prop drive washer	Fig. 8-19 #1 Same as Model 10-1
Prop front washer	Fig. 8-20 #1 Same as Model 10-1
Prop nut	Fig. 8-21 #1 Same as Model 10-1
Timer	Fig. 10-2 #1 Same as Model 10-1 offset cast timer
Timer arm	Fig. 10-2 #2 Same as Model 10-1 production change, but used on all engines
Timer cam	Same as Model 10-1
Intake manifold	New — Formed brass tube, cadmium plated
Carburetor	Fig. 9-13 New — Straight intake tube, square-top air filter, tank top with Gits filler, cad plated
Needle valve	Fig. 9-23 Same as Model 10-1
Fuel tank	Fig. 9-13 New — Celluloid, 1/4 ounce, 1.05" diameter, no center post, crimped to top with gasket

Production Changes

Carburetor	Date Unknown— a few engines made with swivel filler cap on tank top.
Prop drive washer	Fig. 8-19 #2 Date Unknown — Beveled rear surface

Options/Accessories

Twin coil	Fig. 10-9
Condenser	Fig 10-13
Motor mounts	Fig. 12-7
Exhaust pipe	Fig. 12-3 #1
Propeller	Fig. 13-4

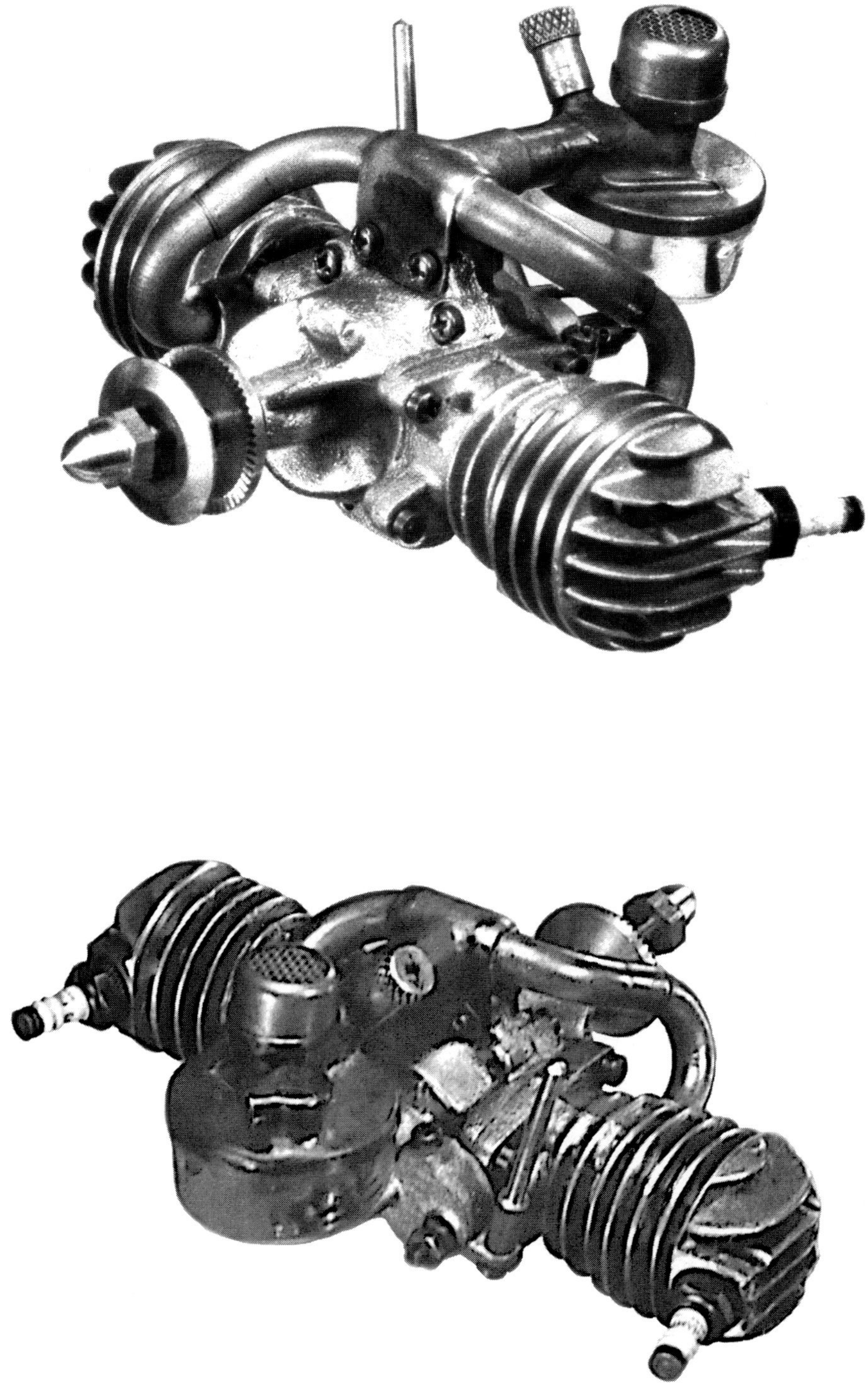

Figure 6-12 Model 20, Series 2 #260

ELF TWIN, MODEL 20, Series 2 ENGINES

Side Port, Unmachined Fins, WEBBED NOSE, Offset Cast Timer

Specifications

Bore	0.468 inches
Stroke	0.564 inches
Displacement	0.198 cubic inches
Type	Two cylinder, two-stroke, change from four-port to three-port
Overall height	2 1/16 inches
Overall length	4.0 inches
Overall width	4 1/8 inches
RPM	8000 using a 10x6 propeller
Horsepower	1/6
Weight	5 Ounces
Production period	October 31, 1940 to April 25, 1942 and June 3, 1944 to August 30, 1945
Quantity produced	254
Spark plug	Champion V-3
Cylinder head	Fig. 7-5,6 Same as Model 10-1
Cylinder	Fig. 7-7,8 #2 New — Same as Model 10-2 — Bypass enlarged, fins shallower in bypass, 3 flange holes
Cylinder liner	Fig. 7-9 #2 New — Same as Model 10-2 — Transfer port eliminated
Piston	Fig. 7-11 #2 New — Same as Model 10-2 — Cast aluminum, transfer slot eliminated
Wrist pin	Fig. 7-13 Same as Model 10-1
Connecting rod	Fig. 7-13 Same as Model 10-1
Crankcase, front	Fig. 8-8 New — Four reinforcing webs added
Crankcase, rear	Fig. 8-9 Same as Model 20-1
Crankshaft	Fig. 8-14 Same as Model 20-1
Bearings	Fig. 8-17 Same as Model 20-1
Prop drive washer	Fig. 8-19 #2 New — Same as Model 10-2
Prop front washer	Fig. 8-20 #1 Same as Model 10-1
Prop nut	Fig. 8-21 #1 Same as Model 10-1
Timer	Fig. 10-1 #2 Same as Model 10-1 Offset cast timer
Timer arm	Fig. 10-2 #2 Same as Model 10-1 production change — Straight, stepped arm, no knurl
Timer cam	Same as Model 10-1
Intake manifold	Same as Model 20-1
Carburetor assembly	Fig. 9-13 Same as Model 20-1 — #301 Gits cap filler
Needle valve	Fig. 9-23 Same as Model 10-1
Fuel tank	Fig. 9-13 Same as Model 20-1 — Center post glued-in, screw under air filter

Production Changes

Piston	Fig. 7-11 #3 Same as Model 10-1 production change June 3, 1944 — 2-piece screwed together
Wrist pin	Fig. 7-13 Same as Model 10-1 production change June 3, 1944 — Shortened for new piston
Carburetor	Fig. 9-25 Same as Model 10-1 production change 1941 — Rounded top on air filter, both types used until production stopped in 1942
Carburetor	Fig. 9-25 Same as Model 10-2 production change — July 3, 1944 — Only round-top air filter used

Options/Accessories

Twin coil	Fig. 10-9,11
Condenser	Fig. 10-13
Motor mounts	Fig. 12-7
Exhaust pipe	Fig. 12-3 #2
Exhaust pipe	Fig 12-3 #3 From June 23, 1944
Propeller	Fig. 13-3,4

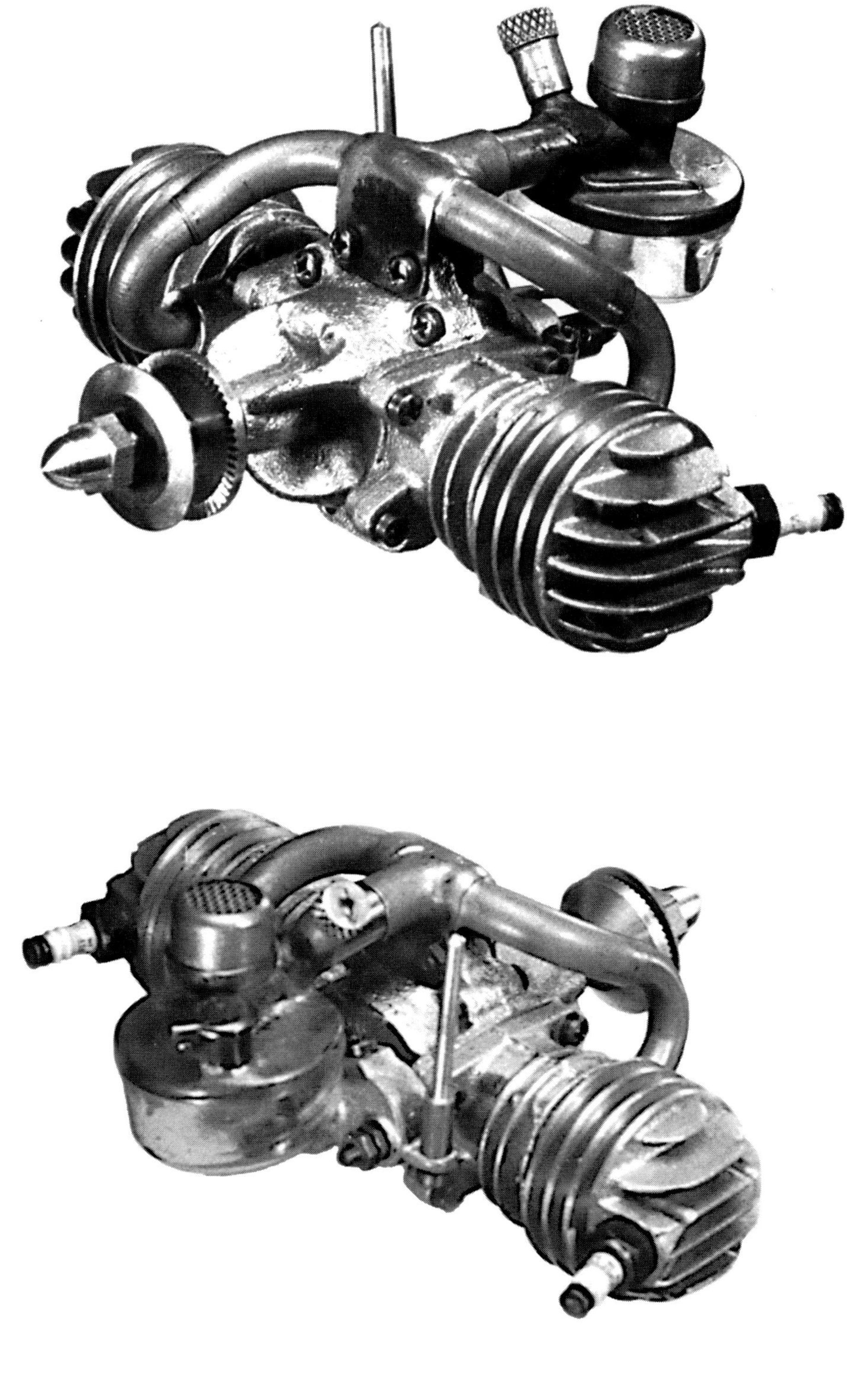

Figure 6-13 Model 20, Series 3A #666

ELF TWIN, MODEL 20, Series 3A ENGINES

Side Port, MACHINED FINS, Webbed Nose, Offset Cast Timer

Specifications

Bore	0.468 inches
Stroke	0.564 inches
Displacement	0.198 cubic inches
Type	Two cylinder, two-stroke, three-port
Overall height	2 1/16 inches
Overall length	4.0 inches
Overall width	4 1/8 inches
RPM	8500
Horsepower	1/6+
Weight	5 ounces
Production period	December 18, 1945 to September 12, 1946
Quantity produced	142
Spark plug	Champion V-3
Cylinder head	Fig. 7-5,6 New — Same as Model 10-3A — Slightly enlarged
Cylinder	Fig. 7-7,8 #3 New — Same as Model 10-3A — Slightly enlarged and machined (turned) top three fins
Cylinder liner	Fig. 7-9 #2 Same as Model 10-2
Piston	Fig. 7-11 #3 Same as Model 10-2 production change — 2-piece, screwed together
Wrist pin	Fig. 7-13 Same as Model 10-2 production change — Shortened for new 2-piece piston
Connecting rod	Fig. 7-13 Same as Model 10-1
Crankcase, front	Fig. 8-8 Same as Model 20-2 — Four reinforcing webs added
Crankcase, rear	Fig. 8-9 Same as Model 20-1
Crankshaft	Fig. 8-14 Same as Model 20-1
Bearings	Fig. 8-17 Same as Model 20-1
Prop drive washer	Fig. 8-19 #2 Same as Model 10-2
Prop front washer	Fig. 8-20 #1 Same as Model 10-1
Prop nut	Fig. 8-21 #1 Same as Model 10-1
Timer	Fig. 10-2 #2 Same as Model 10-1 Offset cast timer
Timer arm	Fig. 10-2 #2 Same as Model 10-1 production change — Straight stepped arm, no knurl
Timer cam	Same as Model 10-1
Intake manifold	Same as Model 20-1
Carburetor	Fig. 9-13 Same as Model 20-1 production change — #301 Gits cap, rounded-top air filter
Needle valve	Fig. 9-23 Same as Model 10-1
Fuel tank	Fig. 9-13 Same as Model 20-1 — Center post glued-in, screw under air filter

Production Changes

Connecting rod	Fig. 7-13 March 1946 — Thickness increased, centered exactly
Cylinder liner	Fig. 7-9 #3 May 1, 1946 — Same as Model 10-3A — Slotted exhaust, 2 bars removed
Crankshaft	Fig. 8-14 August 5, 1946 — Shortened crankshaft for new prop hardware
Prop drive washer	Fig. 8-19 #3 August 5, 1946 — Same as Model 10-3A — 5/8 inch diameter
Prop front washer	Fig. 8-20 #2 August 5, 1948 — Same as Model 10-3A — 5/8 inch diameter
Prop nut	Fig. 8-21 #2 August 5, 1946 — Same as Model 10-3A — Threaded sleeve extends behind prop nut

Options/Accessories

Twin coil	Fig. 10-11
Condenser	Fig. 10-13
Motor mount	Fig. 12-7
Exhaust pipe	Fig. 12-3 #3
Propeller	Fig. 13-5
Extension shaft	Fig. 12-8 New

Figure 6-14 Model 20, Series 3B #931

ELF TWIN, MODEL 20, Series 3B ENGINES

Side Port, MACHINED FINS, Webbed Nose, ROUND TIMER

Specifications

Bore, Stroke	Bore 0.468 inches, Stroke 0.564 inches
Displacement	0.196 inches
Type	Two cylinder, two-stroke, three-port
Dimensions	Height 2 1/16 inches, Length 4.0 inches, Width 4 1/8 inches
RPM	8500
Horsepower	1/6+
Weight	5 ounces
Production period	October 11, 1946 to April 24, 1950
Quantity produced	112
Spark plug	Champion V-3
Cylinder head	Fig. 7-5,6 Same as Model 10-3A
Cylinder	Fig. 7-7,8 #3 Same as Model 10-3A — slightly enlarged, machined (turned) fins & exhaust
Cylinder liner	Fig. 7-9 #3 New — Same as Model 10-3A production change — Slotted exhaust
Piston	Fig. 7-11 #4 Same as Model 10-3B production change — 2-piece, pressed together
Wrist pin	Fig. 7-13 Same as Model 10-2 production change — Shortened for new 2-piece piston
Connecting rod	Fig. 7-13 Same as Model 10-3A production change — Thickness increased, centered exactly
Crankcase, front	Fig. 8-8 Same as Model 20-2 — Four reinforcing webs added
Crankcase, rear	Fig. 8-9 Same as Model 20-1
Crankshaft	Fig. 8-14 Same as Model 20-3A production change — Shortened for longer prop nut
Bearings	Fig. 8-17 Same as Model 20-1
Prop drive washer	Fig. 8-19 #3 Same as Model 10-3A production change — 5/8 inch diameter
Prop front washer	Fig. 8-20 #2 Same as Model 10-3A production change — 5/8 inch diameter
Prop nut	Fig. 8-21 #2 Same as Model 10-3A production change — Extended with threaded sleeve
Timer	Fig. 10-4 New — Same as Model 10-3B — Round, machined aluminum timer
Timer arm	Fig. 10-4 New — Same as Model 10-3B — Cast arm
Timer cam	Fig. New — Same as Model 10-3B — Machined, split, pressedon to crankshaft
Intake manifold	Same as Model 20-1
Carburetor	Fig. 9-13 Same as Model 20-1 — #301 Gits cap filler
Needle valve	Fig. 9-23 Same as Model 10-1
Fuel tank	Fig. 9-13 Same as Model 20-1 — center post glued-in, screw under air filter

Production Changes

Carburetor, manifold	April 1947 — Change from cadmium plating to tin plating
Crankshaft	Fig. 8-14 Date unknown — Holes for wire cam eliminated
Cylinder	Fig. 7-7,8 #3 January 1, 1948 — Cylinder fins milled instead of turned
Cylinder	Fig. 7-7,8 #3 June 15, 1948 — Larger 0.275 inch diameter exhaust port
Piston	Fig. 7-11 #4 April 14, 1949 — Two-piece pressed-together
Cylinder	Fig. 7-7,8 #3 October 5, 1949 — Rounded fins
Timer	Fig. 10-4 October 19, 1949 — Steel case instead of aluminum
Fuel tank	Fig. 9-14 February 18, 1947 — Same as Model 10-3B prod. change, double beveled acetate tank
Crankshaft	Fig. 8-14 June 15, 1948 — Counterweights no longer used on shaft

Options/Accessories

Twin coil	Fig. 10-11
Condenser	Fig. 10-13
Motor mounts	Fig. 12-7
Exhaust pipe	Fig. 12-3 #4
Propeller	Fig. 13-5
Metal tank	Fig. 9-15 Metal tank replacement for acetate tank
Extension shaft	Fig. 12-8
Carburetor	Fig. 9-16 Tankless w/ air filter
Carburetor	Fig. 9-17 Tankless, bell-mouthed intake
Double timer (2 speed)	No photo available
Fuel pump	Fig. 12-10
Large tank carburetor	Figure 9-20, 3-20, 3-11— Used with intake manifold extension from Model 40.
Large metal tank	Replacement for above tank

Figure 6-15 Model 20, Series 3G #908

ELF TWIN, MODEL 20, Series 3G ENGINES

Side Port, Machined Fins, Webbed Nose, NO TIMER (GLOW)

Specifications

Bore	0.468 inches
Stroke	0.564 inches
Displacement	0.198 cubic inches
Type	Two cylinder, two-stroke, three-port
Overall height	2 1/16 inches
Overall length	4.0 inches
Overall width	4 1/8 inches
RPM	8500
Horsepower	1/6+
Weight	5 ounces
Production period	February 15, 1950 to March 4, 1950
Quantity produced	3 (#942, 947, 4--?)
Glow plug	Champion VG-3
Cylinder head	Fig. 7-5,6 Same as Model 10-3A
Cylinder	Fig. 7-7, 8 #3 New — Same as Model 10-3B prod. change — Rounded, milled fins, 0.275" diameter exhaust
Cylinder liner	Fig. 7-9 #3 Same as Model 10--3C production change — Slotted exhaust, hardened
Piston	Fig. 7-11 #4 Same as Model 10-3B production change — 2-piece, pressed-together
Wrist pin	Fig. 7-13 Same as Model 10-2 production change — Shortened for 2-piece piston
Connecting rod	Fig. 7-13 Same as Model 10-3A production change — Thickness increased, centered exactly
Crankcase, front	Fig. 8-8 Same as Model 20-2 — Four reinforcing webs added
Crankcase, rear	Fig. 8-9 Same as Model 20-1
Crankshaft	Fig. 8-14 Continue Model 20-3B prod. change — Shortened for longer prop nut, no counter-weights
Bearings	Fig. 8-17 Same as Model 20-1
Prop drive washer	Fig. 8-19 #4 Same as Model 10-3A production change — 5/8 inch diameter
Prop front washer	Fig. 8-20 #2 Same as Model 10-3A production change — 5/8 inch diameter
Prop nut	Fig. 8-21 #2 Same as Model 10-3A production change — Extended with threaded sleeve
Timer	None
Timer arm	None
Timer cam	None
Carburetor	Fig. 9-15, 9-16, 9-16 See Options
Needle valve	Fig. 9-23 Same as Model 10-1

Production Changes

None

Options/Accessories

Motor mount	Fig. 12-7
Carburetor/metal tank	Fig. 9-15
Carburetor	Fig. 9-16 Tankless, with air filter
Carburetor	Fig. 9-17 Tankless, bell-mouthed intake
Exhaust pipe	Fig. 12-3 #4
Propeller	Fig. 13-5
Extension shaft	Fig. 12-8
Fuel pump	Fig. 12-10 Round timer case needed for mounting
Large tank carburetor	Figure 9-20, 3-20, 3-11 Used with intake manifold extension from Model 40.
Large metal tank	Replacement for above tank

Figure 6-16 Model 20, Series 4B #1006. Notice the plugged intake port on the cylinders.

ELF TWIN, MODEL 20, Series 4B ENGINES (No Series 4A Twins made)

REED VALVE, Machined Fins, Webbed Nose, ROUND TIMER

Specifications

Bore	0.468 inches
Stroke	0.564 inches
Displacement	0.198 cubic inches
Type	Two cylinder, two-stroke, change from three-port to two-port
Overall height	2 1/16 inches
Overall length	3 5/16 inches
Overall width	4 1/8 inches
RPM	9000
Horsepower	Unknown
Weight	4 ounces
Production period	October 16, 1950 to April 7, 1952
Quantity produced	21
Spark plug	Champion V-3, AC 4H or Auto-lite M4
Cylinder head	Fig. 7-5,6 Same as Model 10-3A
Cylinder	Fig. 7-7,8 #3 New, Same as Model 10-3B prod. change, but intake port plugged on #1000, 1002, 1005, 1006, 1008, 1011, 1016
Cylinder liner	Fig. 7-9 #4 New, Same as Model 10-4A — Two-port liner
Piston	Fig. 7-11 #4 Same as Model 10-3B production change — 2-piece, pressed-together
Wrist pin	Fig. 7-13 Same as Model 10-2 production change — Shortened for new piston
Connecting rod	Fig. 7-13 Same as Model 10-3A production change — Thickness increased, centered exactly
Crankcase, front	Fig. 8-8 Same as Model 20-2
Crankcase, rear	Fig. 8-9 New — Houses near vertical intake tube and reed valve
Crankshaft	Fig. 8-14 Same as Model 20-3A prod. change — Shortened for longer prop nut, no counterweight
Bearings	Fig. 8-17 Same as Model 20-1
Prop drive washer	Fig. 8-19 #3 Same as Model 10-3A production change — 5/8 inch diameter
Prop front washer	Fig. 8-20 #2 Same as Model 10-3A production change — 5/8 inch diameter
Prop nut	Fig. 8-21 #2 Same as Model 10-3A production change — Extended with threaded sleeve
Timer	Fig. 10-4 Same as Model 10-3B — Round, machined aluminum timer
Timer arm	Fig. 10-4 Same as Model 10-3B
Timer cam	Same as Model 10-3B
Carburetor	Fig. 9-21 New — Straight tube with needle valve
Needle valve	Fig. 9-24 New — Second style, longer body

Production Changes

Cylinder	Fig. 7-7,8 #4 March 6, 1951 New — 2 port cylinder, no intake (Same as Model 10-4B prod change)

Options/Accessories

Twin coil	Fig. 10-11
Condenser	Fig. 10-13
Motor mounts	Fig. 12-7
Exhaust pipe	Fig. 12-3 #5 Same as Model 10-4B — Long, straight pipe
Propeller	Fig. 13-5
Extension shaft	Fig. 12-8
Throttle for carburetor	Fig. 12-11
Metal tank, half-round	Fig. 9-18 Firewall mount tank

Figure 6-17 Model 20, Series 4C #1532

ELF TWIN, MODEL 20, Series 4C ENGINES

REED VALVE, Machined Fins, Webbed Nose, OPEN CAST TIMER

Specifications

Bore	0.468 inches
Stroke	0.564 inches
Displacement	0.198 cubic inches
Type	Two cylinder, two-stroke, change from three-port to two-port
Overall height	2 1/16 inches
Overall length	3 5/16 inches
Overall width	4 1/8 inches
RPM	9000
Horsepower	Unknown
Weight	4.5 ounces
Production period	May 19, 1952 to November 14, 1955
Quantity produced	62
Spark plug	Auto-lite M4, Champion after October, 1955
Cylinder head	Fig. 7-5,6 Same as Model 10-3A
Cylinder	Fig. 7-7,8 #4 Same as Model 10-4B production change — Intake port eliminated
Cylinder liner	Fig. 7-9 #4 Same as Model 10-4B — 2-port liner
Piston	Fig. 7-11 #4 Same as Model 10-3B production change — 2-piece, pressed-together piston
Wrist pin	Fig. 7-13 Same as Model 10-2 production change — Shortened for 2-piece piston
Connecting rod	Fig. 7-13 Same as Model 10-3A production change — Thickness increased, centered exactly
Crankcase, front	Fig. 8-8 Same as Model 20-2
Crankcase, rear	Fig. 8-9 Same as Model 20-4B — Houses near vertical intake tube and reed valve
Crankshaft	Fig. 8-14 Same as Model 20-3A prod. change — Shortened for longer prop nut, no counterweight
Bearings	Fig. 8-17 Same as Model 20-1
Prop drive washer	Fig. 8-19 #3 Same as Model 10-3A production change — 5/8 inch diameter
Prop front washer	Fig. 8-20 #2 Same as Model 10-3A production change — 5/8 inch diameter
Prop nut	Fig. 8-21 #2 Same as Model 10-3A production change — Extended with threaded sleeve
Timer	Fig. 10-5 New — Open cast timer
Timer cam	Same as Model 10-3B
Carburetor	Fig. 9-21 Same as Model 20-4B — Straight tube with needle valve
Needle valve	Fig. 9-24 Same as Model 20-4B — Second style, longer body

Production Changes

Gasket	Fig. 7-13 January 28, 1954 — between cylinders and crankcase

Options/Accessories

Twin coil	Fig. 10-11
Condenser	Fig. 10-13
Motor mounts	Fig. 12-7
Exhaust pipe	Fig. 12-3 #5 Same as Model 10-4B — Long, straight pipe
Propeller	Fig. 13-5
Extension shaft	Fig. 12-8
Throttle for carburetor	Fig. 12-11
Metal tank, half-round	Fig. 9-18 Firewall mount

Figure 6-18 Model 20, Series 4G: Left:: Phil Riner's #1108 with Champion glow plugs, Right: #1534 with AC Glow Plugs.

ELF TWIN, MODEL 20, Series 4G ENGINES

REED VALVE, Machined Fins, Webbed Nose, NO TIMER (GLOW)

Specifications

Bore	0.468 inches
Stroke	0.564 inches
Displacement	0.198 cubic inches
Type	Two cylinder, two-stroke, change from three-port to two-port
Overall height	2 1/16 inches
Overall length	3 5/16 inches
Overall width	4 1/8 inches
RPM	9000
Horsepower	Unknown
Weight	4.3 ounces
Production period	August 26, 1950 to March 6, 1956
Quantity produced	182
Glow plug	Champion V-G3, AC 4G or Auto-lite M4G
Cylinder head	Fig. 7-5,6 Same as Model 10-3A
Cylinder	Fig. 7-7,8 #3 New — Same as Model 10-4B — Intake port plugged on #996, 998, 1003, 1007, 1009, 1010
Cylinder liner	Fig. 7-9 #4 Same as Model 10-4B — 2-port liner
Piston	Fig. 7-11 #4 Same as Model 10-3B production change — 2-piece, pressed-together
Wrist pin	Fig. 7-13 Same as Model 10-2 production change — Shortened for 2-piece pistons
Connecting rod	Fig. 7-13 Same as Model 10-3A production change — Thickness increased, centered exactly
Crankcase, front	Fig. 8-8 Same as Model 20-2
Crankcase, rear	Fig. 8-9 Same as Model 20-4B — Houses near vertical intake tube and reed valve
Crankshaft	Fig. 8-14 Same as Model 20-3A prod. change — Shortened for longer prop nut, no counterweight
Bearings	Fig. 8-17 Same as Model 20-1
Prop drive washer	Fig. 8-19 #3 Same as Model 10-3A production change — 5/8 inch diameter
Prop front washer	Fig. 8-20 #2 Same as Model 10-3A production change — 5/8 inch diameter
Prop nut	Fig. 8-21 #2 Same as Model 10-3A production change — Extended with threaded sleeve
Carburetor	Fig. 9-21 Same as Model 20-4B — Straight tube with needle valve
Needle valve	Fig. 9-24 Same as Model 20-4B — Second style, longer body

Production Changes

Cylinder	Fig. 7-7,8 #4 March 6, 1951 — 2-port cylinder, no intake (Same as Model 10-4A prod change)
Gasket	Fig. 7-13 January 28, 1954 — Same as Model 20-4B — Between cylinders and crankcase

Options/Accessories

Motor mounts	Fig. 12-7
Exhaust pipe	Fig. 12-3 #5 Same as Model 10-4B — Long, straight pipe
Propeller	Fig. 13-5
Extension shaft	Fig. 12-8
Throttle for carburetor	Fig. 12-11
Half-round metal tank	Fig. 9-18

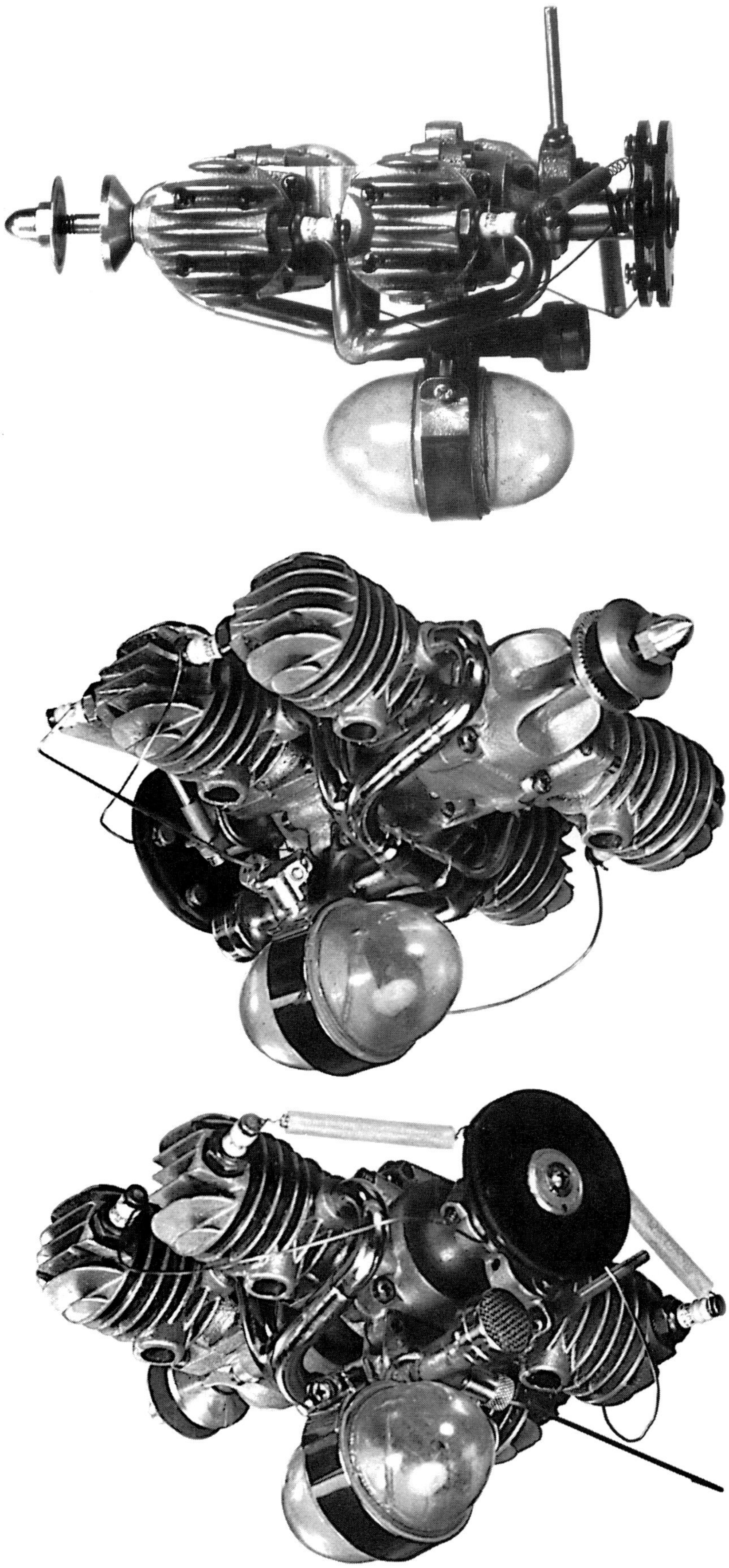

Figure 6-19 Model 40, Series 2. Carl VanCourt's #375

ELF FOUR, MODEL 40, Series 2 ENGINES (No Series 1 Fours Made)

Side Port, Machined Fins, Webbed Nose, Offset Cast Timer/Distributor

Specifications

Bore	0.468 inches
Stroke	0.564 inches
Displacement	0.396 cubic inches
Type	Four cylinder, two-stroke, three-port
Overall height	3 1/16 inches
Overall length	5 1/8 inches
Overall width	4 1/8 inches
RPM	8000
Horsepower	1/3
Weight	9 ounces
Production period	May 30, 1941 to November 12, 1942
Quantity produced	55
Spark plug	Champion V-3
Cylinder head, front	Fig. 7-5,6 Same as Model 10-1
Cylinder head, rear	Fig. 7-5,6 New — Same as Model 10-1 except front of head relieved for plug clearance
Cylinder	Fig. 7-7,8 #2 New — Same as Model 10-2
Cylinder liner	Fig. 7-9 #2 New — Same as Model 10-2
Piston	Fig. 7-11 #2 New — Same as Model 10-2
Wrist pin	Fig. 7-13 New — Same as Model 10-1
Connecting rod	Fig. 7-13 Same as Model 10-1
Crankcase, front	Fig. 8-8 New — Same as Model 20-2 except radial mounts removed from top and bottom
Crankcase, rear	Fig. 8-9 Same as Model 20-1
Crankcase extension	Fig. 8-10 New — Two piece center section, machined aluminum castings
Crankshaft	Fig. 8-15 New — One piece steel crank, no counter-weights,with rear extender, piano wire cam
Bearings	Fig. 8-17 New — Three main bearings with 12-ball thrust bearing
Prop drive washer	Fig. 8-19 #2 Same as Model 10-2
Prop front washer	Fig. 8-20 #1 Same as Model 10-1
Prop nut	Fig. 8-21 #3 Same as Model 10-1
Timer	Fig. 10-3 New — Same as Model 20-2, except mounting hole enlarged, used with distributor
Timer arm	Fig. 10-3 Same as Model 10-1 production change — Straight, stepped arm, no knurl
Timer cam	Fig. 8-15 New — Double formed piano wire
Distributor	Fig. 10-3 New — Phenolic discs with small contacts
Intake manifold	Underslung manifold
Carburetor	Fig. 9-19 New — Houses an egg shaped tank and intake tube/needle valve, mounted under engine
Needle valve	Fig. 9-23 #3 New — Same as Model 10-1 but with wire extender
Fuel tank	Fig. 9-19 New — 3-piece egg shaped celluloid tank

Production Changes

None

Options/Accessories

Coil, single	Fig. 10-9 One required
Condenser	Fig. 10-13 One required
Exhaust pipe	Fig. 12-3 #2
Motor mounts	Fig. 12-7 Supplied with each engine
Propeller	Fig. 13-3

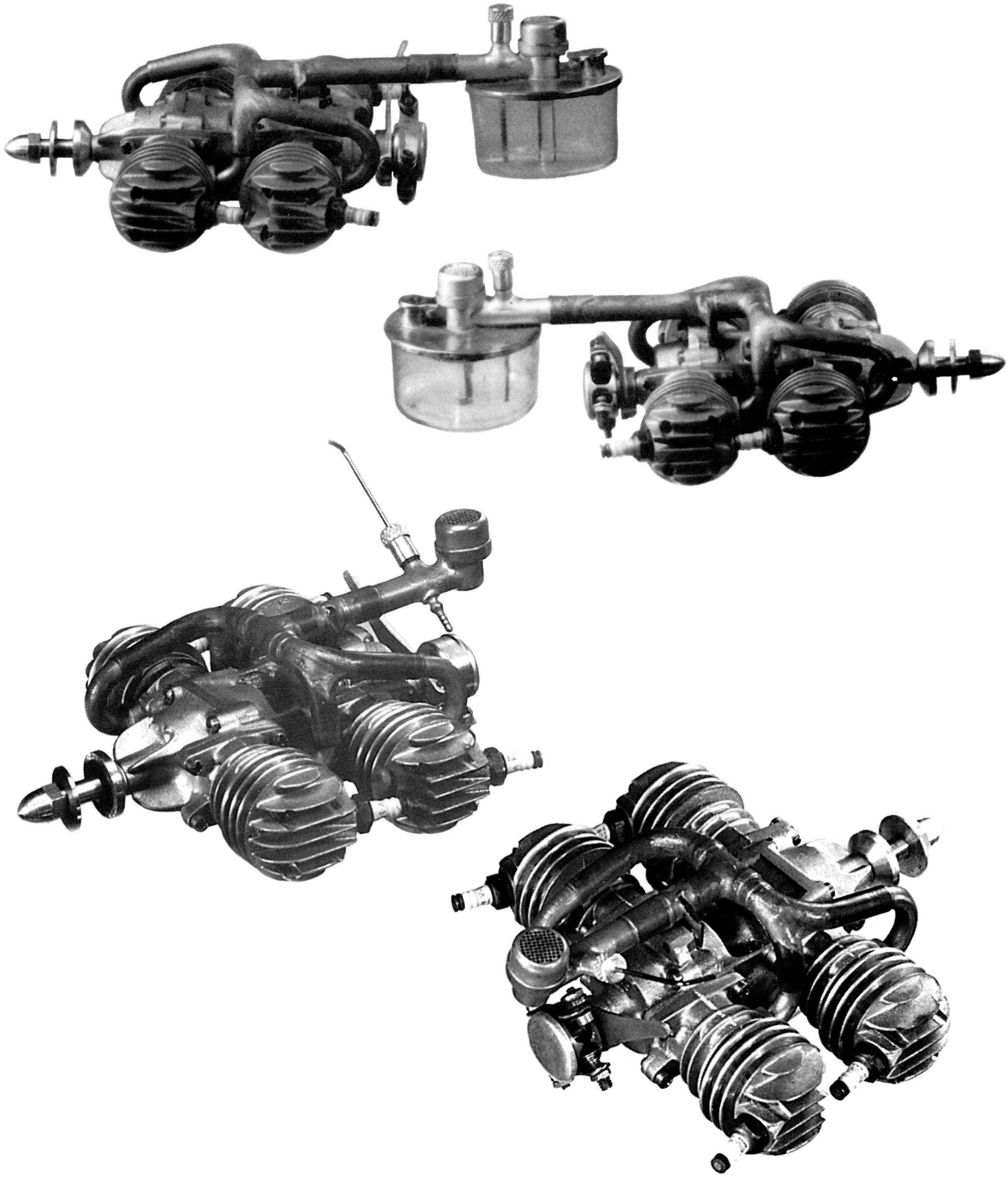

Figure 6-20 Model 40, Series 3B Top, #919 with celluloid tank. Bottom, #950 with tankless carburetor.

ELF FOUR, MODEL 40, Series 3B ENGINES (No Series 3A Fours Made)

Side port, Machined Fins, Webbed Nose, ROUND TIMER

Specifications

Bore	0.468 inches
Stroke	0.564 inches
Displacement	0.396 cubic inches
Type	Four cylinder, two-stroke, three-port
Overall height	3 1/16 inches
Overall length	5 1/8 inches
Overall width	4 1/8 inches
RPM	8000-15,000
Horsepower	1/3
Weight	9 ounces
Production period	May 9, 1947 to June 29, 1950
Quantity produced	50
Spark plug	Champion V-3
Cylinder head, front	Fig. 7-5,6 New — Same as Model 10-3A
Cylinder head, rear	Fig. 7-5,6 New — Same as 10-3A, except front of heads relieved for plug clearance
Cylinder	Fig. 7-7,8 #3 New — Same as Model 10-3A — Slightly enlarged, machined (turned) top 3 fins
Cylinder liner	Fig. 7-9 #3 New — Same as Model 10-3A production change — Slotted exhaust, hardened
Piston	Fig. 7-11 #3 New — Same as Model 10-2 production change — Two-piece screwed-together
Wrist pin	Fig. 7-13 New — Same as Model 10-2 — Shortened length for 2-piece piston
Connecting rod	Fig. 7-13 New — Same as Model 10-3A — Thickness increased, centered exactly
Crankcase, front	Fig. 8-8 Same as Model 40-2
Crankcase, rear	Fig. 8-9 Same as Model 20-2
Crankcase extension	Fig. 8-10 Same as Model 40-2
Crankshaft	Fig. 8-15 New — Shortened front end, no rear extender
Bearings	Fig. 8-17 Same as Model 40-2
Prop drive washer	Fig. 8-19 #3 New — Same as Model 10-3A production change — 5/8 inch diameter
Prop front washer	Fig. 8-20 #2 New — Same as Model 10-3A production change — 5/8 inch diameter
Prop nut	Fig. 8-21 #2 New — Same as Model 10-3A production change — Extended with threaded sleeve
Timer	Fig. 10-4 New — Round, machined aluminum case with two sets of breaker points
Timer arm	Fig. 10-4 New — Same as Model 10-3B — Cast aluminum arm
Timer cam	New — Same as Model 10-3B. Machined, split cam, pressed-on crankshaft
Manifold	New — Mounted above engine rather than below, tin plated
Carburetor	Fig. 9-20 New — Long intake with air cleaner, tank top and Gits cap filler, tin plated, celluloid tank
Needle valve	Fig. 9-23 First style, same as Model 10-1, no wire extender

Production Changes

Cylinder	Fig. 7-7,8 #3 January 1, 1948 — Cylinder fins milled, instead of turned
Cylinder	Fig. 7-7,8 #3 April 19, 1948 — Exhaust diameter increased to 0.275 inch
Piston	Fig. 7-11 #4 October 18, 1948 — Two-piece pressed-together piston
Timer	Fig. 10-4 October 5, 1949 — Case material changed to steel
Cylinder	Fig. 7-7,8 #3 October 5, 1949 — Rounded Fins

Accessories/Options

Twin coils	Fig. 10-11 Two required
Condensers	Fig. 10-13 Two required
Motor mounts	Fig. 12-7 Supplied with each engine until July 1948
Exhaust pipe	Fig. 12-3 #4
Propeller	Fig. 13-5,6
Extension shaft	Fig. 12-8 New
Metal tank	Fig. 9-20 September 1, 1948 — Replacement for celluloid tank for alcohol fuels
Carburetor	Fig. 9-16 October 5 , 1948 — Tankless carburetor with air cleaner
Carburetor	Fig. 9-17 Tankless, bell-mouthed
Fuel pump	Fig. 12-10

Figure 6-21 Model 40, Series 3G engines

ELF FOUR, MODEL 40, Series 3G ENGINES

Side Port, Machined Fins, Webbed Nose, NO TIMER (GLOW)

Specifications

Bore	0.468 inches
Stroke	0.564 inches
Displacement	0.396 cubic inches
Type	Four cylinder, two-stroke, three-port
Overall height	3 1/16 inches
Overall length	5 1/8 inches
Overall width	4 1/8 inches
RPM	8000-15,000
Horsepower	1/3
Weight	9 ounces
Production period	December 2, 1949 to May 2, 1950
Quantity produced	8 (#933, 948, 957, 967, 971, 973, 975, 976)
Glow plug	Champion VG-3
Cylinder head, front	Fig. 7-5,6 Same as Model 10-3A
Cylinder head, rear	Fig. 7-5,6 Same as Model 40-3B — Front of head relieved for plug clearance
Cylinder	Fig. 7-7,8 #3 Same as Model 10-3B prod. change — Rounded, milled fins, 0.275 exhaust
Cylinder liner	Fig. 7-9 #3 Same as Model 10-3A production change — Slotted exhaust, hardened
Piston	Fig. 7-11 #4 Same as Model 10-3B production change — 2-piece pressed-together
Wrist pin	Fig. 7-13 Same as Model 10-2 production change — Shortened for 2-piece piston
Connecting rod	Fig. 7-13 Same as Model 10-3A — Thickness increased, aligned perfectly
Crankcase, front	Fig. 8-8 Same as Model 40-2
Crankcase, rear	Fig. 8-9 Same as Model 20-1
Crankcase, extension	Fig. 8-10 Same as Model 40-2
Crankshaft	Fig. 8-15 Same as Model 40-3B — Shortened, no rear extender
Bearings	Fig. 8-17 Same as Model 40-2
Prop drive washer	Fig. 8-19 #3 Same as Model 10-3A production change — 5/8 inch diameter
Prop front washer	Fig. 8-20 #2 Same as Model 10-3A production change — 5/8 inch diameter
Prop nut	Fig. 8-21 #2 Same as Model 10-3A production change — Extended with threaded sleeve
Carburetor	Fig. 9-20, but with metal tank, 9-16 or 9-17 Same as Model 40-3B options

Production Changes

Cylinder head, rear	Fig. 7-5,6 Date unknown — Rear heads not relieved if sold using AC 4G, or Auto-lite M4G plugs

Options/Accessories

Motor mount	Fig. 12-7
Exhaust pipe	Fig. 12-3 #4
Propeller	Fig. 13-5,6
Extension shaft	Fig. 12-8
Metal tank carburetor	No photo available — Replacement for celluloid tank in Fig 9-20
Carburetor	Fig. 9-16 Tankless carburetor with air cleaner
Carburetor	Fig. 9-17 Tankless carburetor w/bell-mouth
Fuel pump	Fig. 12-10 Required round timer case for mounting

Figure 6-22 Model 40, Series 4B #1379

ELF FOUR, MODEL 40, Series 4B ENGINES (No Series 4A Fours made)

REED VALVE, Machined Fins, Webbed Nose, ROUND TIMER

Specifications

Bore	0.468 inches
Stroke	0.564 inches
Displacement	0.396 cubic inches
Type	Four cylinder, two-stroke, Change from three-port to two-port
Overall height	3 1/16 inches
Overall length	5 1/8 inches
Overall width	4 1/8 inches
RPM	8000
Horsepower	1/3+
Weight	8.5 ounces
Production period	August 23, 1950 to April 7, 1952
Quantity produced	46
Spark plug	Champion V-3, AC 4H or Auto-lite M4
Cylinder head, front	Fig. 7-5,6 Same as Model 10-3A
Cylinder head, rear	Fig. 7-5,6 Same as Model 40-3B — Front of head relieved for plug clearance
Cylinder	Fig. 7-7,8 #3 Same as Model 10-3B prod. change, but with intake port plugged on #983, 984, 989, 990, 1004, 1012, 1027, 1035, 1040, 1042, 1047, 1048, 1051, 1052
Cylinder liner	Fig. 7-9 #4 New — Same as Model 10-4B — Two-port liner, hardened
Piston	Fig. 7-11 #4 Same as Model 10-3B production change — 2-piece, pressed-together
Wrist pin	Fig. 7-13 Same as Model 10-2 production change — Shortened for 2-piece piston
Connecting rod	Fig. 7-13 Same as Model 10-3A production change — Thickness increases, centered exactly
Crankcase, front	Fig. 8-8 Same as Model 40-2
Crankcase, rear	Fig. 8-9 Same as Model 20-1
Crankcase extension	Fig. 8-11 New — Houses reed valve with vertical intake tube, machined for reed valves
Crankshaft	Fig. 8-15 Same as Model 40-3B — Shortened, no extender
Bearings	Fig. 8-17 Same as 40-2 but using new center sections
Prop drive washer	Fig. 9-19 #3 Same as Model 10-3A production change — 5/8 inch diameter
Prop front washer	Fig. 9-20 #2 Same as Model 10-3A production change — 5/8 inch diameter
Prop nut	Fig. 9-21 #2 Same as Model 10-3A production change — Extended with threaded sleeve
Timer	Fig. 10-4 Continue Model 40-3B production change — Round, steel, machined case
Timer cam	Same as Model 10-3B
Carburetor	Fig. 9-21 New — Same as Model 20-4B — No manifold, straight tube with needle valve
Needle valve	Fig. 9-24 Second style, without wire extender

Production Changes

Cylinder	Fig. 7-7,8 #4 February 16, 1951— Same as Model 10-4B — no intake port
Cylinder head, rear	Fig. 7-5,6 Date unknown — Rear heads not relieved if sold using AC 4G, or Auto-lite M4G plugs

Options/Accessories

Twin Coils	Fig. 10-11 Two required
Condensers	Fig. 10-13 Two required
Motor mounts	Fig. 12-7
Exhaust pipes	Fig. 12-3 #5
Propeller	Fig. 13-5,6
Extension shaft	Fig. 12-8
Throttle for carburetor	Fig. 12-11
Half-round fuel tank	Fig. 9-18

Figure 6-23 Model 40, Series 4C #1385

ELF FOUR, MODEL 40, Series 4C ENGINES

REED VALVE, Machined Fins, Webbed Nose, OPEN CAST TIMER

Specifications

Bore	0.468 inches
Stroke	0.564 inches
Displacement	0.396 cubic inches
Type	Four cylinder, two-stroke, change from three-port to two-port
Overall height	3 1/16 inches
Overall length	5 1/8 inches
Overall width	4 1/8 inches
RPM	8000
Horsepower	1/3+
Weight	8.5 ounces
Production period	November 7, 1952 to December 20, 1955
Quantity produced	66
Spark plug	Champion V-3 or Auto-lite M4
Cylinder head	Fig. 7-5,6 Same as Model 10-3A
Cylinder head, rear	Fig. 7-5,6 Same as Model 40-3B, except front of head relieved for plug clearance
Cylinder	Fig. 7-7,8 #4 Same as Model 10-4B production change
Cylinder liner	Fig. 7-9 #4 Same as Model 10-4B — Two-port liner
Piston	Fig. 7-11 #4 Same as Model 10-3B production change — 2-piece, pressed-together
Wrist pin	Fig. 7-13 Same as Model 10-2 production change — Shortened for 2-piece pistons
Connecting rod	Fig. 7-13 Same as Model 10-3A production change — Thickness increased, centered exactly
Crankcase, front	Fig. 8-8 Same as Model 40-2
Crankcase, rear	Fig. 8-9 Same as Model 20-1
Crankcase extension	Fig. 8-11 Same as Model 40-4B
Crankshaft	Fig. 8-15 Same as Model 40-3B — Shortened, no extender
Bearings	Fig. 8-11 Same as 40-4A
Prop drive washer	Fig. 8-19 #3 Same as Model 10-3A production change — 5/8 inch diameter
Prop front washer	Fig. 8-20 #2 Same as Model 10-3A production change — 5/8 inch diameter
Prop nut	Fig. 8-21 #2 Same as Model 10-3A production change — Extended with threaded sleeve
Timer	Fig. 10-5 New — Open cast timer
Timer cam	Fig. Same as Model 40-3B
Carburetor	Fig. 9-21 Same as Model 20-4B
Needle valve	Fig. 9-24 Same as Model 10-4B, second style, without wire extender

Production Changes

Gaskets	Fig. 7-13 February 18, 1954 — Same as Model 20-4B — Between cylinders and crankcase
Cylinder head, rear	Fig. 7-5,6 Date unknown — Rear heads not relieved if sold using AC 4G, or Auto-lite M4G plugs

Options/Accessories

Twin Coils	Fig. 10-11 Two required
Condensers	Fig. 10-13 Two required
Motor mounts	Fig. 12-7
Exhaust pipes	Fig. 12-3 #5
Propeller	Fig. 13-5,6
Extension shaft	Fig. 12-8
Throttle for carburetor	Fig. 12-11
Half-round fuel tank	Fig. 9-18

Figure 6-24 Model 40, Series 4G #1416

ELF FOUR, MODEL 40, Series 4G ENGINES

REED VALVE, Machined Fins, Webbed Nose, NO TIMER (GLOW)

Specifications

Bore	0.468 inches
Stroke	0.564 inches
Displacement	0.396 cubic inches
Type	Four cylinder, two-stroke, change from three-port to two-port
Overall height	3 1/16 inches
Overall length	5 1/8 inches
Overall width	4 1/8 inches
RPM	8000
Horsepower	1/3+
Weight	8.2 ounces
Production period	August 23, 1950 to April 11, 1956
Quantity produced	53
Glow plug	Champion V-G3 or AC 4G or Auto-lite M4G
Cylinder head, front	Fig. 7-5,6 Same as Model 10-3A
Cylinder head, rear	Fig. 7-5,6 Same as Model 40-3C — Some heads with AC or Auto-lite glow plugs not relieved
Cylinder	Fig. 7-7,8 #3 Same as Model 10-3C, but with intake port plugged on #985, 1020, 1032
Cylinder liner	Fig. 7-9 #4 Same as Model 10-4B — two-port liner
Piston	Fig. 7-11 #4 Same as Model 10-3B production change — 2-piece, pressed-together
Wrist pin	Fig. 7-13 Same as Model 10-2 production change
Connecting rod	Fig. 7-13 Same as Model 10-3A production change
Crankcase, front	Fig. 8-8 Same as Model 40-2
Crankcase, rear	Fig. 8-9 Same as Model 20-2
Crankcase extension	Fig. 8-11 Same as Model 40-4B
Crankshaft	Fig. 8-15 Same as Model 40-3B production change
Bearings	Fig. 8-17 Same as 40-4B
Prop drive washer	Fig. 8-19 #3 Same as Model 10-3A production change — 5/8 inch diameter
Prop front washer	Fig. 8-20 #2 Same as Model 10-3A production change — 5/8 inch diameter
Prop nut	Fig. 8-21 #2 Same as Model 10-3A production change — Extended with threaded sleeve
Carburetor	Fig. 9-21 Same as Model 20-4B, Straight tube with needle valve
Needle valve	Fig. 9-24 Same as Model 10-4B, second style,without wire extender

Production Changes

Cylinder head, rear	Fig. 7-5,6 Date unknown — Rear heads not relieved if sold using AC 4G, or Auto-lite M4G
Cylinder	7-7,8 #4 March 8, 1951— 2 port cylinder without intake port
Gaskets	7-13 February 18, 1954 — Same as Model 20-4C — Between cylinders and crankcase

Options/Accessories

Motor mounts	Fig. 12-7
Exhaust pipes	Fig. 12-3 #5
Propeller	Fig. 13-5,6
Extension shaft	Fig. 12-8
Throttle for carburetor	Fig. 12-11
Half-round fuel tank	Fig. 9-18

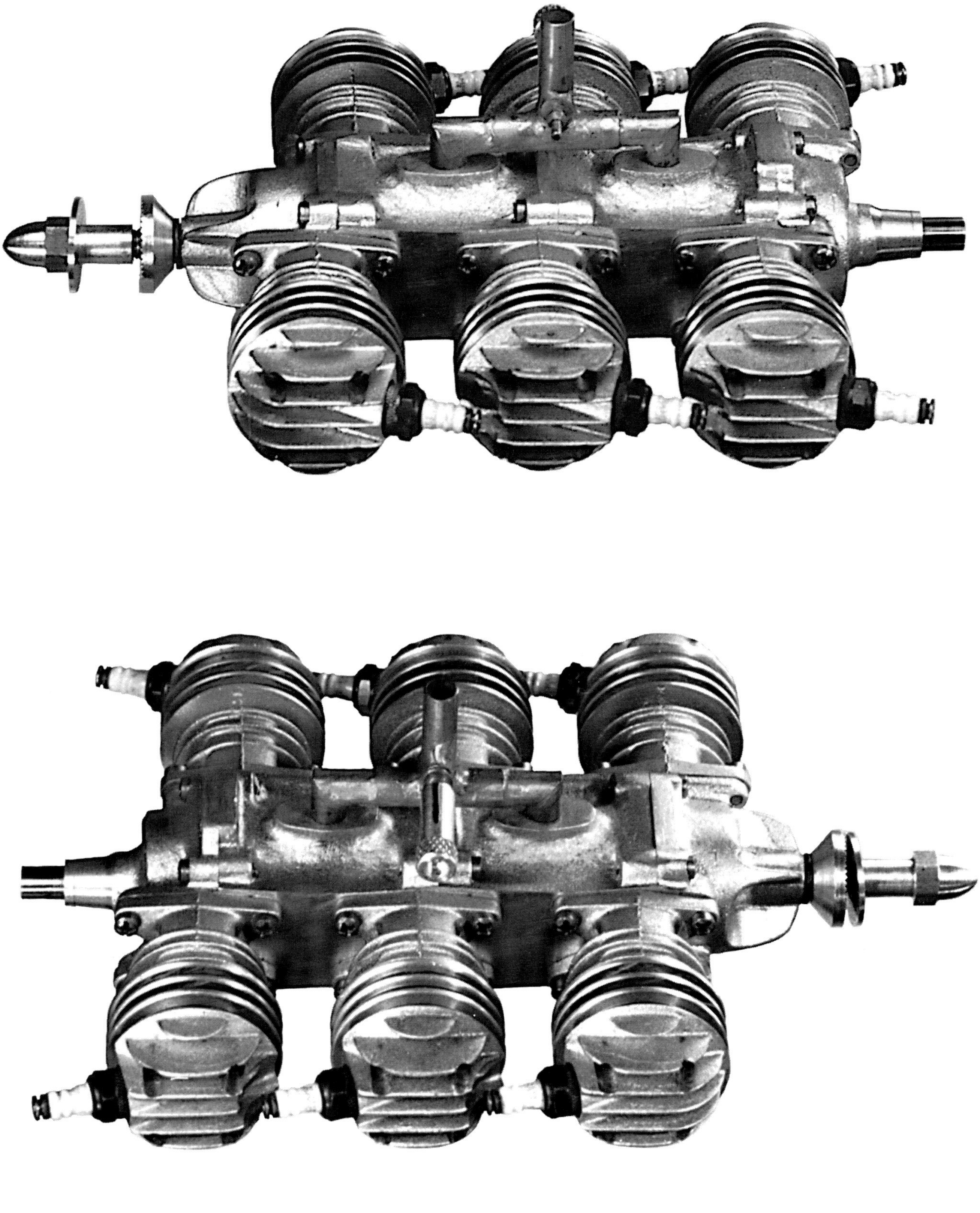

Figure 6-25 Model 60, Series 4G #1369

ELF SIX, MODEL 60, Series 4G ENGINES

REED VALVE, Machined Fins, Webbed Nose, NO TIMER (GLOW)

Specifications

Bore	0.468 inches
Stroke	0.564 inches
Displacement	0.582 cubic inches
Type	Six cylinder, two-stroke, two-port
Overall height	2 1/4 inches
Overall length	6 1/8 inches
Overall width	4 1/8 inches
RPM	8500
Horsepower	Unknown
Weight	12.2 ounces
Production period	March 11, 1951 to September 20, 1954
Quantity produced	12
Glow plug	Champion V-G3, AC 4G or Auto-lite M4G
Cylinder head, front	Fig. 7-5,6 Same as Model 10-3A
Cylinder head, rear	Fig. 7-5,6 Same as Model 40-2 — Front of rear heads relieved
Cylinder	Fig. 7-7,8 #3 Same as Model 10-3G, but intake un-drilled or plugged on #986 only
Cylinder liner	Fig. 7-9 #4 Same as Model 10-4B
Piston	Fig. 7-11 #4 Same as Model 10-2 production change — 2-piece pressed-together
Wrist pin	Fig. 7-13 Same as Model 10-2
Connecting rod	Fig. 7-13 Same as Model 10-2
Crankcase, front	Fig. 7-8 Same as Model 40-2
Crankcase, rear	Fig. 7-9 Same as Model 20-2
Crankcase extension	Fig. 8-11 Same as Model 40-4B, Two required
Crankshaft	Fig. 8-16 New — Crank throws 120 degrees apart
Bearings	Fig. 8-17 New — Four main plain bearings, 12-ball thrust bearing
Prop drive washer	Fig. 8-19 #4 New — Larger center hole
Prop front washer	Fig. 8-20 #4 New — Larger center hole
Prop nut	Fig. 8-21 #4 New — 10-32 thread, larger diameter spool
Carburetor	Fig. 9-21 Same as Model 20-4B
Needle valve	Fig. 9-24 Same as Model 10-4B

Production Changes

Cylinder	Fig. 7-7,8 #4 March 1951 — 2 port cylinder without intake port
Cylinder heads	Fig. 7-5,6 Date unknown — Rear heads not relieved if sold using AC 4G, or Auto-lite M4G plugs

Options/Accessories

Motor mounts	Fig. 12-7
Exhaust pipes	Fig. 12-3 #5
Half-round fuel tank	Fig. 9-18
RC throttle	Fig. 12-11
Extension shaft	Fig. 12-8
Propeller	Fig. 13-5,6

Chapter 7 ELF ENGINES, TOP END

ALL THE CHANGES, ALL THE PARTS

This chapter attempts to guide the reader through all the significant changes made to the cylinder heads, cylinders, cylinder liners, pistons, piston rings, wrist pins, connecting rods and gaskets.

MODEL 1

Cylinder/Head

The Model 1 had a one-piece cylinder and head. There was but one version of the cylinder casting used on the Model 1 during its production period from 1935 to 1939. A very few pre-production engines had a cylinder with no ELF logo, but these are quite rare and, as has been stated, they were not production engines. The standard cylinder with the ELF logo is shown in Figure 7-1.

Dan Calkin did not machine the cylinder fins on the Model 1 ELFs, although a few engines are known to exist having machined fins.

For a look at the interior of the machined casting, refer to Figure 7-2. Note the domed combustion chamber. The shape was produced by the core during the casting process and was not machined in.

Figure 7-1 A Model 1 ELF cylinder.

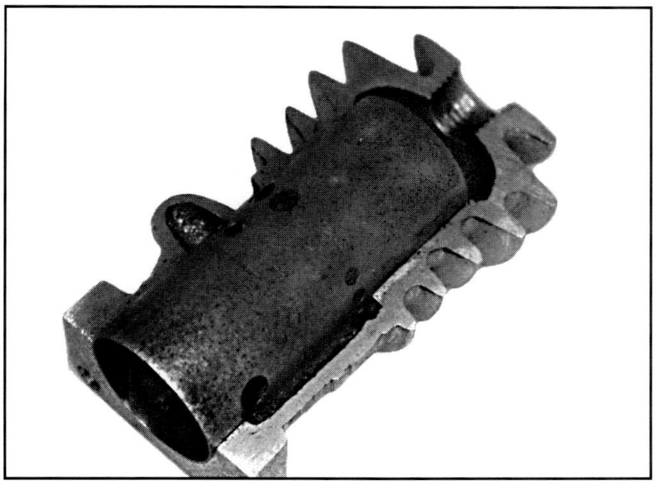

Figure 7-2 A cutaway Model 1 cylinder with liner. (Jim Gerard photo)

Figure 7-3 The cast aluminum piston, cast iron rings and wrist pin used on the Model 1 ELF. Note: the projection on the top of the piston was the sprue for casting and center for machining. It would normally be machined off.

Figure 7-4 The connecting rod, strap and rivet for the Model 1 ELF. The rivet in this photo has not been set. (Dick Tichenor photo)

Note particularly how thin the fins are and how delicate the logo is. Many reproduction engines have been made, but they invariably have thicker fins and a heavier-looking ELF logo. Some of these have been passed off as original ELF engines, so beware.

Finally, if you are lucky enough to own a Model 1 ELF, look at the serial number of your engine. The known existing range of numbers is from 6 to 432. Calkin used a set of number punches that was rather uncommon, therefore, reproduction engines usually have no serial number.

Cylinder Liner

Only one version of the cylinder liner was used. It was machined from seamless, mild steel, shelby tubing. Porting was through round drilled holes.

Piston

A cast aluminum piston with a cored interior was used. Three piston-ring grooves were machined into it. Figure 7-3

The piston was advertised as being forged. The method of forging the piston casting is unknown.

Piston Rings

Dan Calkin made his own cast iron rings. A set is shown, also, in figure 7-3. The rings originally were cut at an angle at the opening. Near the end of production the mating surfaces were changed so that they met squarely.

Wrist Pin

A hollow wrist pin was used and it was pinned in place to the piston. Figure 7-3.

Connecting Rod/Brass Strap

A sand cast connecting rod was used in conjunction with a riveted-on brass strap. Figure 7-4

Gaskets

No gaskets of any kind were used on the Model 1 ELF engines.

MODEL 10, 20, 40 and 60

Cylinder Heads

Two slightly different heads were used on these engines. They were machined to make two versions of each head. The early heads (until November 1945) were a little smaller in diameter than the heads used with the machined fin cylinders after that date. The diameter of the heads varied slightly. The diameter was not a critical dimension and, even on a four cylinder ELF, you may find heads varying in size by as much as plus or minus 0.010 inch from the nominal size.

The circumference of the heads made before November 1945 were left "as cast", but all heads made after that date were machined.

A modified version of the standard head was used on the rear heads of all the spark-ignition ELF Fours to provide clearance for the front spark plugs. (Figure 7-5). Many of the glow plug-equipped engines also used the modified rear heads, except some engines made after March, 1950, which used the smaller AC or Auto-lite glow plugs. The AC and Auto-lite glow plugs supplied with these engines were small enough so as to not require that the front of the rear heads be relieved.

The exception to the standard head was the high compression head that Dan tried when he changed to the 0.275 inch diameter exhaust. The bottom of a high compression head was contoured to match the piston. (Figure 7-6). Between May 25, and September, 1, 1948, Dan shipped two Singles (#1568 and #1565), one Twin (#896), and five Fours (#899, #900, #910, #907 and #921) before realizing that the high compression head created more problems than it solved. For example, it required a much hotter spark. Calkin subsequently returned to using the former lower compression head.

The surface quality of the heads, as with all sand castings, varied somewhat due to the variations in the sand casting operation. But, overall, the surface quality of the heads and other castings continued to improve throughout the life of the product.

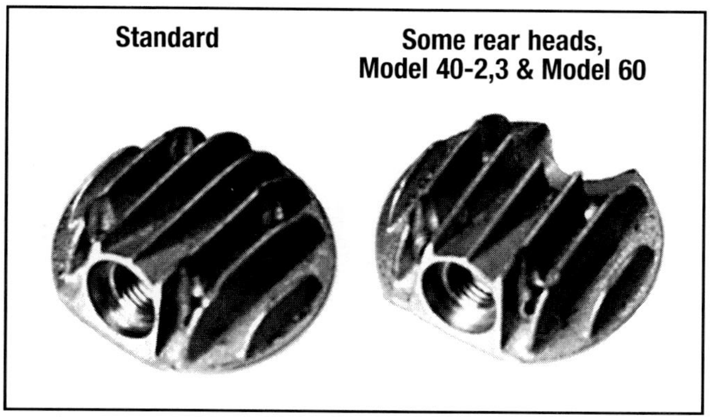

Figure 7-5 Heads for slant plug ELFs. Left is a standard head. Right is a rear head machined for plug clearance on some ELF Four or Six engines. (Dick Tichenor photo)

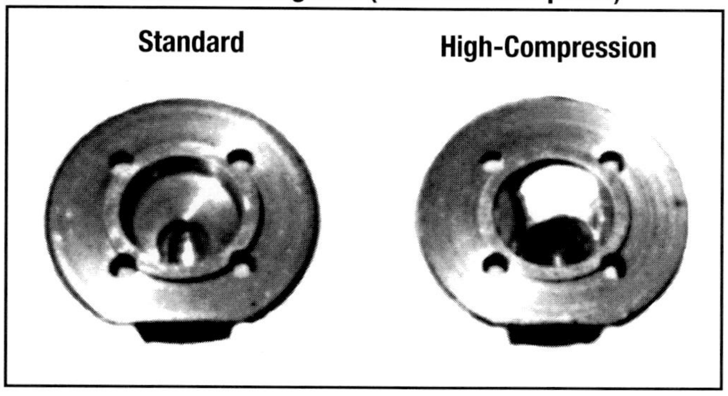

Figure 7-6 The bottom side of two heads, standard and high compression. The bottom of the high compression head was machined to match the contour of the piston.

Cylinders

For the slant plug engines, there were eight somewhat different cylinders. They were machined from three different castings. Not all versions are shown here because some of the changes cannot be readily shown in photographs.

The eight versions of the machined cylinder castings were as follows:

1. The first version was used from October, 1939-October, 1940, on the Series 1 engines. It had a small bypass. The area above the flange was lightly machined, but the cylinder fins were unmachined. The exhaust port diameter was 0.250 inches. See Figure 7-7 #1, 7-8 #1.

Figure 7-7 Two views of four of the eight slightly different cylinders used on the slant plug engines. Top row shows the front of the cylinders with the exhaust port to right. The bottom row shows exhaust port facing you with intake port to right.

#1 Series 1 had unmachined fins, machined top of flange, 0.250 inch exhaust port. #2: Series 2 engines had shallower bypass fins, unmachined flange, 0.250 inch exhaust port. Cylinder for the Series 3 engines had the top three cylinder fins and exhaust machined flush. Cylinder for Series 4 engines (#4) had milled and rounded fins, no intake port on cylinder.

Figure 7-8 These are the same cylinders shown above, but now looking at the bottom surface. #1: Note that this one doesn't have the three small holes for porting that the other three have. This is because the Series 1 engines used a transfer port cut in the piston to deliver the fuel/air charge to the cylinder. The larger machined fins are visible on the two on the right.

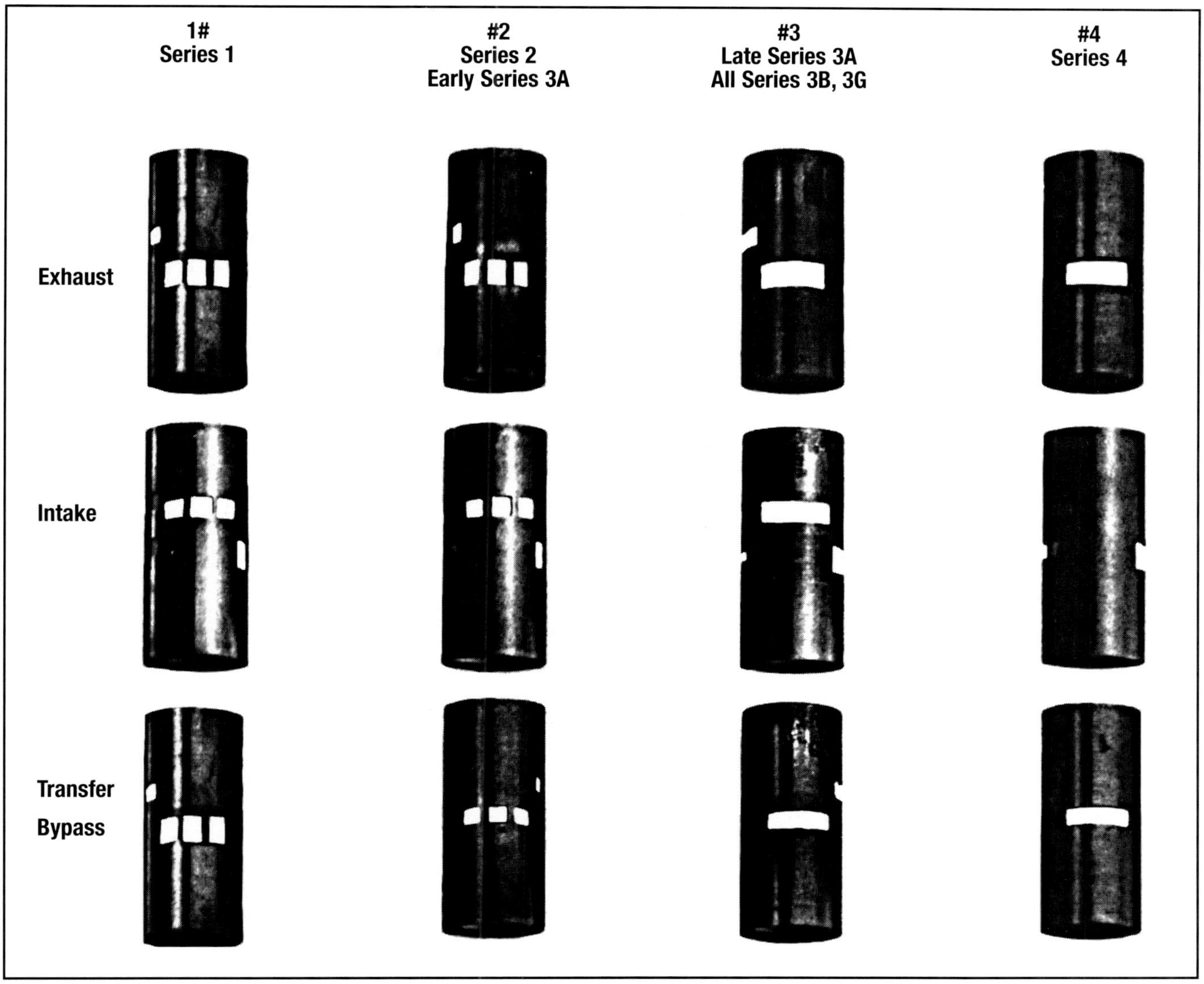

Figure 7-9 The cylinder liners used in all slant plug engines. Top row is looking at the exhaust ports, center row looks at intake port. bottom row shows the transfer port. Only the Series 1 engines had the transfer port.

2. A new casting was used from October, 1940-October, 1945, for the Series 2 engines with an enlarged internal bypass area. Enlarging the bypass area resulted in the fins in the bypass area being shallower than the previous style. Three holes were drilled in the base of the flange into the bypass area. This allowed the flow of the fuel/air mixture from the crankcase to the cylinder, replacing the transfer port in the side of the Series 1 pistons. See Figure 7-7 #2 and 7-8 #2.

3. Starting in November 1945, The outside diameter of the machined fins was increased slightly and the top three cylinder fins and the exhaust port were turned on a lathe. These cylinders were used on the early Series 3 engines. The exhaust port was still 0.250 inches. This cylinder's appearance is similar to 7-7 #3 and 7-8 #3.

4. Beginning about January, 1948, on the Series 3 engines with 0.250 inch exhaust port, the fins were milled instead of turned. The outside diameter of the cylinder was the same as the previous version. With careful examination it is usually possible to see the circular mill marks between the fins and on the top surface of the top fin. See Chapter 15 for a picture of the tooling used to perform this operation. Dan continued to mill the cylinder fins until the end of ELF production in 1956. See Figure 7-7 #3 and 7-8 #3.

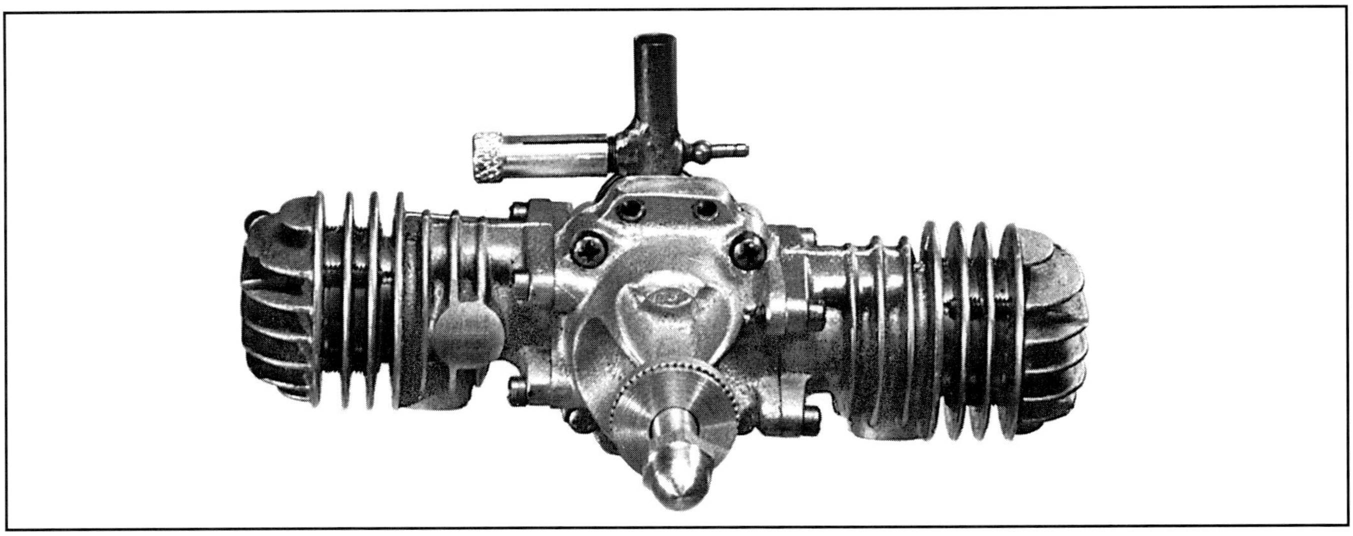

Figure 7-10 An early Series 4 Twin (#1006) made with Series 3 Cylinders. Note the plugged intake port.

5. In August of 1948, the cylinders were modified to enlarge the exhaust port from 0.250 inch to 0.275 inch. This size exhaust was continued until ELF engine production ceased. This cylinder has the characteristics of #3, but the larger exhaust of #4.

6. Starting about March, 1950, Dan started rounding the edges of the top three milled fins. The rounded-edge look continued until the end of production. This cylinder looks like #3, but with the larger exhaust and rounded fins of #4.

7. The first reed valve (Series 4) engines made, starting in March, 1950, used the previous side port type cylinder with 0.275 inch exhaust port and milled and rounded fins. The intake port on the cylinder was either plugged or left un-drilled, the intake tube being moved to the rear crankcase. See Figure 7-10.

8. Beginning in March, 1951, a new two-port cylinder casting was made for the reed valve (Series 4) engines. These were made without the intake port. This cylinder, with milled and rounded fins and 0.275 inch exhaust, was retained until production ended. Refer to Figure 7-7 #4 and 7-8 #4.

Cylinder Liners

Five different cylinder liners were used during the production period of the slant plug engines.

1. A four-port (intake, transfer, bypass and exhaust) liner was used only for the Series 1 engines. (October, 1939-October, 1940). These had square holes for the porting. Figure 7-9 #1.

2. A three-port (intake, bypass, exhaust) liner with square ports was used for the Series 2 engines and the early Series 3 engines. (October, 1940-May, 1946) Figure 7-9 #2.

3. From May, 1946 to September, 1946, Calkin used a three-port liner made with slotted ports instead of the square holes. The liners were first made with three square holes, then the punch used to make the square holes was repositioned to remove the two bridges between the holes. Figure 7-9 #3.

4. The next liners were the same as the previous type except that they were all hardened. These were used from September, 1946 to October, 1950.

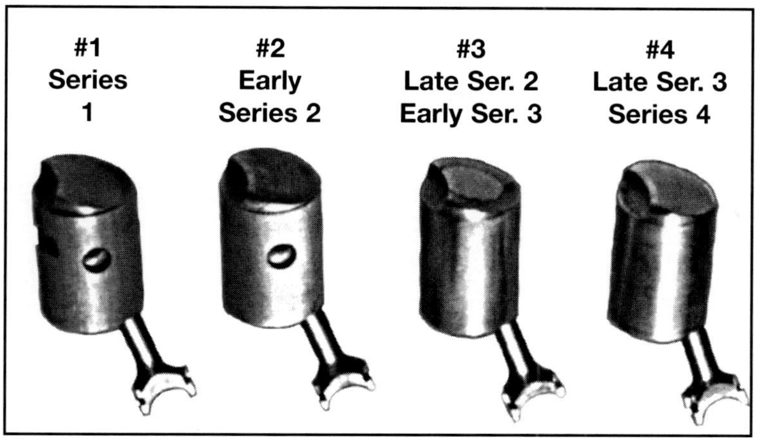

Figure 7-11 The four styles of pistons used on the slant plug engines.

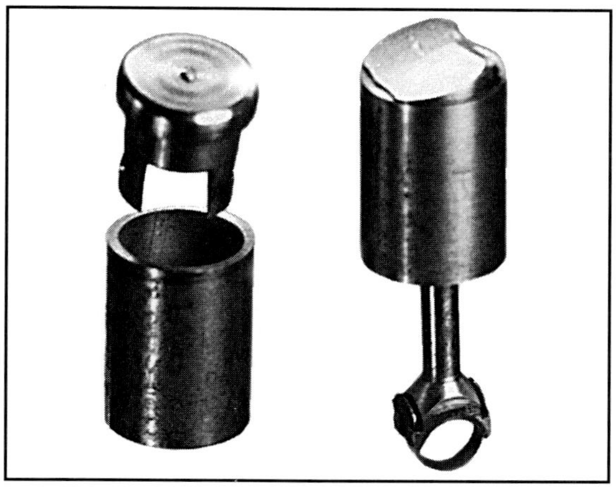

Figure 7-12 A #4 piston before machining of the top, and a finished piston with rod and strap.

5. Finally, all reed valve engines used a two-port hardened liner with slotted ports. (October, 1950-April, 1956) Figure 7-9 #4.

Pistons

Calkin used four different pistons throughout the life of the slant plug engines. Figure 7-11.

1. The Series 1 engines all used the cast aluminum piston (7-11 #1) with a cored interior. These had a slot for fuel transfer cut through the side of the piston. (October, 1939-October, 1940)

2. This cast aluminum piston was the same as the previous piston, but without the transfer slot. See Figure 7-11 #2. It was used on all Series 2 engines from October, 1940, until the wartime shutdown in November 1942.

3. When production resumed in June, 1944, the Series 2 engines used the two-piece piston shown in Figure 7-11 #3. The two pieces were screwed together. This piston can be identified by the serrated circle around the top, about 1/8 inch inside the edge. This piston skirt concealed the ends of the wrist pin.

4. Beginning in November 1950, until April, 1956, the new two-piece pressed-together piston in Figure 7-11 #4 was used. This simplified the machining process by eliminating the need to cut threads on the two parts. On these pistons the ends of the wrist pin were concealed by the piston skirt.

Piston Rings

No piston rings were ever used on the production slant plug engines.

Wrist pins

Two wrist pins were used, cut from 9/32 inch diameter drill rod, differing only in length.

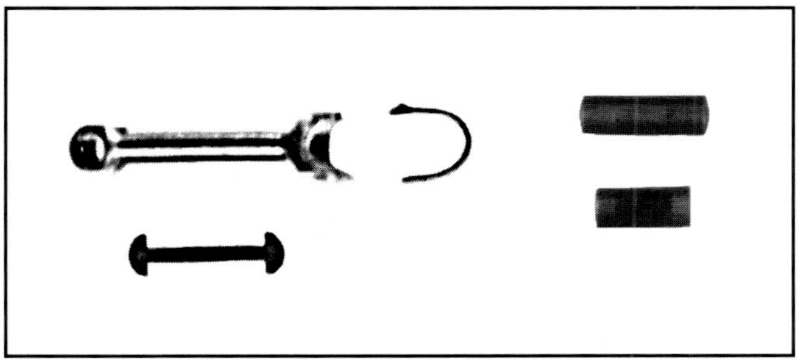

Figure 7-13 The connecting rod, strap and both sizes of wrist pins for for the slant plug engines.

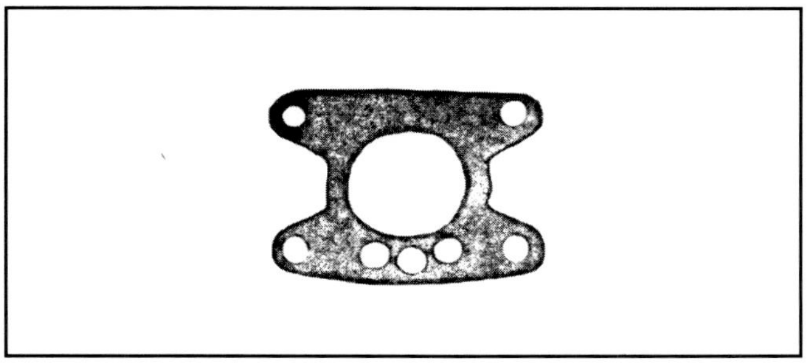

Figure 7-14 The paper gaskets used between the cylinder and crankcase from 1954 to 1956.

1. From October, 1939, until June, 1942, the pin went through the cast-aluminum pistons (Figure 7-11 #1 and #2) and the ends of the pin were visible on the sides of the piston. This wrist pin was press-fit into one side of the piston to prevent wrist pin float that might scuff the cylinder wall.

2. With the introduction of the two-piece pistons in June, 1944, (Figure 7-11 #3 and #4) the wrist pin was shortened and its ends were concealed inside the piston. Figure 7-13.

Connecting Rods

Only two connecting rods are known to have been used. The first rod was used until about March, 1946. (Figure 7-13) At that time Dan told Carl VanCourt that he had increased the diameter of the rods and corrected a slight misalignment. The difference cannot be shown with conventional photographs. The diameter of the early rods measured about 0.120 inch, while the later ones were about 0.126 inch.

Connecting Rod Straps

The phosphor-bronze strap that secures the connecting rod to the crankshaft was originally made of a material 0.006 inches thick. In mid-1945, the thickness of the material was increased to 0.010 inches. The strap is shown with the connecting rod in Figure 7-13.

Gaskets

It was always a matter of pride with Dan that the ELF engines were machined to such close tolerances that no gaskets were needed. But, beginning in January, 1954, a few engines were experiencing problems due to a slight leakage between the cylinder and the crankcase. To resolve the problem, Dan made paper gaskets and used them on all subsequent engines. The gasket is shown in Figure 7-14.

While on the subject of gaskets, all Series 1, 2 and 3 engines with celluloid fuel tanks had a paper gasket between the tank and the tank top.

Chapter 8 ELF ENGINES, BOTTOM END

ALL THE PARTS, ALL THE CHANGES

This chapter examines the progression of important changes to the crankcases, crankshafts, bearings, propeller hardware, reed valves and shims.

MODEL 1

Crankcase

Dan made two identical crankcase castings, then machined the front and rear halves from them. Figure 8-1. The rear case half was machined for the timer assembly. Oil holes were drilled in both halves of the case for improved lubrication of the crankshaft and Calkin cut grooves in the crankshaft journals for extra lubrication of the crank.

Bearings

The crankshaft was supported on both ends by the crankcase halves providing excellent support for the shaft. The aluminum used for the castings was high in copper content, thus providing a good bearing surface. A bronze thrust washer was used inside the crankcase, between each end of the shaft and the crankcase. Figure 8-2.

Crankshaft

The Model 1 ELF had a crankshaft turned from a piece of drill rod with two pressed-on steel counterweights. Figure 8-2. The front end was threaded for an 8-32 prop nut. The rear end had a pressed-on hardened cam to operate the breaker points.

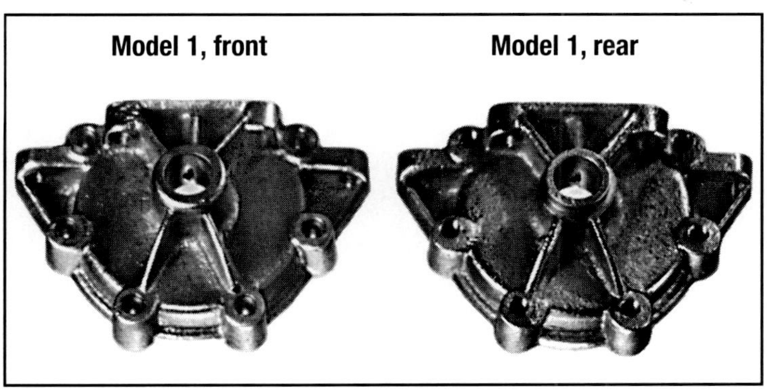

Figure 8-1 Crankcase used on the Model 1 ELFs. The right case has been machined for the timer assembly. (Dick Tichenor photo)

Figure 8-2 Crankshaft with prop hardware used on the Model 1 ELFs. Thrust washers inset. (Dick Tichenor photo)

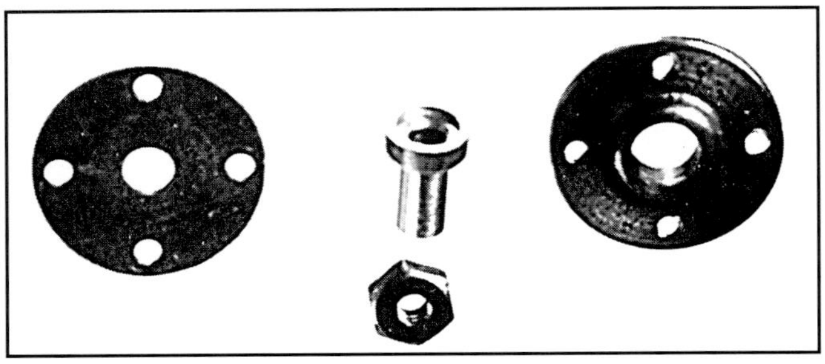

Figure 8-3 The Model 1 propeller hardware Left: Front propeller washer, Center: prop bushing and prop nut, Right: Prop drive washer. (Dick Tichenor photo).

Propeller Hardware

Propeller hardware consisted of an aluminum front prop washer and prop bushing and a steel drive washer and an unplated 8-32 hex nut. Four unplated 3-48 x $\frac{1}{2}$ inch round head screws went through the prop and both prop washers. (Figures 8-3).

MODELS 10, 20, 40 and 60

Model 10 Crankcases

Two different front cases and three different rear cases were used over the lifetime of the Model 10.

Front Crankcases

The two different front crankcases for the Model 10 engines are shown in Figure 8-4 and 8-5.

The Series 1 engines used a crankcase without reinforcing webs. This was the only slant plug production ELF with beam mounts. The production period was from October, 1939 to December, 1940.

The Series 2 and 3 engines used a completely redesigned case with four reinforcing webs added. The new crankcase was similar in appearance to the Twin crankcases. The beam mounts were removed and tabs for radial mounting added. Production was from March, 1941 to October, 1946.

For the Series 3 engines made after August, 1946, the machining of the internal dimensions was changed slightly to accommodate the six-ball thrust bearing. The outside of the crankcase remained the same. This case was also used for the Series 4 engines until the last Single sold on August, 6, 1951.

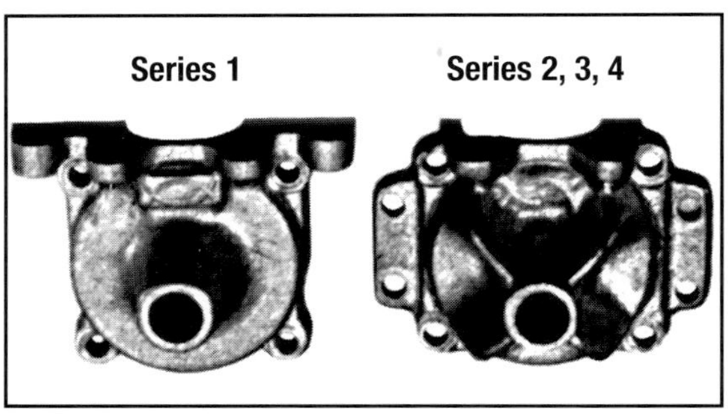

Figure 8-4 The two front crankcase halves used for the Model 10 ELFs.

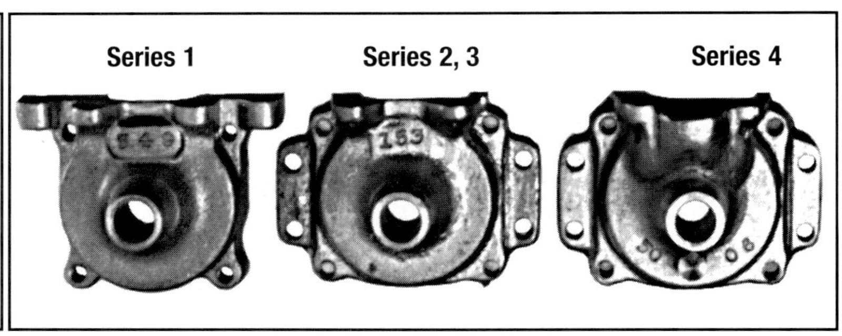

Figure 8-6 The rear crankcases for the Model 10 ELFs.

Figure 8-5 A look at the inside of the Model 10 front crankcases.

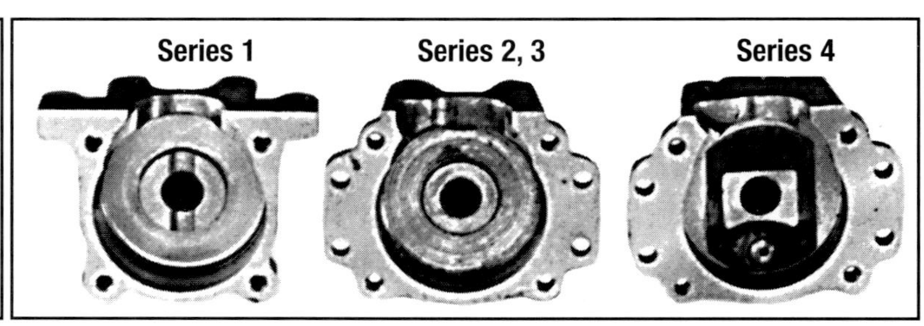

Figure 8-7 The inside of the rear crankcase for the Model 10 engines. Left: Series 1, Middle: Series 2, 3, Right: Series 4 with reed valve installed.

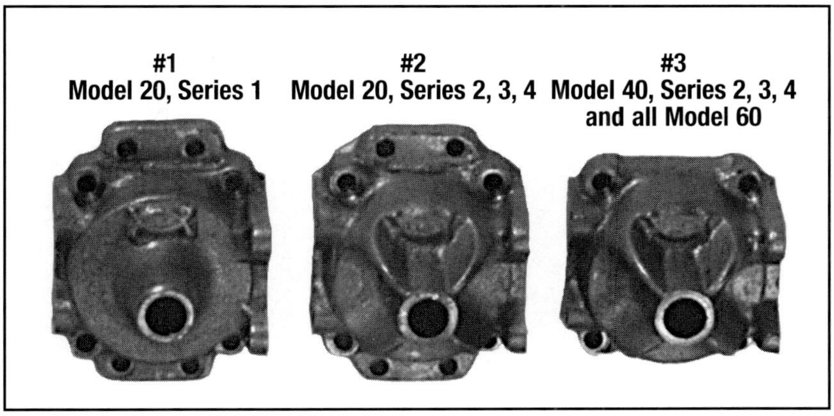

Figure 8-8 The front crankcases used on the ELF multi-cylinder engines.

Rear Crankcases

The Series 1, Model 10 rear crankcase, as with the front case, had provisions for beam mounting. Production was from October, 1939 to January, 1940.

For the Series 2 and 3 engines, the crankcase was redesigned and the mounting was changed to radial, allowing the use of the same metal motor mounts as the Model 20 engines. A noticeable difference was that the mounts were attached to the left and right sides of the crankcase on the Model 10s, instead of to the top and bottom, as was the practice on the multi-cylinder engines. This crankcase was used from March, 1940 to April, 1950.

The Series 4 Singles needed a new rear case to allow the intake tube to enter the top rear of the case. This was done by adding wax to the rear of the pattern thus adding material to the casting for the intake tube. After casting, the case was machined to house the reed valve. The reed valve inside this case can be seen in Figure 8-7 Right. Production period was from March, 1950 to August, 1951.

MODEL 20, 40, and 60

Multi-cylinder Front Crankcases

Two different castings in three configurations were used for the front crankcase on the multi-cylinder engines. Figure 8-8 shows them all.

1. The Series 1, Model 20 engines used a front case that lacked reinforcing webs. March, 1940-October, 1940. (Figure 8-8 #1)

2. The second front case had four reinforcing webs added. This was used on all subsequent Model 20 engines. October, 1940-March, 1956. (Figure 8-8 #2)

3. A slightly modified version of the second front case was used on all Model 40 and 60 engines by cutting off the top and bottom radial mounts. May, 1941-April, 1946. (Figure 8-8 #3)

Multi-cylinder Rear Crankcases

Two different rear crankcases were used on the multi-cylinder ELFs. (Figure 8-9)

1. Rear crankcase #1 was used on all Series 1, 2, and 3 Twins and all Fours and Sixes. May, 1941-April, 1946.

2. As with the Model 10, the #1 crankcase pattern was modified to become crankcase #2 by adding wax to the pattern for a rear intake tube. This crankcase was used only on the Series 4 Twins. The inside of the case was machined for a reed valve. October, 1950-March, 1956.

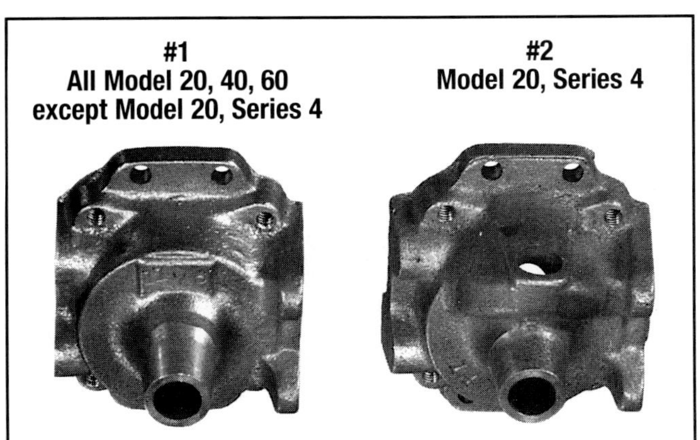

Figure 8-9 The rear cases used on all ELF multi-cylinder engines Left is #1176, Right is #1499.

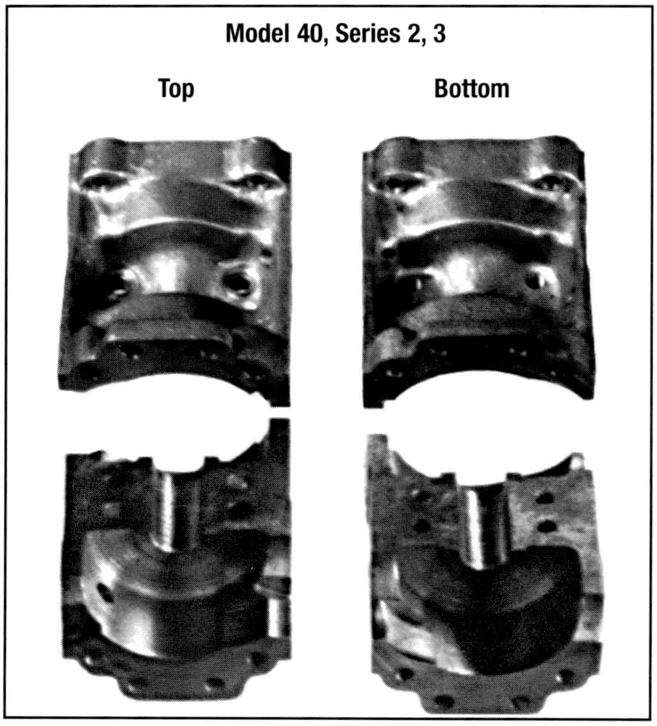

Figure 8-10 The center crankcase extender used on the Model 40, Series 2 and 3 engines

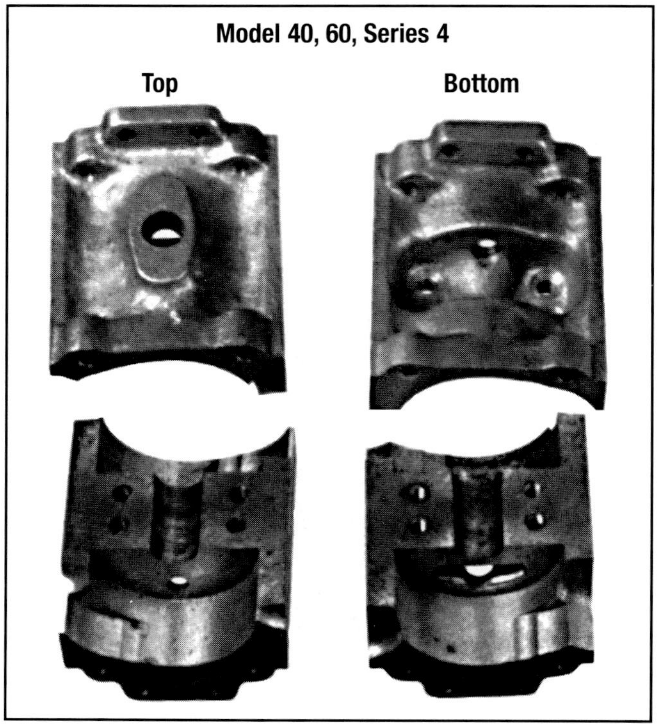

Figure 8-11 The center crankcase extender for the Series 4, Model 40 and 60 engines.

Center Crankcase Extender

The ELF Four and Six cylinder engines used a center crankcase extender casting to separate the front and rear crankcases, thus allowing the addition of another row(s) of cylinders. Only one bottom case casting was used, with two different top castings.

The Series 2 and 3 ELF Fours had identical top and bottom crankcase castings. May, 1941-April, 1952. Figure 8-10

The Series 4 (reed valve) Fours and Sixes used a new top case with an intake port added. (Figure 8-11). The casting for the bottom of the case was the same as in Figure 8-11. It was machined internally for the reed valves. August, 1950-April, 1956.

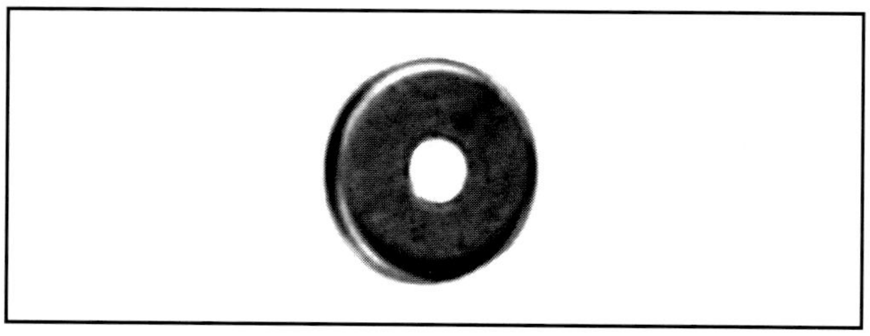

Figure 8-12 This is the stamped steel shroud that was soldered to the crankshaft with a crescent shaped piece of lead to form the counterbalance.

Model 10 Crankshaft

The Model 10 ELFs used four similar crankshafts. All are shown in Figure 8-13.

The Series 1 engines had a crankshaft machined from $\frac{1}{4}$ inch thick, steel plate. Small holes were drilled in the rear end of the shaft for the piano wire cam. Some engines had two sets of holes so that the timer could be rotated 180 degrees.

After machining, two lead counter-weights and steel shrouds were soldered onto the crank throws for counterbalance. All Model 10 engines had the counter-weights and shrouds.

Series 2 engines had a crankshaft similar to the previous crankshaft, but the lead counter-weights were a little thinner and thus lighter.

Series 3 engines initially had the same shaft as the Series 2s but, in late May, 1946, the front end

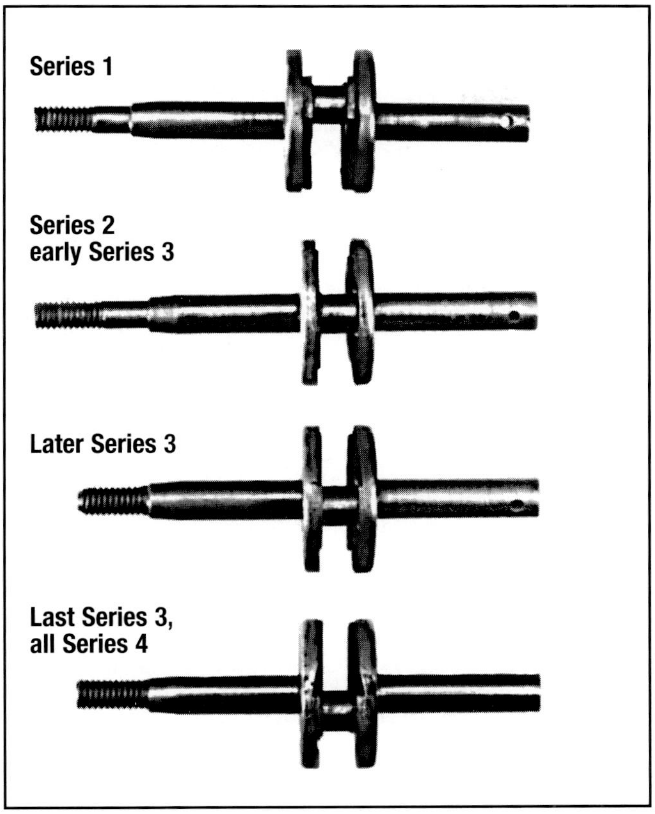

Figure 8-13 The crankshafts for the Model 10 engines.

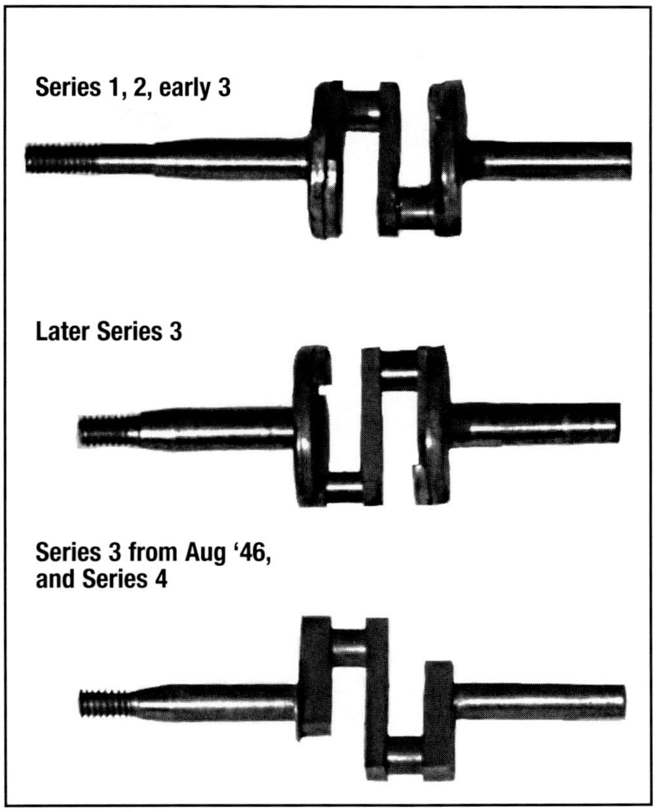

Figure 8-14 Crankshafts for the Model 20 engines

of the shaft was shortened to be used with a new prop nut having a threaded sleeve and 5/8 inch diameter prop washers.

The Series 3 engines, made starting in late August, 1946, used the round timer with a pressed-on cam. Thus it was no longer necessary to have the small holes in the rear of the shaft for the wire cam. As the remaining stock of shafts was used up, the new shafts were machined without the two small holes in the rear end, so we find that the first engines with the round timer have holes for the wire cam, the later ones do not, but very few were made without the small holes.

The Series 3G (glow plug models) and all Series 4 Singles continued using the same shortened Series 3 shafts with the wire-cam holes removed.

Model 20 Crankshafts

The progression of Model 20 crankshafts was similar to that of the Model 10. Three slightly different crankshafts were used.

The crankshafts for the Series 1 and 2 Twins were cut from 1/4 inch steel plate and had two counter-weights and shrouds soldered-on. Two holes on the rear were used for the piano wire cam.

Series 3 engines used the shaft from the Series 1 and 2 engines until August, 1946. At that time, the front end of the crankshaft was shortened to accommodate the new propeller hardware.

As nearly as can be determined, in June, 1948, when the exhaust size was increased to 0.275 inch, the counter-weights were discontinued. The counter-weights on the Twins were never really as necessary as they were on a Single, anyway. In October, 1946, the round timer was introduced on the Twins. The unneeded holes in the rear of the shaft were eliminated sometime after this date as the supply of the older shafts was exhausted.

Series 4 engines used the same Series 3 shaft without the counterweight. No other significant changes were made. All Model 20 crankshafts are illustrated in Figure 8-14.

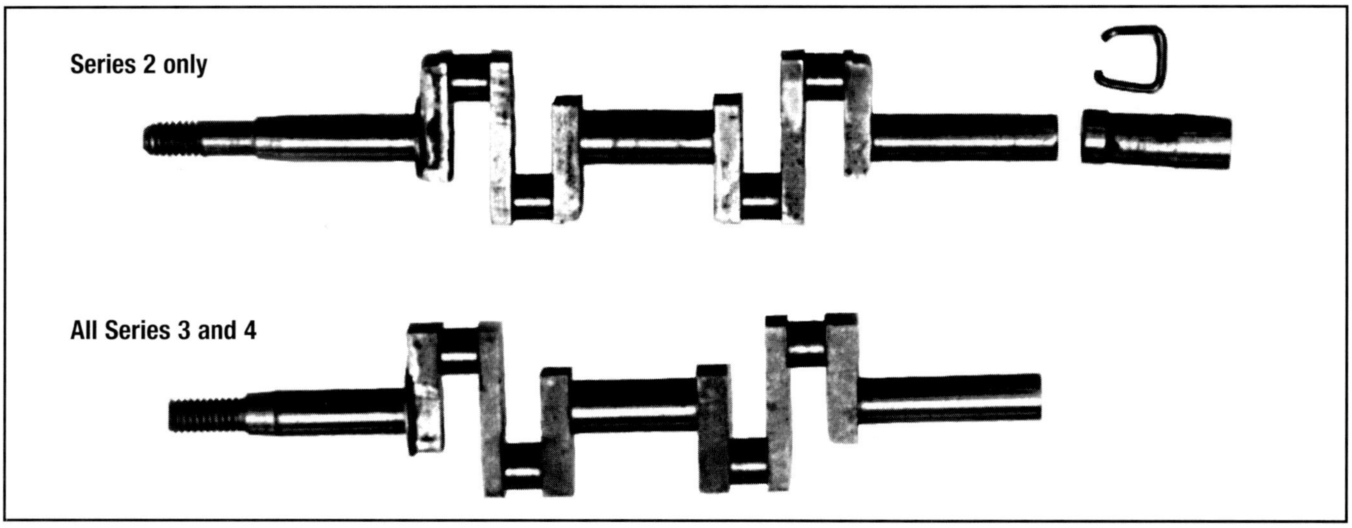

Figure 8-15 The Model 40 crankshafts. The top shaft was used on all pre-war Fours. The bottom shaft was used on all post-war ELF Fours.

Model 40 Crankshafts

Only two different crankshafts were used on the ELF Fours. The first ELF Fours, the Series 2 engines, used the crankshaft in Figure 8-15, top, from April, 1941 until November, 1942. The rear crankshaft extender, used to drive the distributor, was secured to the shaft via a double, formed piano wire cam.

When production of ELF Fours (Series 3B) resumed in 1947, a second version of the shaft was used without the extender. This shaft's front end was shortened to use the $\frac{5}{8}$ inch diameter prop hardware. All subsequent ELF Four production used this shaft. Figure 8-15, bottom.

A counterweight was never necessary on the Fours and consequently never used.

Model 60 Crankshafts

The Model 60 had a complicated and difficult-to-build shaft. The throws are 120 degrees apart. See Figure 8-16. This was the only ELF production engine to use a 10-32 thread instead of the 8-32 thread that the other ELFs used.

Bearings

The Model 10 engines had, as a thrust bearing, a steel washer soldered on the front of the shaft throw until August, 1946, when a six-ball bearing with race became the standard for Singles. Figure 8-17.

All the multi-cylinder engines used the same twelve-ball thrust bearing with a race. Also Figure 8-17.

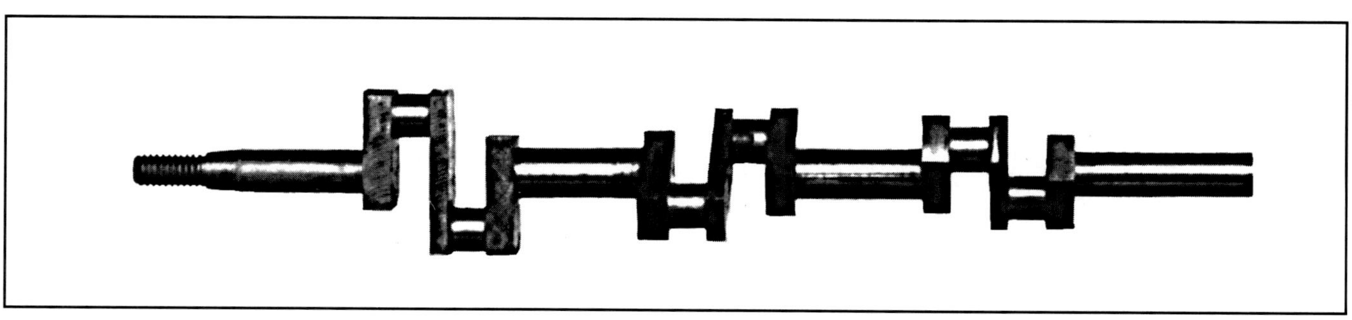

Figure 8-16 The complex crankshaft for the Model 60 ELF. (This reproduction was machined by Rich Clymer)

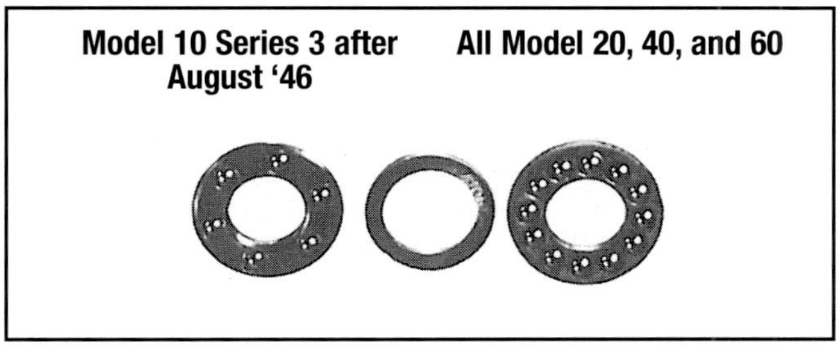

Figure 8-17 The six-ball bearing was used on the Model 10 engines. The twelve-ball bearing was used on all multi-cylinder engines. The race both used was in the center.

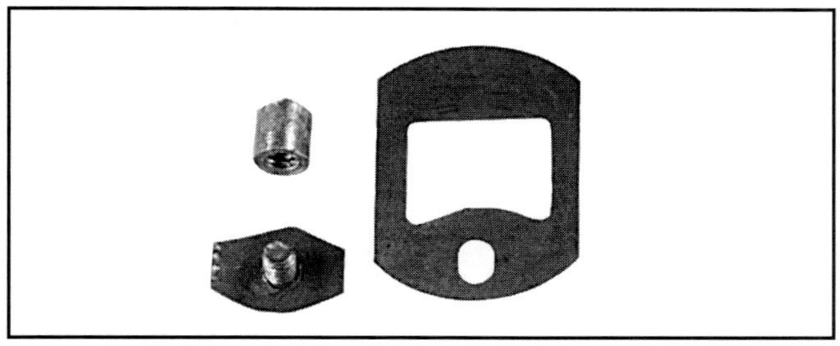

Figure 8-18 The reed valve with the special screw and nut used to secure it in the crankcase.

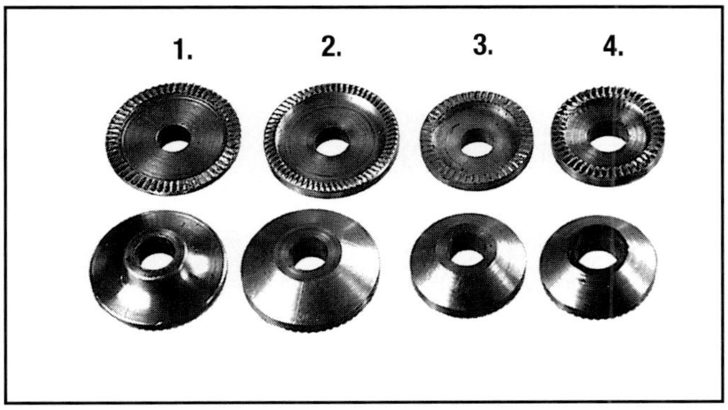

Figure 8-19 The prop drive washers. The larger washers are 3/4 inch and the smaller are 5/8 inch diameter. 1: Used from 1939-1940, 2: used from 1940-1946, 3: used from 1946 to 1956, 4: used only on the ELF Six, 1951-1954.

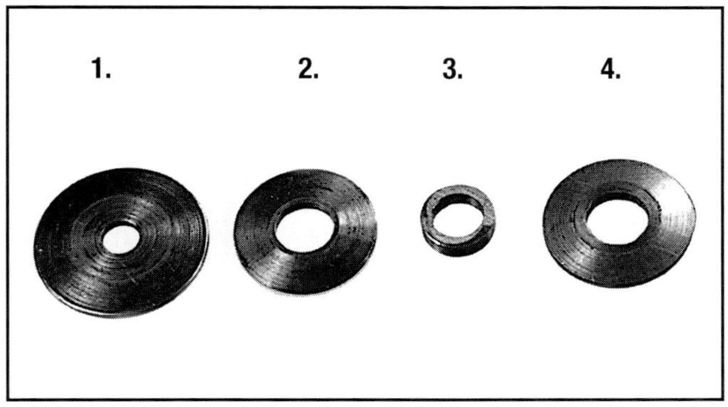

Figure 8-20 The front prop washers. 1: large washer was used until 1944. 2: used from 1944-1956. 3: used with b. only on the reed valve Singles. 4: Used only on the ELF Sixes 1951-1954.

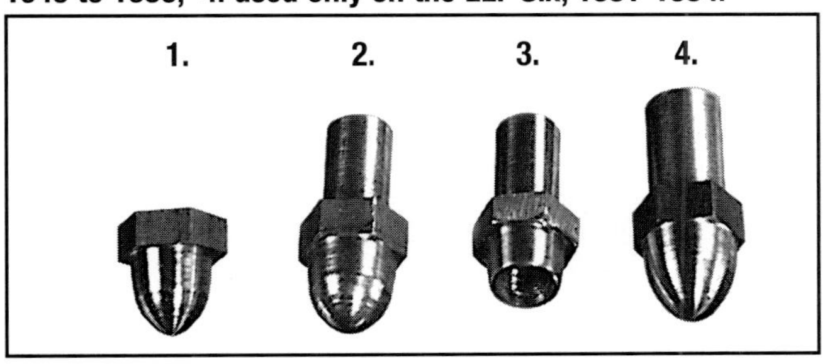

Figure 8-21 The four prop nuts used on the slant plug engines. 1: 1939-1946, 2: 1946-1956, 3: 1950-1952 reed Single only, 4.: ELF Six only.

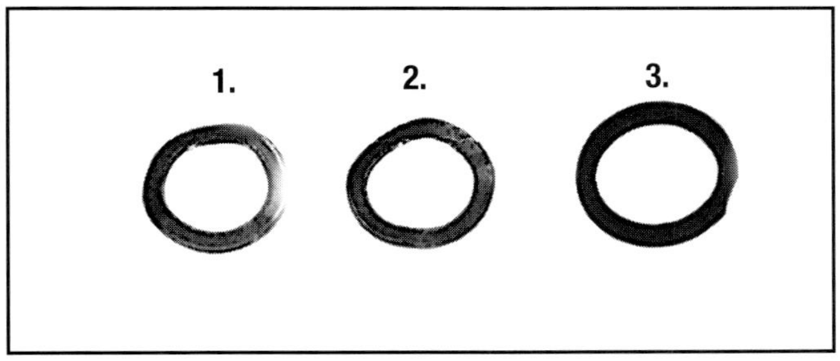

Figure 8-22 The shims used between the drive washer and the front crankcase to reduce crankshaft end play. 1: 0.010", 2: 0.015". and 3: 0.038" thick.

Reed Valves

All Series 4 engines used the same reed valve made of 0.006 inch phosphor bronze. A 3-48 x $\frac{1}{8}$ inch screw and a pointed nut secured the reed valve. The screw was soldered to the specially shaped plate and the screw head was ground down so that it was almost flush with the plate. The ELF Four used two reed valves and the Six used three, one for each pair of cylinders. Figure 8-18

PROPELLER HARDWARE

The propeller hardware consisted of a drive washer behind the prop, a front prop washer, a prop nut and end-play shims.

Prop Drive Washers

Four different drive washers were used as shown in Figure 8-19. The first was $\frac{3}{4}$ inch diameter with a concave rear face and used only on the Series 1 engines. The second, also $\frac{3}{4}$ inch diameter, was strengthened by beveling the rear face. It was used on the Series 2 and 3 engines until May,

1946. After that date the outside diameter was reduced to $\frac{5}{8}$ inch. The ELF Six alone used a drive washer with a larger center hole.

Prop Front Washers

Three front prop washers were used: a $\frac{3}{4}$ inch diameter washer was used first, then a $\frac{5}{8}$ inch, corresponding to the drive washers of the same size. The front surface of all were slightly beveled.

The reed valve Singles used an additional 0.100 inch thick washer between the prop nut and the front prop washer. The extra washer was necessary because the Singles often used propellers with thinner hubs and sometime the prop nut would 'bottom-out' when the nut was tightened. Figure 8-20. The ELF Six used the $\frac{5}{8}$ inch washer with a slightly larger center hole.

Prop Nuts

Four different aluminum prop nuts were used. The first had a simple spinner point and was drilled and tapped 8-32 on the rear side.

The second prop nut was used with the $\frac{5}{8}$ inch prop washers and had a long threaded 8-32 shoulder extending back into the prop.

The third prop nut was a modified version of the second. However, the threaded hole was drilled completely through the prop nut. This one was only used on the Series 4 ELF Singles. The last style, used only by the ELF Six, had a 10-32 thread and was a little larger overall.

Figure 8-21 shows all four prop nuts.

Crankshaft Shims

The use of aluminum shims, to remove crankshaft end-play, probably started in 1944. Dan used a combination of two thicknesses (0.010 inch and 0.015 inch) of shims to get the desired results. Two to four of these shims were installed on the crankshaft between the drive washer and the crankcase. Some ELF Six engines used the thicker shim shown in Figure 8-22.

Chapter 9 ELF INDUCTION SYSTEMS

CARBURETORS, NEEDLE VALVES AND TANKS

Model 1

The Model 1 ELF had a carburetor consisting of a curved intake tube, a needle valve and a float chamber housing a wooden float. A brass tube fuel connection was located on the top of the float chamber. The piano wire clamp that held the top on the float chamber also allowed the float chamber top to rotate so that the fuel line could face any desired direction. Four different carburetors were used. For the first 100 engines, two slightly different carburetors with horizontal needle valves were available, one for upright operation, the other for inverted operation. (Figure 9-1) These engines were shipped with black rubber fuel line.

At about engine number 100, a change was made to the carburetor so that the needle valve was now at a sixty degree angle from the horizontal. Again, the carburetor was available in both upright and inverted styles. (Figure 9-2) These engines were also shipped with black rubber fuel line tubing.

The fuel tank for the Model 1 ELF was a one inch diameter cylinder, one inch long, made of 0.005" brass shim stock, with a wooden filler plug. (Figure 9-3)

Near the end of the production cycle of the Model 1 ELF, a new carburetor with celluloid tank was offered. See Figure 9-4. As you can see, it was for

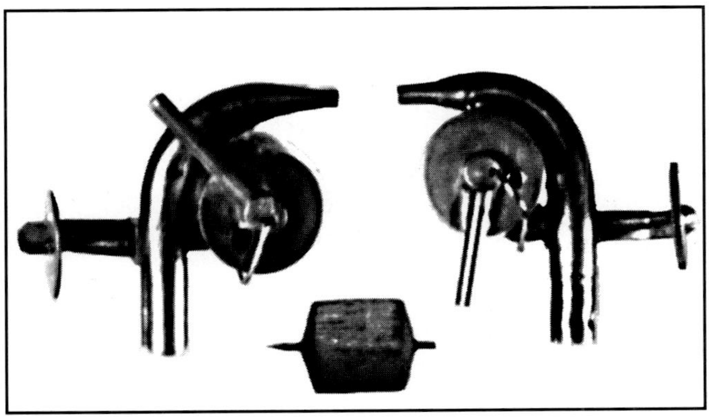

Figure 9-1 The early upright and inverted carburetors for the Model 1 ELF. Inset is the wood float used in the carburetors.

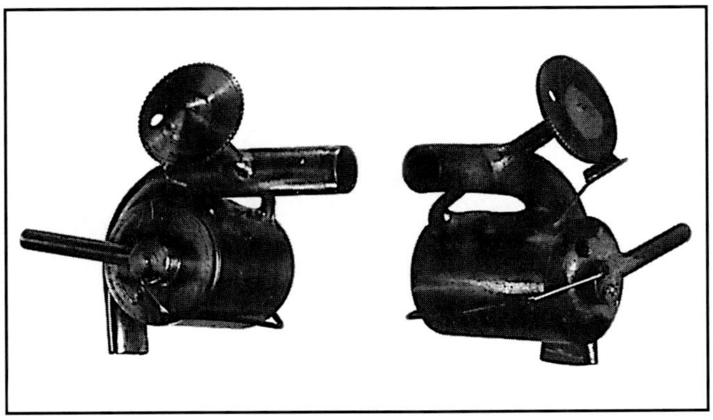

Figure 9-2 The later carburetors used on the Model 1 ELF, both upright and inverted styles.

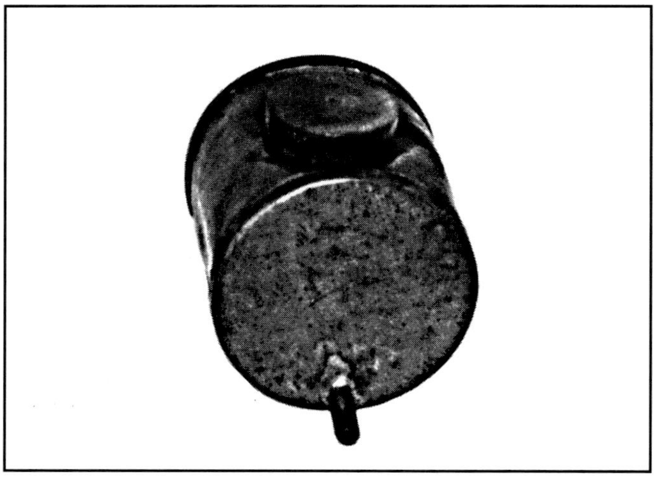

Figure 9-3 The standard gas tank for the Model 1 ELF

Figure 9-4 The carburetor on this Model 1 was made in small quantities near the end of its production.

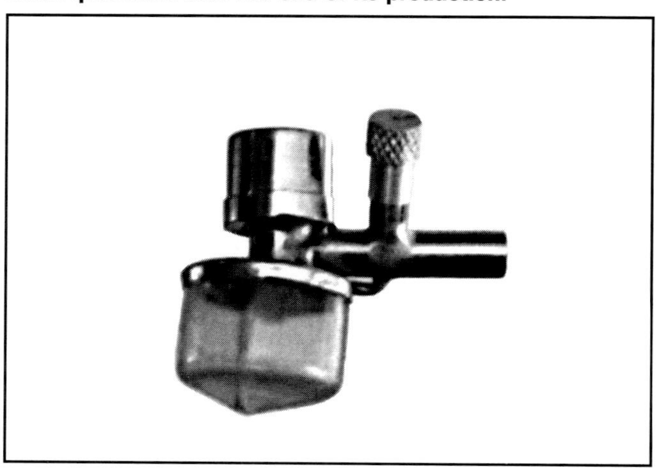

Figure 9-5 Carburetor for the Model 10, Series 1 engines. Tank was crimped on to tank top (no center post).

inverted operation only, consisting of an intake tube, needle valve, tank top and tank. The tank was crimped onto the top. It is unknown if an upright version was also made. Credit Leonard Lucas with bringing this item to our attention.

Model 10

The Model 10 (Series 1) engines initially had a carburetor consisting of a straight intake tube (0.218" diameter) with an air filter on the aft end, a needle valve, fuel tank and tank top. The clear celluloid tank attached to the tank top via a crimped lip on the edge of the tank top with a thin paper gasket between the tank and the tank top. The air filter consisted of two parallel layers of fine brass 30 mesh screen with a little piece of lamb's wool between. The tanks, made from 0.030" celluloid, were approximately $^{3}/_{4}$ inch diameter. (Figure 9-5) The Model 10 Series 2 carburetors were changed in two ways from the Series 1. The fuel filler now was a short piece of brass tube and the fuel tank had a glued-in slightly-offset center post. A screw below the air filter screwed into the tank's center post to secure it. (Figure 9-6)

No further changes were made in the carburetor assembly of the Model 10 until February, 1947, during the early production of the Series 3 engines. To answer the demand for a larger tank, and to avoid interference between the round timer and the tank, a change was made to use

Figure 9-6 Carburetor for the Model 10-2 and 10-3A and early 10-3B engines. Many pre-war Series 2 engines had the square-top air filter.

Figure 9-7 The large, twin size, double-beveled tank for the Model 10-3B engines with the round timer. (#1570)

Figure 9-8 The optional metal tank available for Model 10 engines for use with methanol fuels.

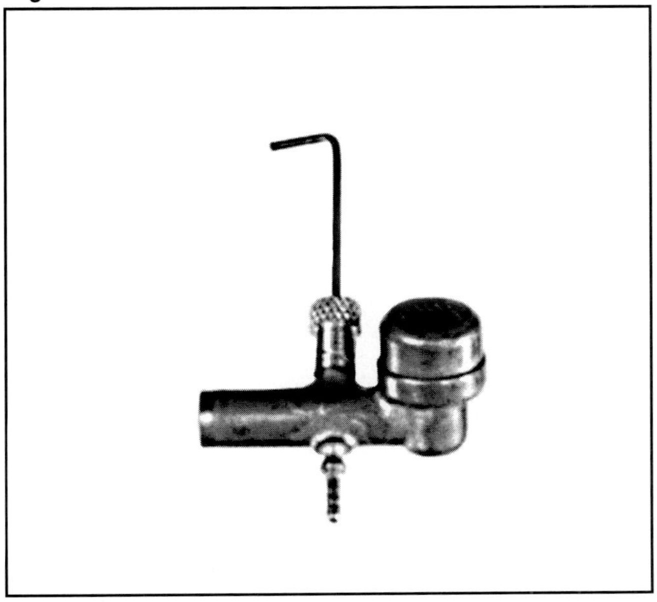

Figure 9-9 Tankless carburetor with air filter for the Model 10, Series 3B and 10-3G.

the new larger double-beveled tank from the Model 20 engines. This, of course, meant that the tank top had to be changed as well. The tank top had a filler tube rather than a Gits cap, although the existence of one engine with a Gits cap is known. The carburetor with one inch tank was retained until the end of the Model 10 Series 3. Only about 45 of the Model 10 engines with the large tank were produced. (Figure 9-7)

Among the options available for the Model 10s in 1949 was a metal tank to replace the celluloid tank. This was for modelers using alcohol based fuels. A smaller metal tank, about $^3/_4$ inch diameter, was available for the earlier Singles. The one inch Twin size tank would fit the few Singles with the round timer and larger tank. Figure 9-8 shows the large metal tank for the Single. No photo of the smaller metal tank for the Single is available.

In December, 1949, near the end of the Model 10, Series 3 production, a new optional tankless carburetor became available. Since sales of Model 10 engines in 1950 were very low, the tankless model is scarce. See Figure 9-9 for the tankless carburetor.

Another optional carburetor was offered with a bell-mouth intake tube. These are also very rare. (Figure 9-10)

In October 1950, the Model 10, Series 4 (Reed valve) engine was introduced. The only carburetor available for it was a simple intake tube with needle valve. (Figure 9-11)

Model 20

The first Model 20s (Series 1, smooth nose models) originally had a carburetor similar to that of the single cylinder engines. Since the Twin was released several months after the Single, Dan learned from the problems of the Single and made changes earlier in the production life of the Twin. The carburetor for the twin cylinder engines originally consisted of a brass intake tube ($^7/_{32}$ inch diameter), a fuel filter, a needle valve set at a 45° angle, a round tank top with a $^1/_8$ inch filler hole

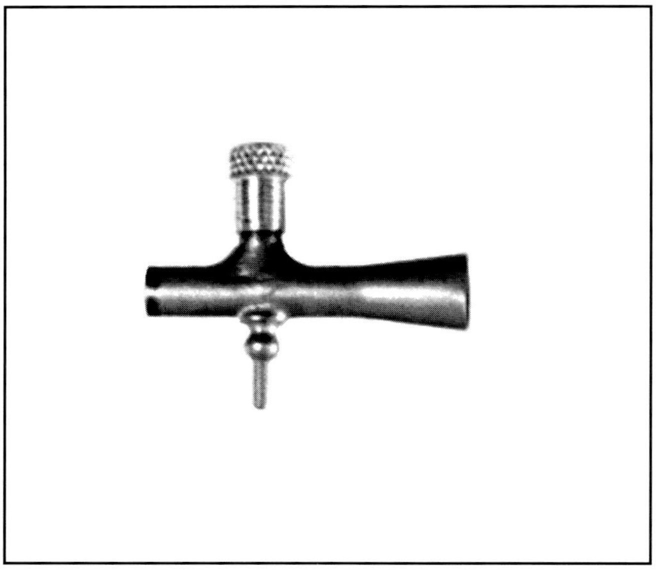

Figure 9-10 Flared carburetor for Model 10, Series 3. It would fit all side port singles.

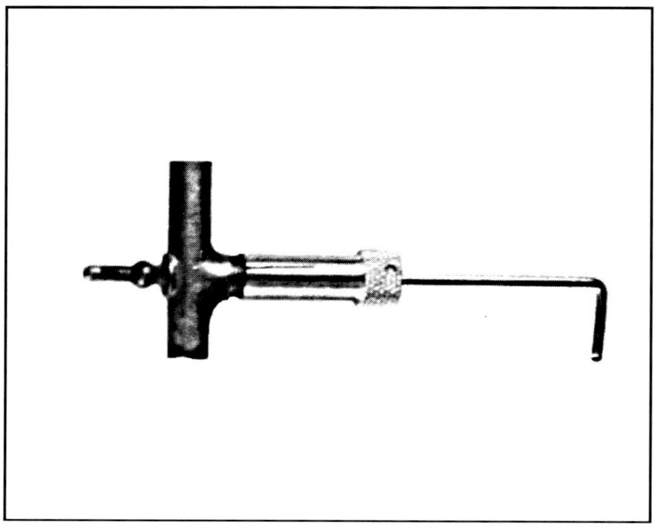

Figure 9-11 Simple carburetor for Model 10, Series 4 engines with the new style needle valve.

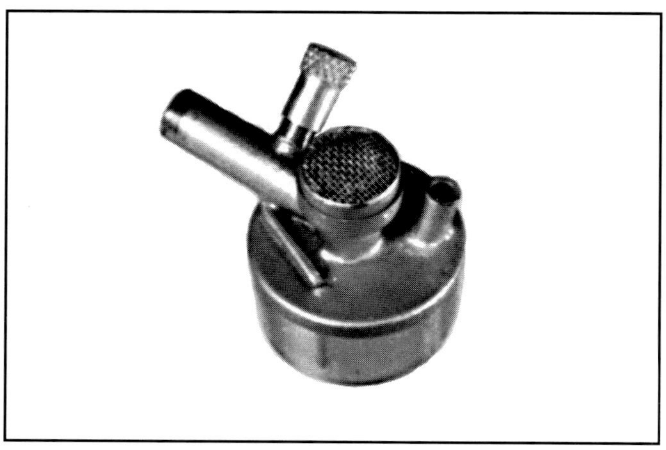

Figure 9-12 The carburetor on the prototype Model 20, Series 1 with the filler tube.

and a one inch diameter celluloid fuel tank with no center post. The tank top was slightly crimped to keep it attached to the tank with a thin paper gasket between. It was found that the fuel would slosh out the filler hole when the engine was running, so a short filler tube was tried. (The filler tube was only used on the prototype #TX, see Figure 9-12.) A swivel filler cap was also tried (no picture available). As with the Model 10 engines, the tank was made from 0.030 inch celluloid. The same square-top air cleaner was used as on the Model 10 engines with two #30 mesh brass filters with lamb's wool between them.

When it was realized that the crimping attachment method wasn't reliable enough, a center post was glued into the tank. During the early production of the Series 1 Twin, a #310 Gits cap filler was added to the tank and became standard. (Figure 9-13)

As with the Model 10 engines, a small screw (3-48 x $\frac{1}{4}$) below the air filter attached the tank to the tank top. Probably most smooth-nose Twins were shipped with a tank having a center post. The center post was slightly off center. All Series 1 and some Series 2 Twins used the air filter with the square top edge. The remainder used the rounded top air filter.

All Model 20, Series 2 engines continued the use of the celluloid fuel tank with center post and Gits cap filler.

The Series 3 Twins continued to use the same fuel tank as the Series 2 engines until the introduction of the round timer (Series 3B engines). As with the Model 10 engines, the new timer interfered with the tank, so a modified celluloid double beveled tank was soon introduced. (Figure 9-14) This is the same tank shown in Figure 9-7 for the Model 10.

The Series 3 Twins could also use the large celluloid, or optional metal tank like the Model 40. (Figure 9-20) To do so required the use of the manifold extension tube listed on the 1949 price list (Figure 3-20). The complete package is shown in Figure 3-11.

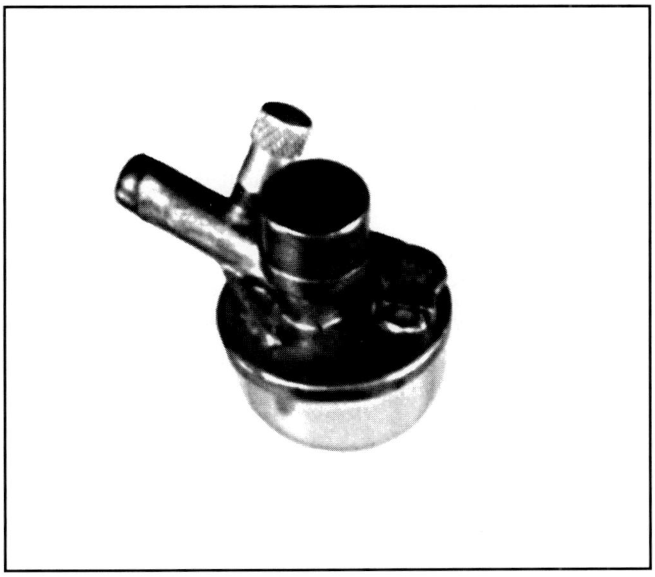

Figure 9-13 The standard carburetor on the Model 20, Series 1, 2, 3A and early 3B engines. During Series 2 production, the air filter was replaced with the rounded top style.

Figure 9-14 The double beveled tank for the Model 20, Series 3B, used with the round timer.

Figure 9-15 The optional metal fuel tank that could be used in most Model 20, Series 2 and 3 engines for methanol fuels.

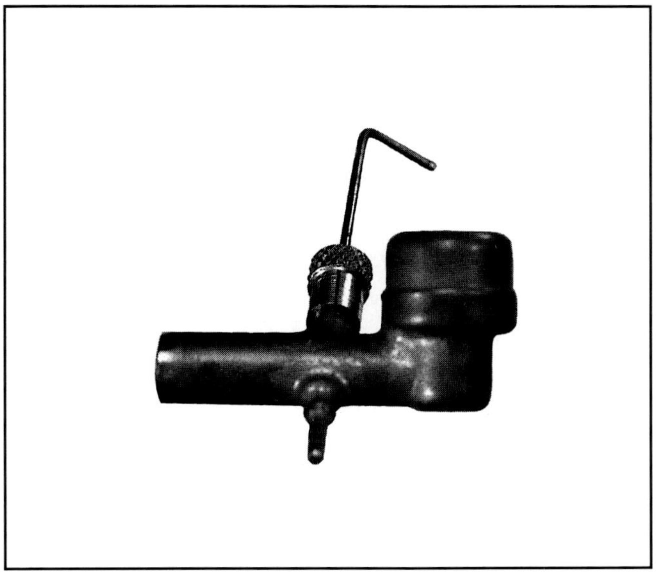

Figure 9-16 The tankless carburetor with air cleaner for some Series 3B and 3C engines, Models 20 and 40.

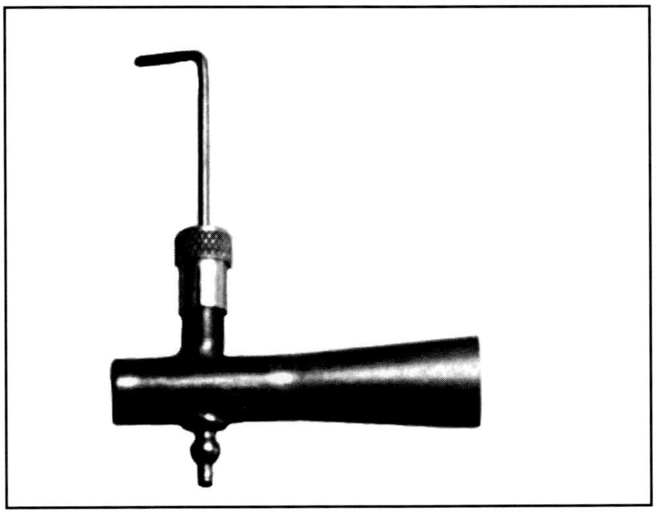

Figure 9-17 The bell-mouthed carburetor for the Model 20 and 40.

Figure 9-18 Optional firewall-mount metal tank for the Model 20, 40 and 60 engines, Series 4.

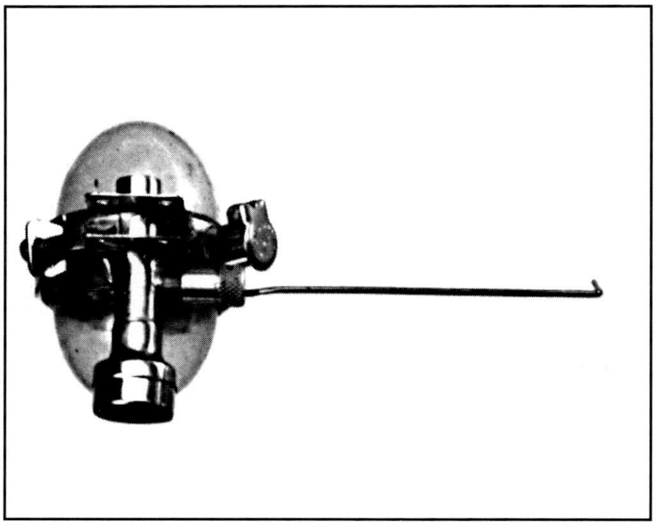

Figure 9-19 The carburetor for the Model 40. Series 2 ELF.

Figure 9-20 The carburetor used for the Model 40, Series 3A engines. It was used with a long manifold extension tube. It was also available for the Model 20 side port engines.

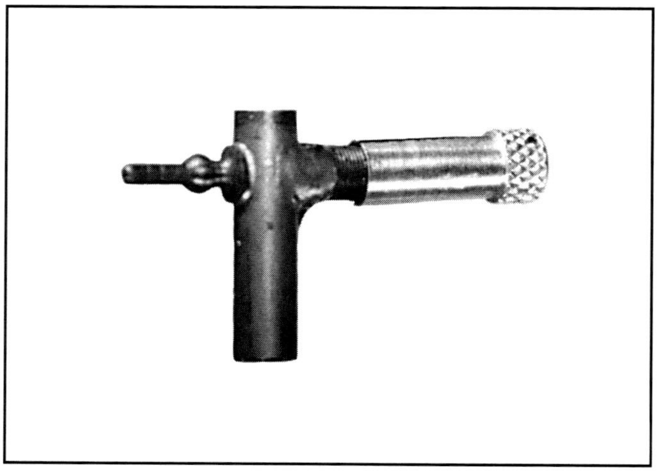

Figure 9-21 The carburetor used for the Model 20, 40, and 60, Series 4 engines.

The 1949 price sheet (Figure 3-20) listed an optional metal fuel tank to replace the celluloid tank. (Figure 9-15) This was the tank sold for the Model 10 engines with the round timer and large tank.

The 1949 ELF parts price list also advertised the optional tankless carburetor. This was for modelers using alcohol based fuels. (See Figure 9-16)

A bell-mouthed carburetor was offered also in 1949. Figure 9-17. As with the bell-mouthed intake for the Model 10 engines, it is quite rare.

In August, 1950, the Model 20, Series 4 (reed valve) engine was introduced. The only carburetor available for it was a simple, straight intake tube with needle valve. (Figure 3-21) The only fuel tank for the reed valve Twin was the half-round metal tank that mounted behind the motor mounts, although it held too much fuel for most ELF Twin uses. (Figure 9-18)

Model 40

The Model 40 engine was first introduced in Series 2 form (reinforced front crankcase). This famous pre-war Four had an egg-shaped celluloid tank mounted below the engine. The carburetor consisted of a brass tube with air filter, a crosswise needle valve having a long wire extender and a three-piece celluloid tank about $1\frac{1}{8}$ inches in diameter and $1\frac{7}{8}$ inches long. A Gits cap filler was attached to the center mounting clamp. (Figure 9-19) This carburetor was attached to the intake manifold underneath the engine. This was the only carburetor made for the Series 2 Four and it was only made until the wartime shutdown in late 1942. As the tank material aged it took on a slight greenish hue. There have been unconfirmed reports that a few engines were made with a red celluloid tank.

It wasn't until May of 1947 that Dan Calkin resumed production of a four cylinder engine. The Model 40 Series 3 came in two different configurations. The celluloid tank model (Figure

9-20) was introduced first, followed by a tankless version in October 1949. (Figure 9-16) The celluloid tank, made from 0.060 inch celluloid, was approximately $1\frac{3}{4}$ inch diameter and was mounted at the aft end of a long manifold extension tube.

The November, 1948 brochure announcing the new ELF Four also mentioned that a metal fuel tank was available for use with alcohol fuels. This tank was a direct replacement for the $1\frac{3}{4}$ inch celluloid tank. (No picture available)

A bell-mouthed intake tube was offered in the 1949 price list. This was the same assembly used with the Model 20 engines. Figure 9-17.

The only fuel tank available for the Model 40, Series 4 engine was the accessory half-round metal tank. Figure 9-18.

Model 60

The Model 60 engine was available only as a Series 4 (reed valve) engine. The only fuel tank ever available for it was the optional half-round metal tank that could be installed on the firewall behind the motor mounts. Figure 9-18.

NEEDLE VALVES

Model 1

One needle valve was used consisting of a threaded, pointed wire with a thin brass knob. The knob had small notches around its circumference. (Figure 9-22)

Model 10, 20, 40 and 60.

Two needle valves were used. The early one was used for all side port engines (Series 1, 2, and 3).

Dan used two knurling tools on the needle valves. A fine knurl was used on the early pre-war engines

Figure 9-22 The needle valve for the Model 1 ELF.

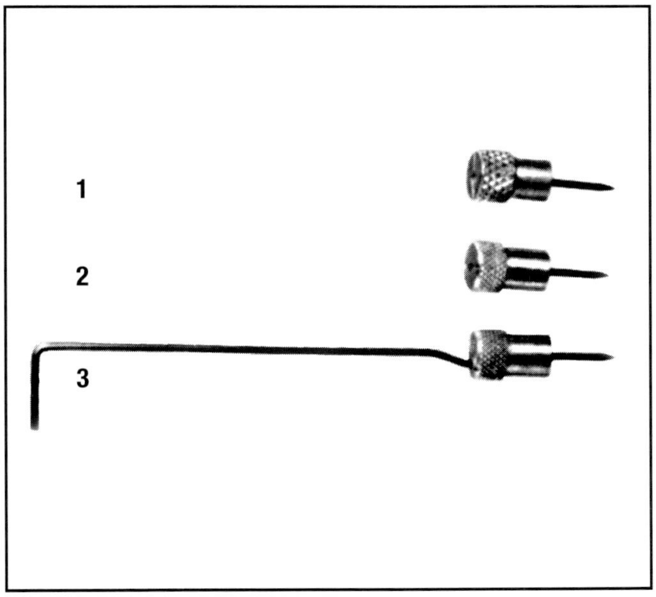

Figure 9-23 The needle valves for the side port (Series 1, 2 and 3) ELFs. Also available with a wire extender, standard for the Model 40, Series 2 and 3. A shorter extension was available for the Model 10 and 20 engines.

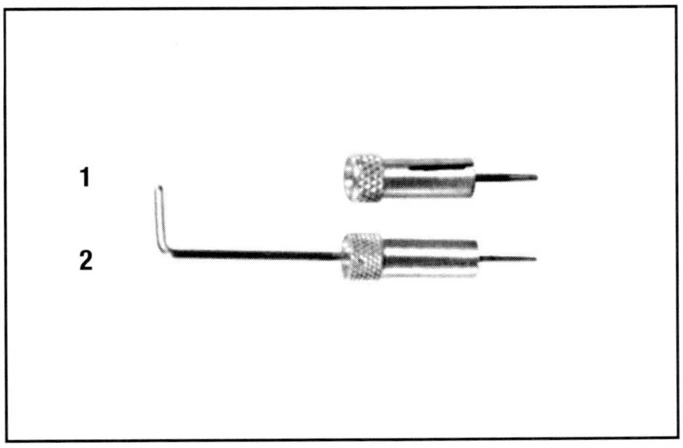

Figure 9-24 The needle valves used on the Series 4 engines.

and both the fine and a coarser knurl were used later (Figure 9-23)

The Model 40, Series 2 and some Model 40 Series 3 engines used a modified needle valve with a wire extension. The wire extender was first used on the Model 40, Series 2. When production resumed in 1944, the wire extender needle valve was available as an option. The length of the wire extension varies, but $2\frac{1}{8}$ inches was the length for the pre-war Four. For the Model 10 and 20 engines, extensions of about 0.9 and 1.5 inches long were available. (Figure 9-23)

With the introduction of the reed valve engines, a new, longer needle valve body was introduced which was used on all subsequent engines. A needle valve with a one inch long wire extender was optionally available for any reed valve engine. (Figure 9-24)

AIR FILTERS

Two air filter caps were used on the slant plug engines. The first air filter was used on all Series 1 engines and some pre-war Series 2 engines and was made of brass tubing. This filter had a square top edge. (Figure 9-25) The height of the air filters was reduced over the years.

The second air filter cap had a rounded top surface and was used on some pre-war Series 2 ELFs and all ELF Series 2 and 3 engines made after 1944. These filter caps were punched out on Dan's kick press.

Inside the air filters was a soldered-on #30 brass screen. Beneath that was some lamb's wool material and, beneath that, another brass screen that was usually not soldered-in.

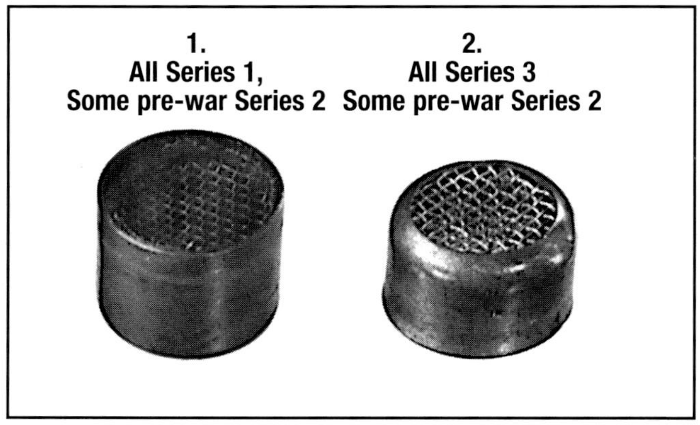

Figure 9-25 The square top and rounded top air filters used on the slant plug engines, Series 1, 2 and 3.

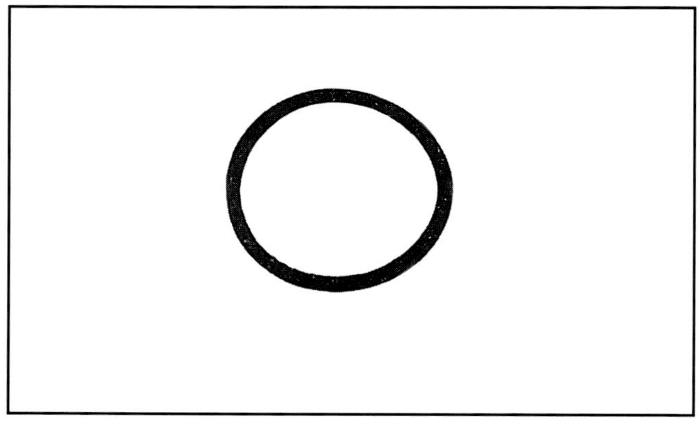

Figure 9-26 The paper gaskets used between the celluloid tank and the tank top.

GASKETS

Dan Calkin used a thin (.015") paper gasket between the celluloid fuel tank and the tank tops. All Series 1, 2 and 3 engines having a metal or celluloid tank attached to the tank top used such a gasket. Figure 9-26 illustrates the gasket used on the Model 20 engines and the Model 10 with the large tank.

Chapter 10 ELF IGNITION DEVICES

TIMERS, COILS AND CONDENSERS

MODEL 1 TIMER ASSEMBLY

Only one timer assembly was used on the Model 1 ELFs. It had a stamped, two-piece cylindrical case housing an automotive style breaker point set. A doubled piece of piano wire formed the spring. One-sixteenth inch diameter tungsten points were used. The cover snapped over the timer assembly. (Figure 10-1)

Figure 10-1 The Model 1 timer assembly with cover removed. (Dick Tichenor photo)

MODEL 10, 20, and 40 TIMERS

Three distinct timer assemblies were used, with variations in the number of breaker point sets in two of them. The timers are the (A) offset cast, (B) round and (C) open cast.

The single and twin cylinder engines used timers with a single set of breaker points. Four cylinder engines had two sets of breaker points, except for the Model 40, Series 2 which had an offset cast timer with a single set of points, used in conjunction with a distributor, serving four high-tension leads.

A. Offset Cast Timer (1940-1946)

The timer assembly used on the early slant plug Singles, Twins and Fours had a offset cast case and a formed wire cam. The Model 10 and 20, Series 1, Series 2 and the Series 3A engines used this timer in three slightly different versions. Figure 10-2 shows three slightly different offset cast timers. The front side of the timer case is not shown, but it has a thin, round aluminum cover pressed into the case. Note the small hole in the bottom of the case for drainage.

Dan Calkin was particularly pleased with the performance of this timer because it presented a very

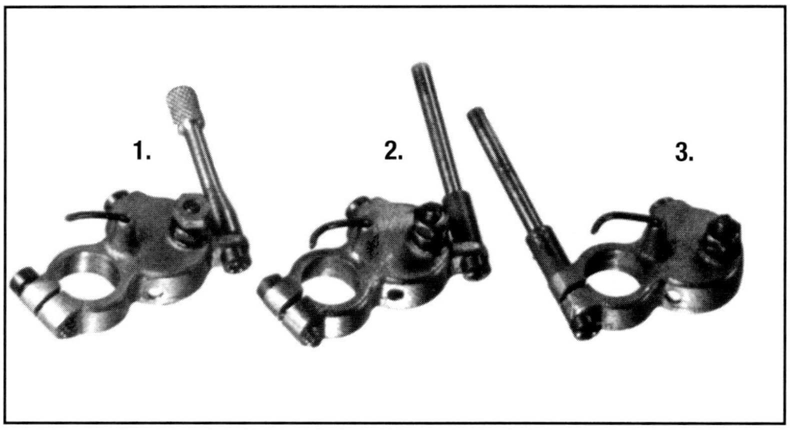

Figure 10-2 The (A) offset cast timer was used on all Series 1, 2, and 3A engines. Timer #1 was used for the Series 1 engines. The knurled timer arm was optional on the Model 10s. The use of a new style timer arm began some time during Series 1 production. The #2 timer was used on the later Model 20, Series 1, Series 2 and 3A engines. The timer #3 was used on the later Model 10, Series 1 and all Model 10, Series 2 and 3A engines. (Dick Tichenor Photo)

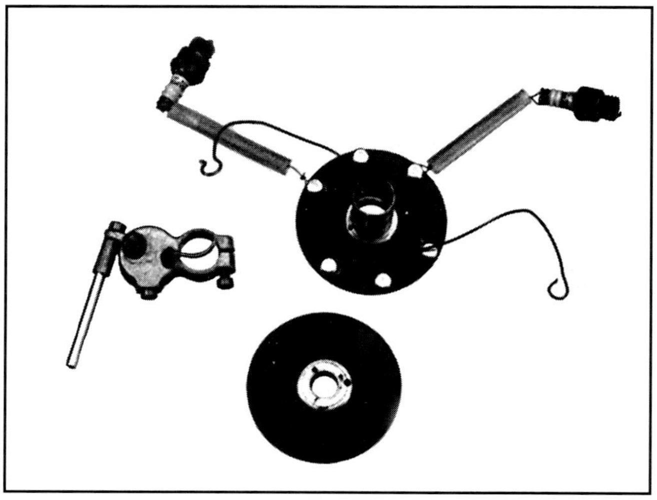

Figure 10-3 The special version of the offset cast (A) timer used for the Model 40, Series 2, with its distributor.

small drag on the engine. The only time there was a load on the engine was when the points were actually triggering the spark to jump across the spark plug contacts.

A modified version of this timer was used on the Model 40, Series 2 engines with a distributor. (Figure 10-3)

B. The Round Timer (1946-1952)

The round timer featured a machined aluminum case with a cast aluminum arm. The case had tabs punched out of the turned case to provide mounting surfaces for the breaker points. The tabs were punched from the inside out with the same device Dan used to punch the ports in the cylinders. The timer case housed one set of points for the Single and Twin, two sets of points for the Fours. This timer was used in conjunction with a machined split cam pressed over the aft end of the crankshaft. Beginning in October of 1949, the timer case material was changed from aluminum to steel. Most steel cases had a natural finish but a few had a black, gun-metal appearance. See Figure 6-16.

Figure 10-4 shows the round timer. On the left side is the cast timer arm that could be used with either the single set of points (center) or with two

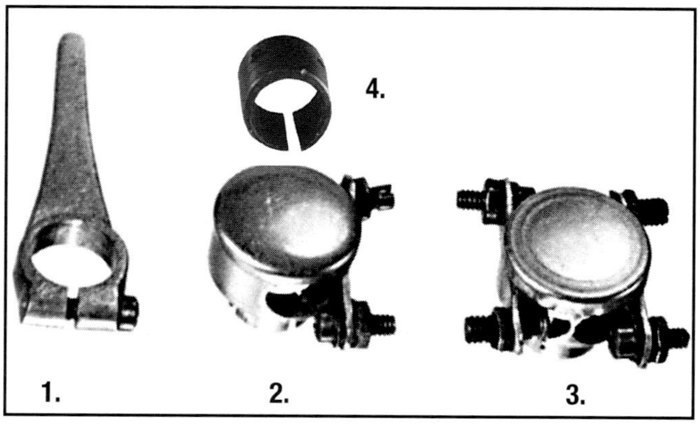

Figure 10-4 The round (B) timers. 1. The timer arm, 2. The timer for the Model 10 and 20, 3. The timer for the Model 40. Inset is an enlarged cam used for the round timers and the later open cast timer.

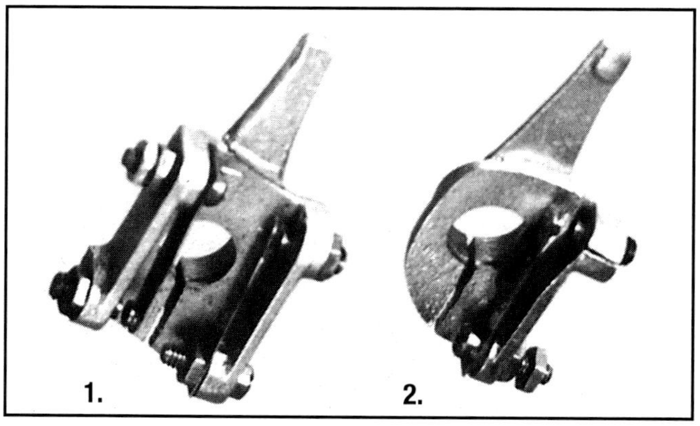

Figure 10-5 The open cast (C) timer was used on the later reed valve engines. 1. was used on the Model 40, Series 4B engines. 2. was used only on the Model 20, Series 4B engines. This open cast timer was never used on the Model 10 engines.

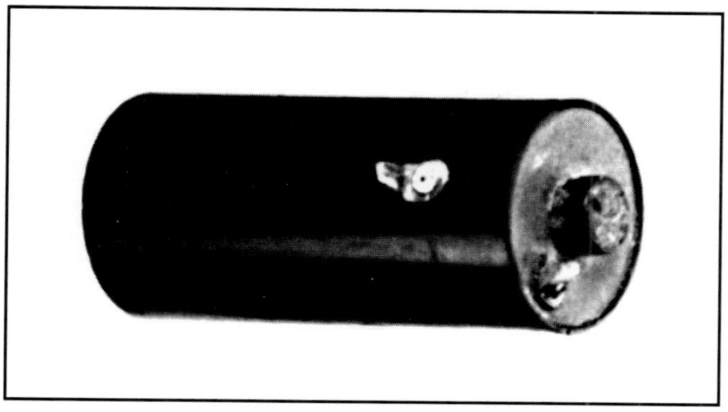

Figure 10-6 First ELF production coil. It was used by the early Model 1 engines. (Dick Tichenor Photo)

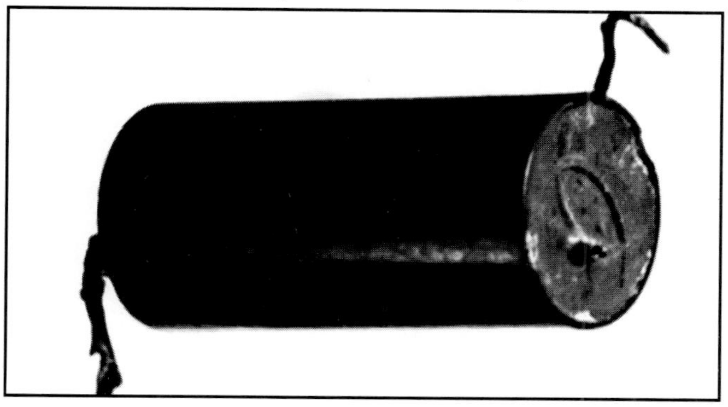

Figure 10-7 The second ELF coil was supplied with the later Model 1 engines. (Dick Tichenor photo)

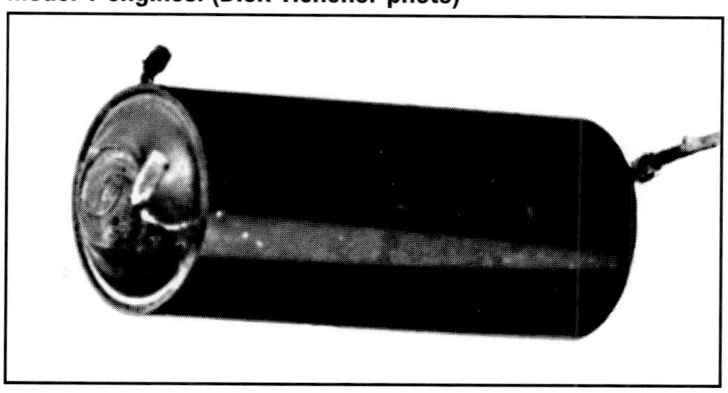

Figure 10-8 The third coil was used with the last Model 1s and pre-war Model 10 and 40 engines. (Dick Tichenor photo)

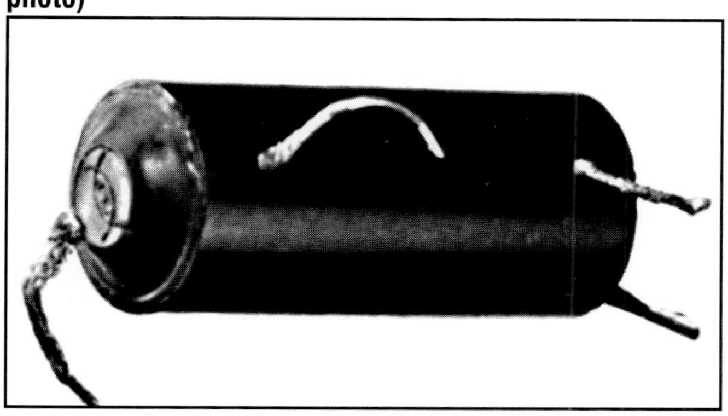

Figure 10-9 The fourth coil, a Twin coil was used by the Model 20 engines until 1942. (Dick Tichenor photo)

sets of points (right). In the center is the timer assembly used on the Singles and Twins On the right is the same timer with two sets of points used by the Model 40, Series 3B and 4B engines.

C. Open Cast Timer (1952-1956)

This sand cast, open style timer was manufactured with both one and two sets of breaker points. The spring, points and cams were the same ones used on the previous round timer. The single-point timer was only used on Model 20, Series 4 engines, never on the Model 10. The dual-point timers were only used on the Model 40, Series 4C engines. Figure 10-5 shows both the single and dual-point timers.

ELF COILS

Six different production ELF coils were produced, in addition to several different experimental units. They were all designed by Dan Calkin and built by ELF. All the coils were designed to operate on one pen cell (1.5 Volts), but could be operated on 3 Volts. The coils for the single cylinder engines have one high-voltage terminal while the twin coils, naturally, have two. The pre-war ELF Four, Series 2 spark ignition engine, because of the distributor mechanism, required only a single coil. The post-war Fours all required two twin coils or four single coils. All ELF coils had a laminated soft iron core and were housed inside a black bakelite case of 15/16 inch diameter (23.8 mm).

First Coil (1935-1937)

The first ELF coils had terminal lugs and red micarta end plates. The bakelite case measured about 2.15 inches long (51.8mm). The overall coil length was 2.5 inches (63.5mm). A bit of beeswax sealed the ends of the iron core. This coil was used only with the early Model 1 ELF engines. Figure 10-6

Second Coil (1938-1939)

This coil is similar to the type 1 except that it had beeswax covering both ends, with the ELF logo

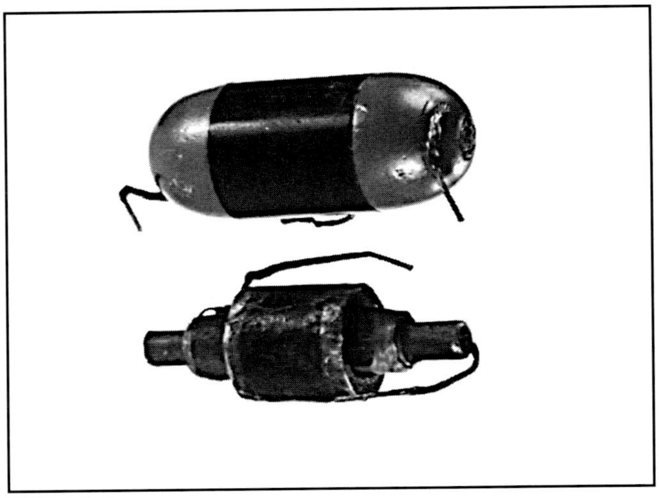

Figure 10-10 The fifth coil, a Single coil used from 1941 until the end of ELF in 1956. Shown both with and without the outside case. (Dick Tichenor photo)

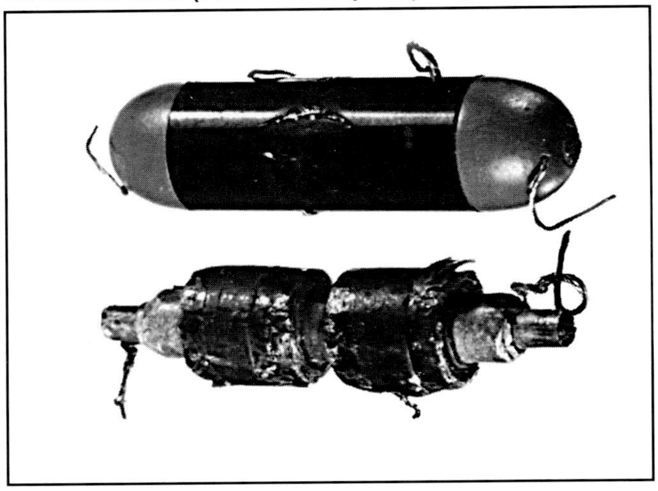

Figure 10-11 The sixth coil, a Twin coil used from 1941 to 1956. Also shown without outside case. (Dick Tichenor photo)

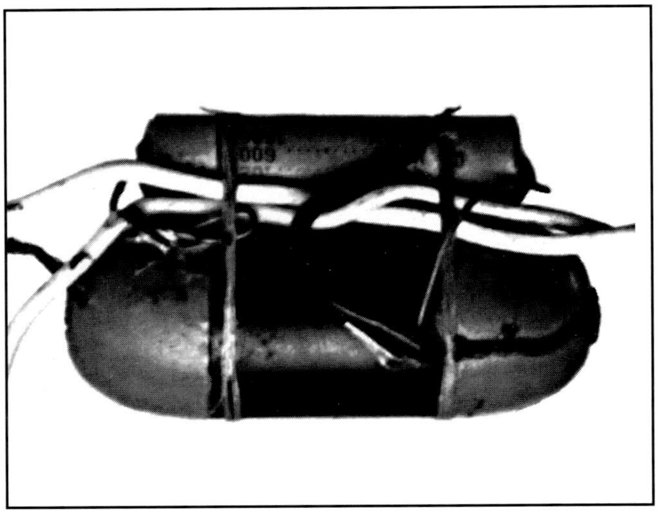

Figure 10-12 Dan Calkin's method for securing the flexible coil leads to prevent breakage.

on each end, instead of micarta. It also had flexible leads instead of solder tabs. This coil was used with some of the Model 1 ELF engines. (Figure 10-7)

Third Coils (1939-1942)

The third type ELF-manufactured coil had a cylindrical case and beeswax insulation on the ends and was the standard coil of this period. This was made in both Single and Twin versions and was intended to be operated on 3 Volts. The Single coil had the ELF logo on one end and the word "ground" stamped on the opposite end. (Figure 10-8) These coils were shipped with the last Model 1 Singles and with the early Model 10, and 40 engines.

Fourth Coil (1939-1942)

This was the Twin version of the previous coil. It was the first twin coil made by ELF and was used for the pre-war Twins. The ELF logo was stamped in the wax on both ends of the coil. (Figures 10-9)

Fifth Coil (1941-1956)

This coil and the Twin version of it were the most common ELF coils. Both had red caps over the ends of the iron core. The coils were known to have been produced starting in 1941 and continued until the end of ELF production in 1956. The Single coil is $1\frac{3}{8}$ inches long. (Figures 10-10)

Sixth Coil (1941-1956)

The Twin version of the previous coil was $2\frac{1}{2}$ inches long. Except for having dual outputs, it was based on the previous coil and was very similar.(Figure 10-11)

CONDENSERS

Five different condensers are know to have been used by ELF, but others may have been used as well. All condensers were standard, off-the-shelf stock as used in radios of that period. ELF Engine Company records do not show any purchases of condensers. The Series 3 and 4 ELF Fours required two condensers, all other ELF spark ignition engines required only one condenser. The source of the ELF condensers remains unknown. (Figure 10-13)

First Condenser (1935-1940)

The original condenser was the flat, olive drab style. The condenser itself was $1\frac{7}{8}$ inch long, 1 inch wide and $\frac{1}{4}$ inch thick. The overall length with the terminals was $2\frac{5}{8}$ inches. This style condenser was used with some prototype engines and during the early Model 1 engine production.

Second Condenser (1939-1942)

The second type condenser used was the square green style. The condenser body was $\frac{3}{4}$ inch square and about $\frac{1}{4}$ inch thick. These were shipped with some later Model 1 engines and some early slant plug engines, and may have been used much later.

Third Condenser (1940-1956)

The third known condenser type, and the one most commonly shipped with the slant plug engines, was the Aerovox brand cylindrical condenser. These Aerovox condensers had a 300 VDC rating. Their body measured about $\frac{3}{8}$ inches in diameter and $1\frac{1}{2}$ inches long. Capacitance was .03 mfd.

Fourth Condenser (1942-1956)

This was another Aerovox condenser similar to the third style, but rated at 400 VDC. The body measured about .410 inches in diameter by $1\frac{1}{2}$ inches long. Capacitance was either .03 or .04 mfd.

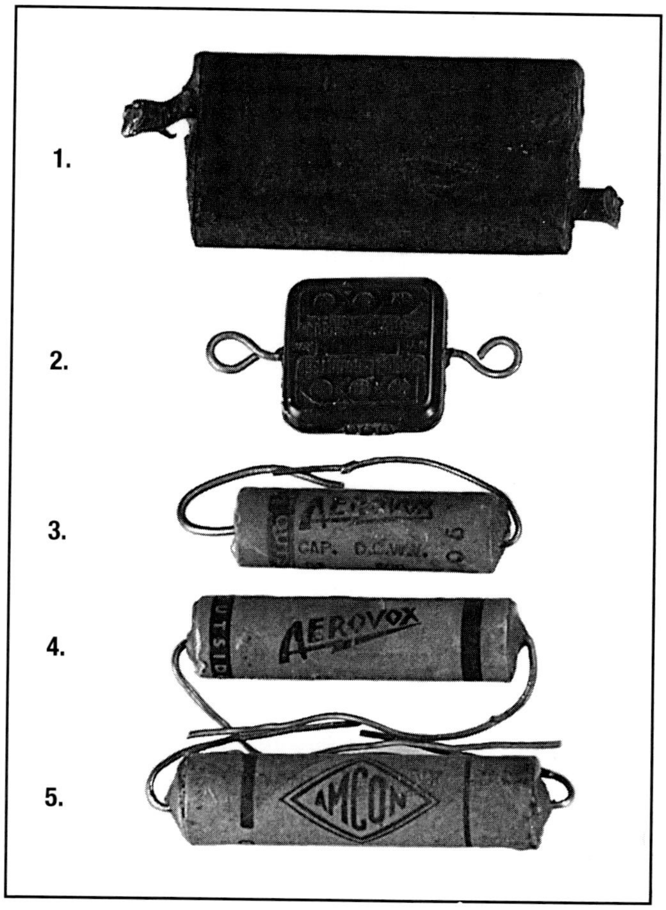

Figure 10-13 The condensers ELF used for the production engines.

Fifth Condenser (1950-1956)

The last known condensers were made by the American Condenser Company. They were similar in appearance to the Aerovox units. They had a yellow waxed-paper-covered, cylindrical body with the AMCON logo inside a diamond. They had a capacitance of 0.02 mfd., rated at 600 VDC. Their diameter was about 0.440 inches with a length of $1\frac{5}{8}$ inches.

Dan Calkin advised the users of ELF coils to tie down the flexible leads to prevent breakage from vibration. Figure 10-12 shows a coil used by Dan showing how it could be done.

Chapter 11 IGNITION PLUGS

SPARK PLUGS AND GLOW PLUGS

THE EARLY SPARK PLUGS

When Dan Calkin first started making his engines, he devoted considerable time and effort to making a spark plug. He made a number of plugs using a variety of insulators. He quickly realized that special equipment would be required to make the insulators. He knew that it would not be practical to obtain such equipment, given the relatively small quantities involved.

Figure 11-1 (left) shows the $^{5}/_{16}$ x 32 spark plug with stacked mica insulator made for his first successful engine, the 1928 boat engine, and (right) the $^{5}/_{16}$ x 48 plug for the 1931 University of Washington engine.

Figure 11-1 The Calkin made spark plugs used in the 1928 boat engine and the 1931 University of Washington engine.

On May 4, 1934, Dan wrote to Champion Spark Plug Company of Toledo, Ohio, to inquire about the availability of a small spark plug. Champion replied by sending him two 6 millimeter spark plugs (Figure 11-2). These plugs were much too long for the ELF, having been designed for much higher voltages than the ELF coil provided. These plugs, having no separate ground electrode, required the spark to jump between the center electrode and the metal case. This is known as an annular gap.

PRODUCTION SPARK PLUGS

The ELF Spark Plug

Dan designed and built a small spark plug of the type he wanted. (See Figure 11-3) Dan's next let-

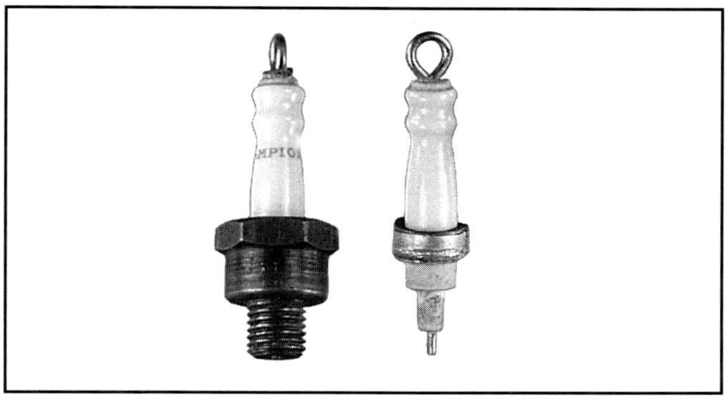

Figure 11-2 The two six millimeter spark plugs that Champion sent Calkin in 1934. (Right hand plug cut away)

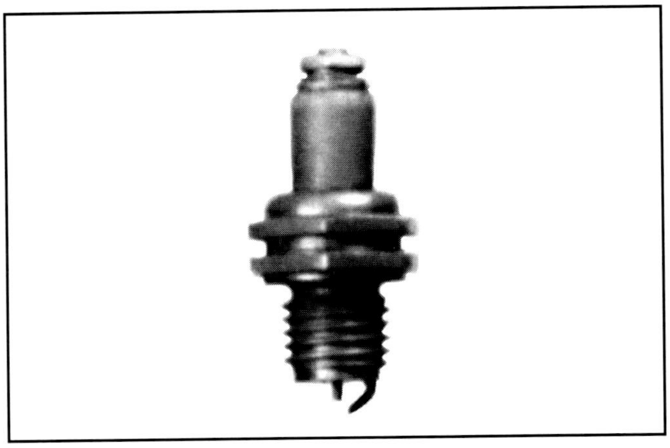

Figure 11-3 The Calkin-developed 1/4-32 spark plug.

Figure 11-4 The ELF spark plug made by Calkin using a Champion insulator. (Enlarged)

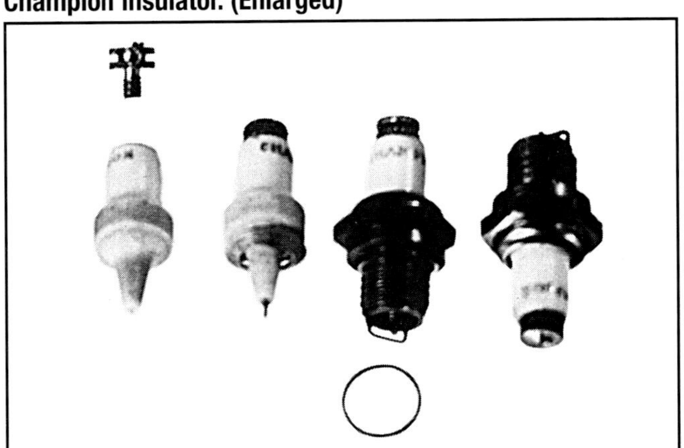

Figure 11-5 Two insulators and two ELF plugs with the top contacts, metal seals and the thin brass washer.

ter of March 13, 1935, included a request for Champion to make only the special insulator of his design. He received a price quotation for both the finished insulators and a supply of the unbaked, raw sillimanite insulator material. Dan considered turning the raw blanks himself, then returning the turned cores to Champion to be glazed, fired and decorated. The price of the finished insulators would be $25.00 for 100 pieces. The price for buying raw insulator material and returning the shaped material to Champion for finishing was $30.00 for a lot of 100 pieces. The obvious choice was to buy the finished insulators and a lot of 100 was ordered. These insulators were used by Calkin to make his famous ELF spark plug, and from the very beginning of production of his Model 1 ELF engines, these spark plugs were used.

Dan shipped a complete, ready-to-run Model 1 ELF engine to Champion to convince them that he had a viable product and a need for plugs. Their reply was:

> I deeply appreciate your sending me the little engine you promised me last year. It arrived in perfect condition, and I am having it set on a little base with the idea of exhibiting it to our engineering staff and to the visitors who, from time to time, go through our Plant in considerable numbers.

In the September 28, 1936, letter from Champion, an order for an additional 100 insulators was acknowledged, and it was stated that 43 pieces had been shipped the previous day and that the remainder would follow soon.

Champion V Spark Plug

Champion notified Calkin via a July 13, 1937, letter that they had recently brought out a new Type V ($^3/_8$ inch – 24) plug and sent him a sample. They hoped that he would be able to use it for his engines. Of course, a large plug such as this was unsuitable in such a small engine.

Figure 11-6 Calkin's display of spark plugs from the middle 1940s. From left to right: boat engine, U of W, first 1/4" plug, ELF with Champion insulator, Champion V-2, Champion V-3.

The next letter from Champion must have come as a shock to Dan. The October 26, 1937, letter states in part:

> We have discontinued the manufacture of any special porcelain parts, such as the small insulator we have furnished you in the past.

The same letter continued with another offer to sell him the Type V plug.

Dan responded with a carefully worded letter explaining again why the Type V plug was unsuitable, and asking if they could recommend someone else who could supply the special insulator.

A surprise turning point occurred with the November 22, 1937, reply from Champion. Hope was offered when the letter said:

> It is quite likely that we may be able to refer you to another source for the parts you need.

Champion V-2

The really startling item was in the post script to the November 22, 1937, letter which read:

> Since writing this letter we find there is a possibility we may manufacture a much smaller plug for model airplanes. One of the men from our Engineering Dept. in Toledo is in the Plant today and he states it is now being discussed. He feels it is almost sure to go through especially since you can order as many as 1,000 at a time.

The next letter from Champion reads:

> Your letter of January 15 (1938) approving our miniature type spark plug, after severe test, is greatly appreciated. In small

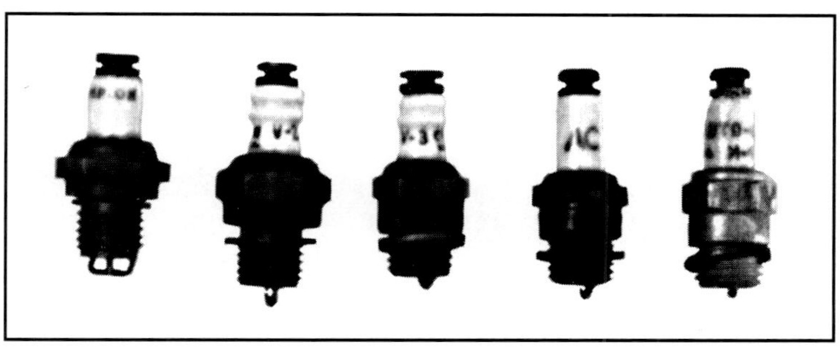

Figure 11-7 All the spark plug types used in the ELF production spark ignition engines.

Figure 11-8 All glow plug types used in the ELF production engines. Left: Champion VG-3, Center: Auto-lite M4G, Right: AC 4G.

Figure 11-9 Model 40, Series 3 ELF #1379 with AC spark plugs .

Figure 11-10 Model 40, Series 3 ELF #1379 with Auto-lite spark plugs .

> quantities we are glad to quote you 31 cents each, F.O.B. Toledo, Ohio.
>
> It might interest you to know that on an average of once a week, for some time now, we have been adding another manufacturer of a miniature type motor to our files of those using Champions for original equipment.

The new plug, of course, was the V-2 spark plug. Except for the ground electrode, this plug was mechanically and electrically very similar to the ELF spark plug.

A few days later, on February 18, 1938, after receiving Champion's latest letter, Dan revised his price list, changing the price of the spark plug from $2.20 to $1.10 each.

Shortly after this date, Dan received a supply of the Champion V-2 spark plugs, and no more ELF spark plugs were ever made after this time. Of the 200 insulators Dan bought from Champion to make the ELF plugs, about 115 still exist. This implies that no more than about 85 ELF spark plugs were made. Since a few engine serial numbers were skipped, due to parts being scrapped during the machining phase, the ELF spark plug could not have been used after engine number 100 of the Model 1 ELF engines. All Elf Model 1 engines after February, 1938, from about serial number 100, would use the new Champion V-2 plug. (See Figure 11-4) Figure 11-5 shows two complete ELF plugs as well as an insulator, internal sealing washer, external sealing washer and top connection. The ground electrode was really a "U" shaped piece of wire inserted into two holes in the base of the spark plug.

Champion V-3

When Dan started producting the ELF slant plug engines in October 29, 1939, the Champion V-3 spark plug was used. Champion introduced this plug in the summer of 1939. The ELF Engine Company used the V-3 spark plug until the wartime shut-down and continued using it when production resumed in 1944. In September, 1945, Dan bought 1,000 V-3 spark plugs in two batches of 500 plugs each.

Figure 11-6 shows a display Dan Calkin made while at the Washington Street shop displaying all the plugs used up to that time. From left to right are the boat engine plug, the U of W engine plug, the Calkin-developed plug, The ELF plug with Champion insulator, the Champion V-2 and the V-3.

AC 4H Spark Plug

Only Champion V-3 spark plugs were used until March, 1950, when Dan bought 50 type 4H spark plugs manufactured by the AC Spark Plug Company of Flint, Michigan. Another shipment of 50 type 4H plugs from AC was received in June, 1951, with 100 more arriving in May of 1951. He probably used these plugs in engines for several months after this date. No additional AC spark plugs were purchased by ELF after the May, 1951, shipment.

Figure 11-11 A Model 40, Series 4 ELF with AC glow plugs #1385. Note that the rear heads are not relieved.

Auto-lite M-4 Spark Plug

In May, 1951, Dan started buying the M4 spark plug from the Electric Auto-lite Company of Toledo, Ohio. The M4, as well as the AC 4H spark plugs, were the electrical and mechanical equivalent of the Champion V-3 spark plug. Four purchases of 100 plugs each took place in May, 1951, July, 1952, January, 1953, and March, 1954. These plugs were used until at least October 1954, when Dan started buying the Champion V-3 again. By that time, the AC and Auto-lite spark plugs were no longer available. Dan bought 100 V-3 plugs in October 1954, another 100 in March, 1955, and a final purchase of 100 in September, 1955. These purchases lasted him until the last ELF spark plug engine was sold in December of 1955.

Figure 11-7 shows all the production spark plugs used in the ELF engines. From left to right are: the ELF, Champion V-2, Champion V-3, AC 4H, and Auto-lite M4.

GLOW PLUGS

Champion VG-3 Glow Plugs

ELF Engine Company records show that Dan bought ten Champion VG-3 glow plugs in October 1949. This was a $\frac{1}{4}$-32 plug with a short reach that looked almost like the V-3 spark plug from the outside. In January, 1950, another 20 units were purchased. The first ELF glow plug equipped engine was shipped in April, 1950.

Figure 11-12 A Model 40, Series 4 ELF with Auto-lite glow plugs #1385. Note that the rear heads were not relieved.

ELF glow ignition engines continued to use the VG-3 plugs until mid 1951.

AC 4G Glow Plugs

In March, 1950, Dan purchased 50 of the 4G glow plugs made by AC. Another fifty were purchased in January, 1951, and a final 100 4G glow plugs were bought in June, 1951.

Auto-lite M4G Glow Plugs

While using AC glow plugs, Dan was also buying the M4G glow plug from Auto-lite. Purchases of 100 were made in May, 1951, July, 1952, and January, 1953. A final purchase of M4G plugs was made in March, 1954.

Both the AC and Auto-lite glow plugs were much shorter than the Champion VG-3, which enabled Dan to forego his usual practice of machining the front of the rear cylinders on the Model 40 and 60 engines for plug clearance. All Model 40 spark ignition engines and all ELF Fours and Sixes equipped with the Champion VG-3 glow plugs had the front of the rear cylinders relieved to make room for the plugs at the rear of the front row.

Champion VG-3 Again

By the middle of 1954, both the AC and Auto-lite glow plugs had been discontinued. Dan returned to the Champion VG-3 plug, buying four orders of 100 plugs each in July and October of 1954, and July and September of 1955. No additional purchases of glow plugs of any kind occurred after September 1955. Apparently, the stock on hand lasted until he sold the last glow plug equipped engine on April, 11, 1956.

In addition to the purchases of both spark plugs and glow plugs recorded above, it is possible, indeed probable, that Calkin made other smaller purchases of plugs that are not reflected in the company records. For instance, it is known that he used glow plugs before the first record of such purchases occurred. They were probably purchased from a local supplier in small quantities.

Figure 11-8 shows all the known glow plugs supplied by ELF as original equipment on their engines.

This information should make it possible for the ELF engine owner to determine which plugs could have been initially shipped on any particular engine. Only reed valve engines used something other than Champion plugs. Simply look up the serial number of the engine in Appendix A to find the date it was first sold, then refer to this chapter to find which plugs were in use at that time.

Chapter 12 ELF ACCESSORIES

MISCELLANEOUS ITEMS

MODEL 1

Flywheel

The 1938 ELF price list had a listing of a flywheel for use with a stationary Model 1 ELF engine. The flywheel had cooling fins to provide a degree of cooling. (No photo is available).

Exhaust Pipe

As far as is known no exhaust pipes were marketed for the Model 1 ELF Singles. However, Dan did have one that he used on his 5 foot airplane model. Refer to Figure 12-1 and 1-20.

Figure 12-1 The exhaust pipe that Dan Calkin used on his own Model 1 ELF prototype.

Figure 12-2 The metal motor mounts suggested for the Model 1 ELF #339. Note the downthrust. (Mounts fabricated by Dan Lutz)

Figure 12-3 The five styles of exhaust pipes used on the slant plug engines. #1—curved exhaust for the Series 1 engines. #2—straight tube for Series 2 engines. #3—flared exhaust for Series 2 engines after 1944 and early Series 3 engines with .250 inch exhaust. #4—flared exhaust for later Series 3 engines with .275 inch exhaust. #5—5 1/2 inch pipe for Series 4 engines (bend to suit model).

Motor Mounts

The Model 1 ELF Singles were beam mounted and Dan had several ways of mounting them to protect the engine from crash damage. Mounting brackets for the Model 1 ELFs were never sold by the ELF Engine Company, but two different mounts were shown on the plans for the ELF three foot airplane. See Figure 12-2 and Figure 2-10.

MODEL 10, 20, 40, 60 ACCESSORIES

Exhaust Pipes

For the slant plug engines, five different exhaust pipes were used. Figure 12-3 shows them all.

Exhaust Pipe #1 (1939-1941)

With the introduction of the Model 10 and 20, Series 1 engines, the so-called smooth nosed Single and Twin, a curved exhaust pipe was standard and was normally shipped with each engine.

Exhaust Pipe #2 (1941-1942)

With the introduction of the Model 10, 20 and 40, Series 2 engines with the four reinforcing gussets on the front crankcase, Dan used a new, short, straight $1\frac{1}{8}$ inch long aluminum exhaust tube. Sometimes they were supplied with each engine, but there was no consistent policy. No other exhaust pipes were made through 1942.

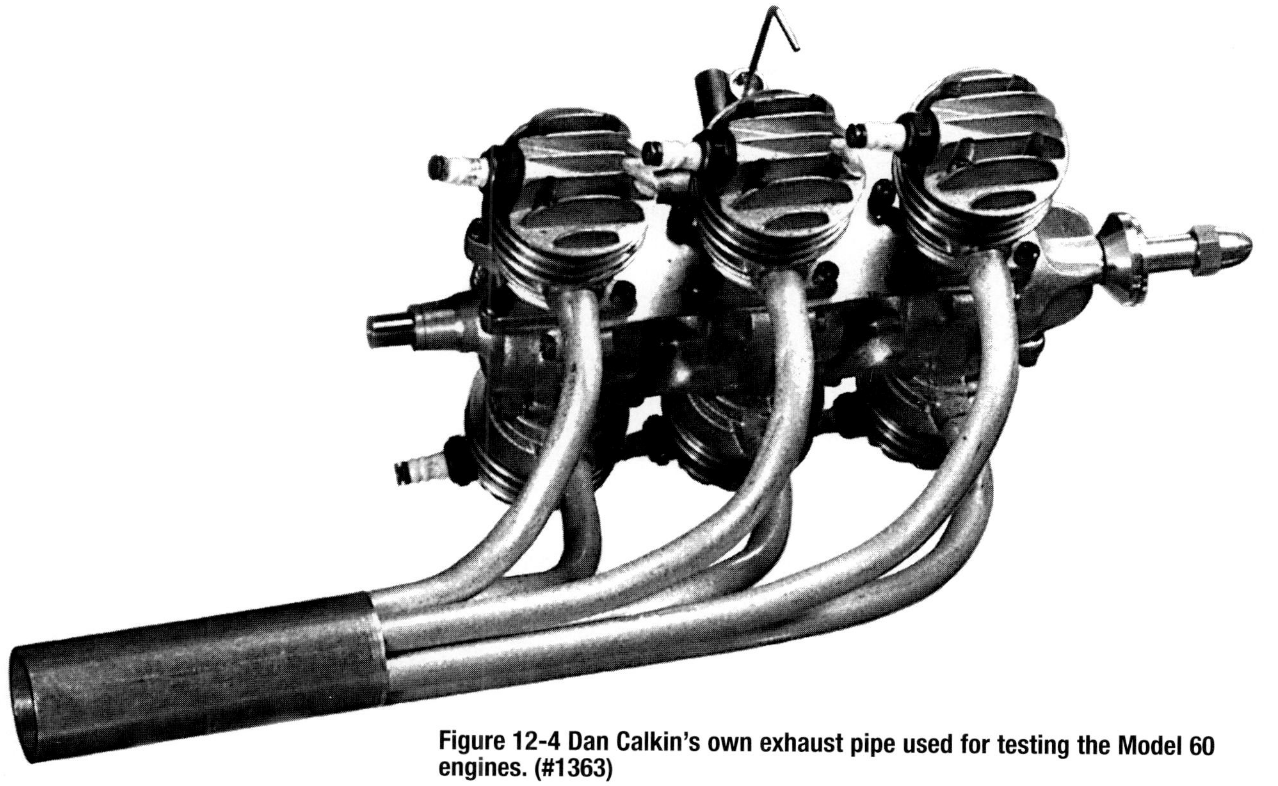

Figure 12-4 Dan Calkin's own exhaust pipe used for testing the Model 60 engines. (#1363)

Exhaust Pipe #3 (1944-1948)

After engine production resumed in 1944, a flared exhaust pipe became available to fit the Singles and Twins. These were about $1\frac{5}{16}$ inches long with an $\frac{11}{32}$ inch diameter at the outer end. The other end would fit in the $\frac{1}{4}$ inch exhaust hole. Naturally they would fit all the pre-war Model 10, 20 and 40 engines.

Exhaust Pipe #4 (1948-1950)

With the increase in the exhaust port diameter to 0.275 inch in August, 1948, a flared exhaust pipe was made to fit these engines. The dimensions were unchanged from the previous exhaust tube except to make the engine end slightly larger to fit the enlarged exhaust port. This item was listed on the 1949 ELF price list.

Exhaust Pipe #5 (1950-1956)

By the time the reed valve engines were brought to market in 1950, Dan seemed to realize that, while the flared exhaust looked attractive, it did little for the performance. Dan found that a much longer exhaust pipe of about $5\frac{1}{2}$ inch length would increase engine speed by 500-1000 RPM. Apparently, Dan had found out about tuned exhausts. This pipe, as supplied, was a straight length of 0.250 aluminum tube with one end of the tube expanded to 0.275 inches to fit the ELF engines made since about mid 1948. The long, soft aluminum exhaust pipe could be bent by the ELF owner to best fit his airplane. This item was listed on the 1950 price list and the flared 0.275 inch stack was crossed out. Notice the use of these long exhaust pipes on Dan's radio plane in Figures 17-13 and 17-14.

For his own bench testing on the ELF six cylinder engines, Dan used the large exhaust manifold shown in Figure 12-4. It consisted of six, type 5 exhaust tubes with a machined collector at its extremity. This item was not offered for sale.

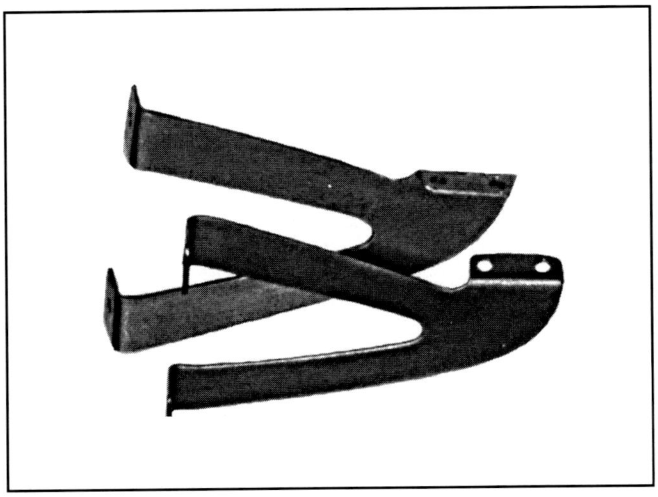

Figure 12-5 The same style metal motor mounts used by the Model 1 ELF engines were used by the early ELF Model 10. (Fabricated by Dan Lutz)

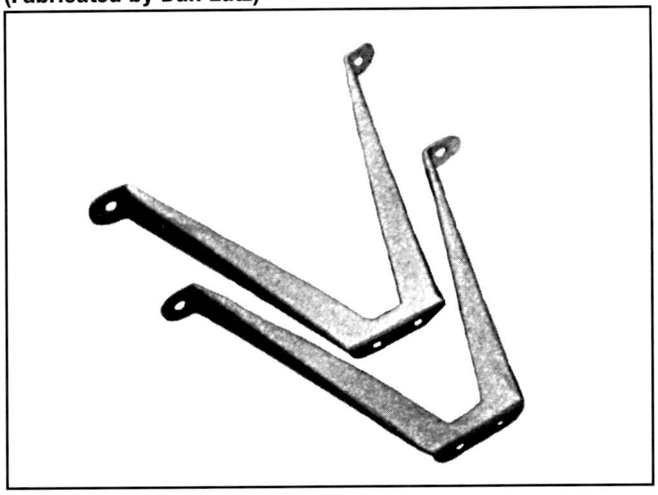

Figure 12-6 The metal motor mounts an owner could fabricate for his radial mount ELF. (Fabricated by Dan Lutz)

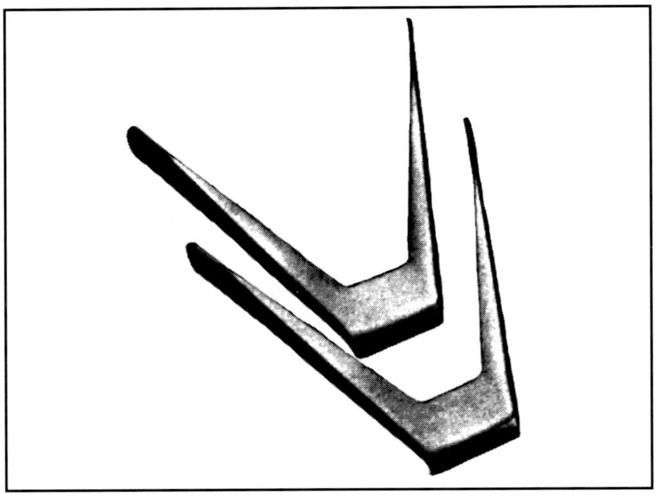

Figure 12-7 The standard metal motor mounts for all ELFs except Model 1 and Model 10, Series 1.

MOTOR MOUNTS

Model 10, Series 1

The Model 10, Series 1 engine utilized a conventional beam mounting so no brackets were required. However, the illustration on the box the engines were shipped in showed the engine mounted on the same style engine mount the Model 1 ELF engine used. Figure 12-5.

Model 10, 20, 40, 60

The earliest instruction sheet for the ELF Model 20 suggests how the modeler could make his own metal mounts. This was before the ELF metal mount became available. Figure 12-6

The Elf mounting bracket (Figure 12-7) was a standard accessory item from about two weeks after the first Twin was shipped. These were supplied with unbent ears for mounting against the airplane's firewall. The ears could be bent in or out, and shortened to suit the airplane. On the Twins, Fours and Sixes, one bracket mounted at the top of the rear crankcase, while the other mounted below. For the pre-war four cylinder ELFs, the brackets had to be bent slightly to clear the carburetor. The Model 10 Singles, from the Series 2 on, used the same mounting bracket as the multi-cylinder engines, but the brackets were located on the left and right rear sides of the engine.

Crankshaft Extension (1947-1956)

An accessory item that was available, although never listed on price sheets, was a crankshaft extension. ELF records list sales of a few extensions in July, 1949. These extensions all seem to vary slightly so they were probably made in small batches, as needed. At least six extensions are known to exist. One is shown in Figure 12-8. Figure 12-9 shows an extension shaft mounted on a reed valve Twin that Dan used in his own airplane. The extension was basically an extended, hollowed-out drive washer with a short shaft extender for the prop nut connection.

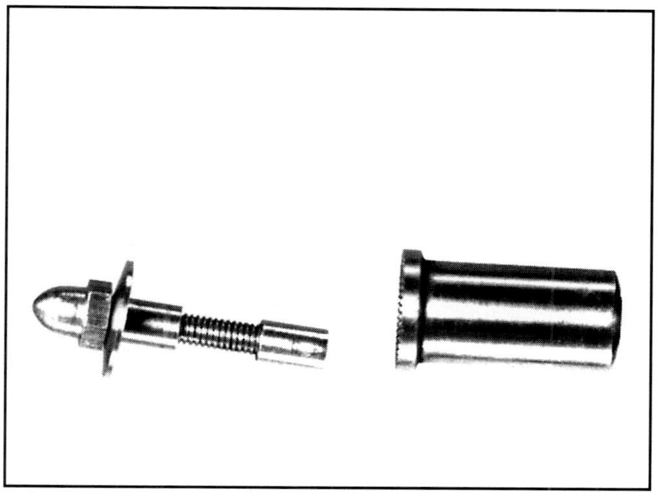

Figure 12-8 The crankshaft extension.

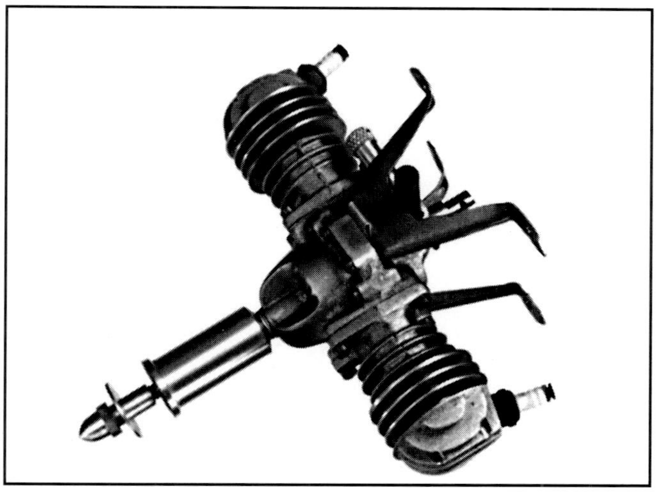

Figure 12-9 The crankshaft extensions on one of Dan's own Twins #1493. Notice the shortened mounts.

Figure 12-10 The little-known fuel pump.

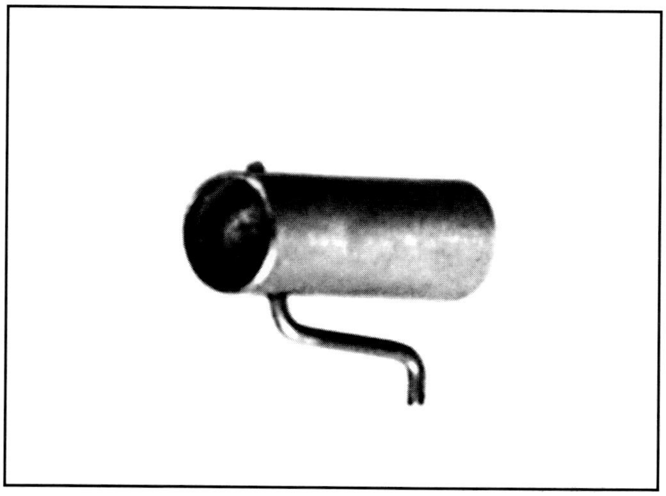

Figure 12-11 Throttle for carburetor.

Double Timer (Two-Speed)

The October 1949 price list offered a double timer for the Model 10, 20 and 40. A double timer was offered to Gordon S. Light on September 22, 1947. It was described this way in a letter to Gordon from Dan:

> We are using a new timer (presumably a round timer) which can be put on your engine. A second similar timer and double cam are added to this for two speed control. The fast and slow speeds are independently adjustable. The new timer price is $3.60. The second timer is $3.20.

The double timer was, most likely, a special version of the round timer. No photo is available.

Fuel Pump (1949-1950)

On the October 1949 price list a fuel pump accessory was offered for sale. The pump could be used on any ELF with the round timer. The pump itself was the same, regardless of whether it was for a Model 10, 20 or 40. See Figure 12-10.

Throttle for Carburetor (1949-1956)

On July 18, 1949, the ELF log book had a notation that a throttle valve was available and was being shipped with some engines at added cost. This consisted of a short brass tube that slipped over the intake tube on the Model 20, 40 and 60 reed valve engines. Inside the tube was a butterfly valve soldered to a piano wire rod bent to provide an attachment place. See Figure 12-11. This item was added to the 1950 price list and was available until the end of ELF production. (Figure 3-23).

Chapter 13 ELF PROPELLERS

ELF, WESTERN MODELCRAFT SUPPLY CO. AND HARRY FOSBURY

PROPELLERS

Model 1 Engines

When Dan first started selling engines in 1935 no other small engines were on the market, and consequently there were no propellers for such small engines.

Dan designed a propeller that he felt was ideal for a situation where fuel economy was the order of the day. One way to achieve fuel economy was to run the engine at a low speed. The propeller that he designed had a 12 inch diameter with a pitch of 6½ inches. This propeller limited the engine speed to about 3600 RPM. Most of his propellers were made of alder, but a few were made of mahogany. See Figure 13-1. Also refer to pages 6 and 7 of Appendix C: *Notes on Model Airplanes and Propeller Design.*

A machine was built to facilitate the rough carving of the props, then hand sanding and finishing was required. Usually, the finishing was done by Rufus Rothrock and his brother Richard, "Dick". (They called themselves Western Modelcraft Supply Co). Rufus, being much older than his brother Dick, was the head of the company.

SLANT PLUG ENGINES

With the introduction of the Model 10 and 20 engines in 1939 and 1940, a new series of propellers was required. Although other small engines were now on the market, such as the *Brat* and *Atom*, props were still not readily available. Again, Dan contracted with the Rothrock brothers to supply the ELF propellers. Usually, Rufus would cut the prop blanks out of alder on a bandsaw and Dick would carve them. Many of the propellers had the name, Western Modelcraft

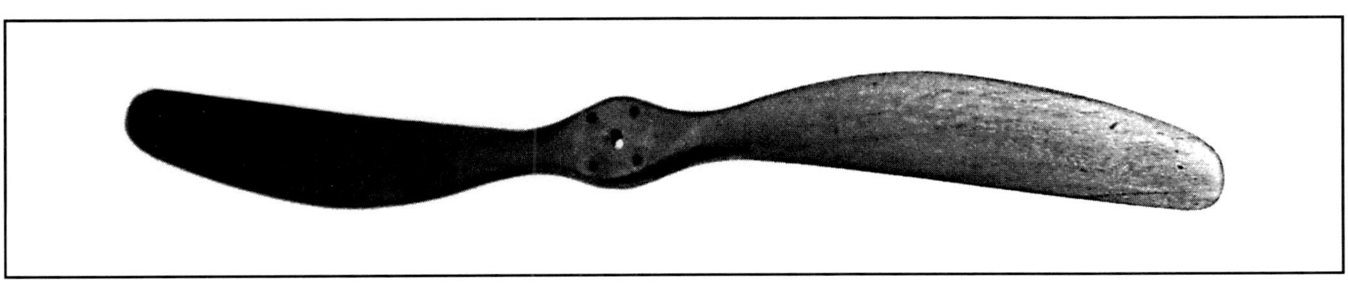

Figure 13-1 The Calkin designed prop for the Model 1 engines. Dan made a machine to do the rough carving, the Rothrock brothers did the rest. (Dick Tichenor photo)

Supply Co., hand written on each blade by Rufus. (See Figure 13-2) They sometimes said 'ELF.' When finished, Rufus would deliver a handful of props, perhaps a dozen or two, to Dan. For his services, Rufus was paid with cash or with ELF merchandise.

Model 10

The standard propeller for an ELF Single Model 10 was an 8 x 5¾ size. For larger, slower models a 9 x 6 prop was recommended.

Model 20

The best all around prop for the ELF Twin was a 9 x 6 or a 10½ x 6, but Dan had other sizes, such as a 10 x 7 and 11½ x 5 prop, for larger and slower models. For control line flying, a 9 x 10 was suggested.

Model 40

For the new Model 40, the recommended prop was a 12 x 6, 12 x 7, or a 13 x 5. For control line flying a 9 x 10 prop was best. See Figure 13-3 for an assortment of ELF propellers.

Dick Rothrock provided me with the original Calkin-made template for the 8 x 6 prop, a carved prop and partly finished prop to show how it was done. (Figure 13-4). He still had the blank and knew how to carve the props, although over 50 years had elapsed. At the top (#1) of Figure 13-4 is a prop blank laid out to begin carving. Next, (#2) is a partly carved prop. Next, (#3) the pattern Dan Calkin supplied for the 8 x 6 prop was traced on each blade. The #4 item is the complete finished prop Dick made for this book. The fifth and sixth items are original props made by the Rothrocks before World War II.

In the days before World War II, model airplane engines were used almost exclusively for free flight, thus relatively low pitch props were needed. Control-line airplanes very often needed higher pitched props of smaller diameter than did the free-flight model planes. Because ELF engines reached their peak torque at relatively low RPMs they required larger propellers than most other comparable sized engines.

HARRY FOSBURY PROPS

When ELF Engine Company resumed production in 1944 after an 18 month shutdown, ELF could no longer supply props, as Rufus and Dick Rothrock were no longer making them. By this time, other companies had sprung up to meet the

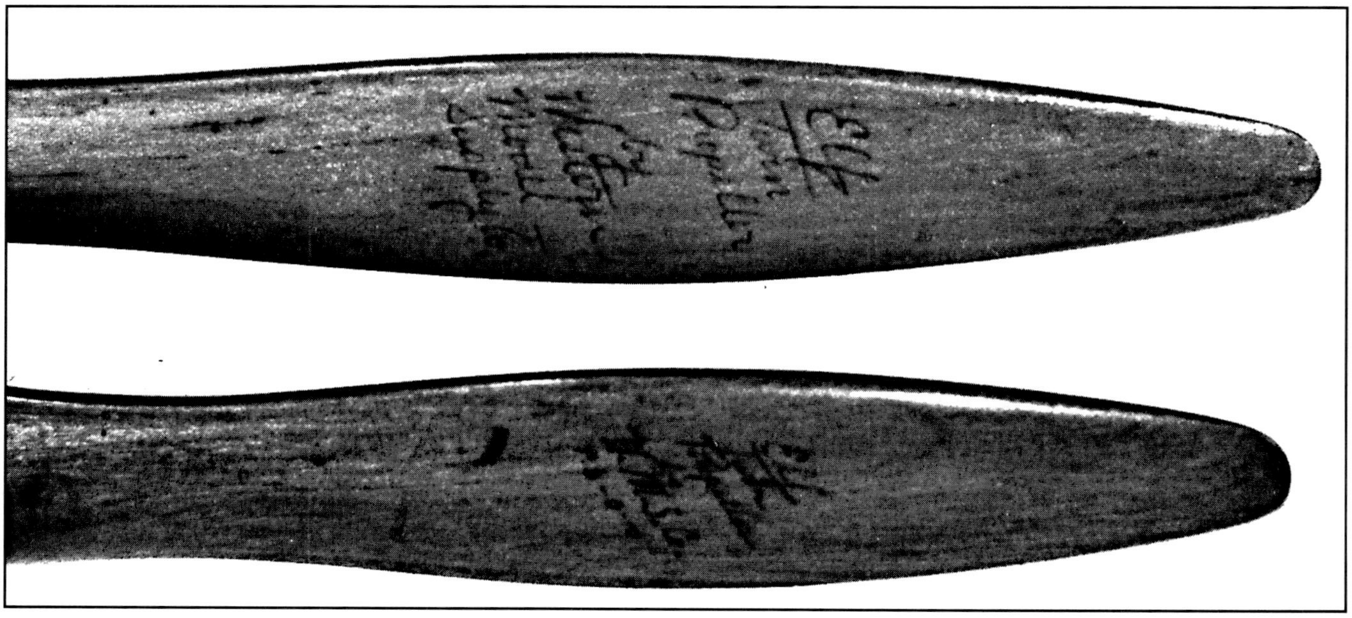

Figure 13-2 Notice the writing on the propeller blades of these Western Modelcraft props.

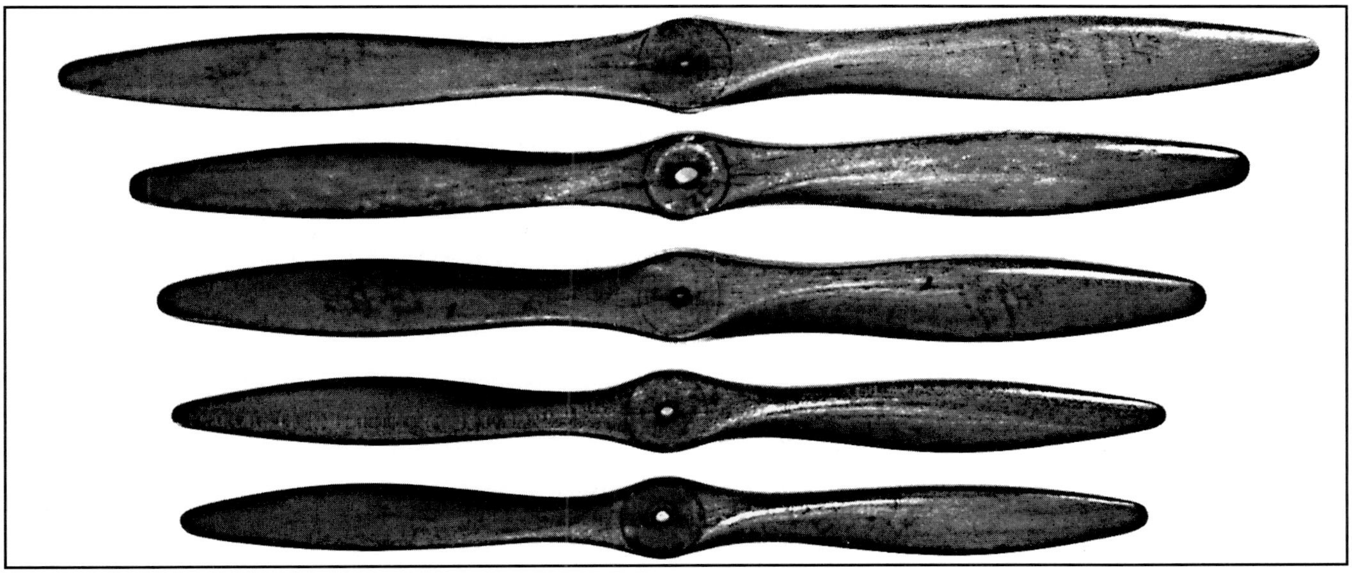

Figure 13-3 A selection of props from Western Modelcraft Supply Co.

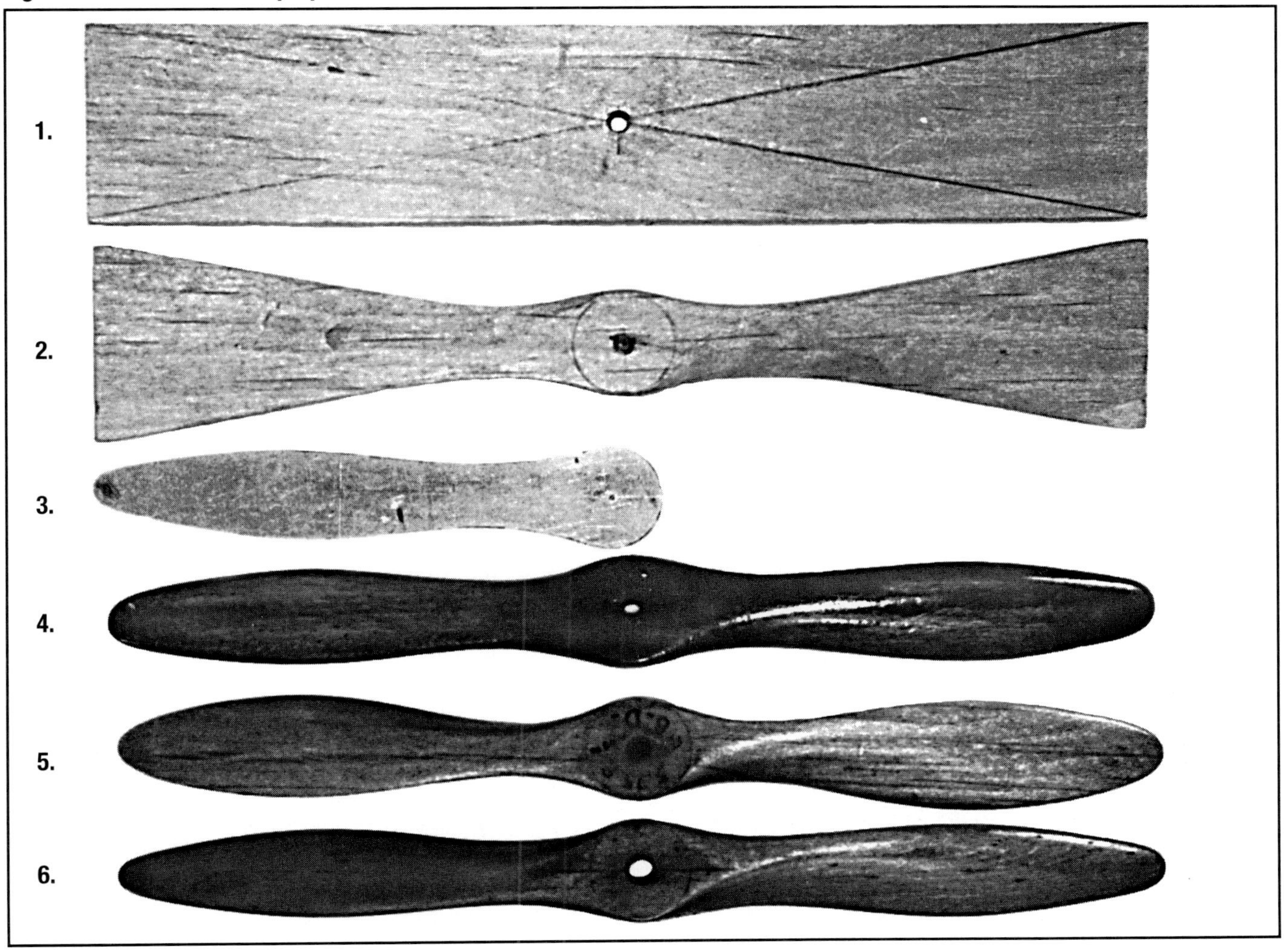

Figure 13-4 The steps Dick Rothrock used in making his props. #1 is the layout of the blank, #2 is a partly carved prop. #3 is the blade pattern given him by Calkin to create the shape of the blade. Under the pattern (#4) is a complete prop Dick made in 1997, bottom two props (#5, 6) were made by Dick in early 1940s.

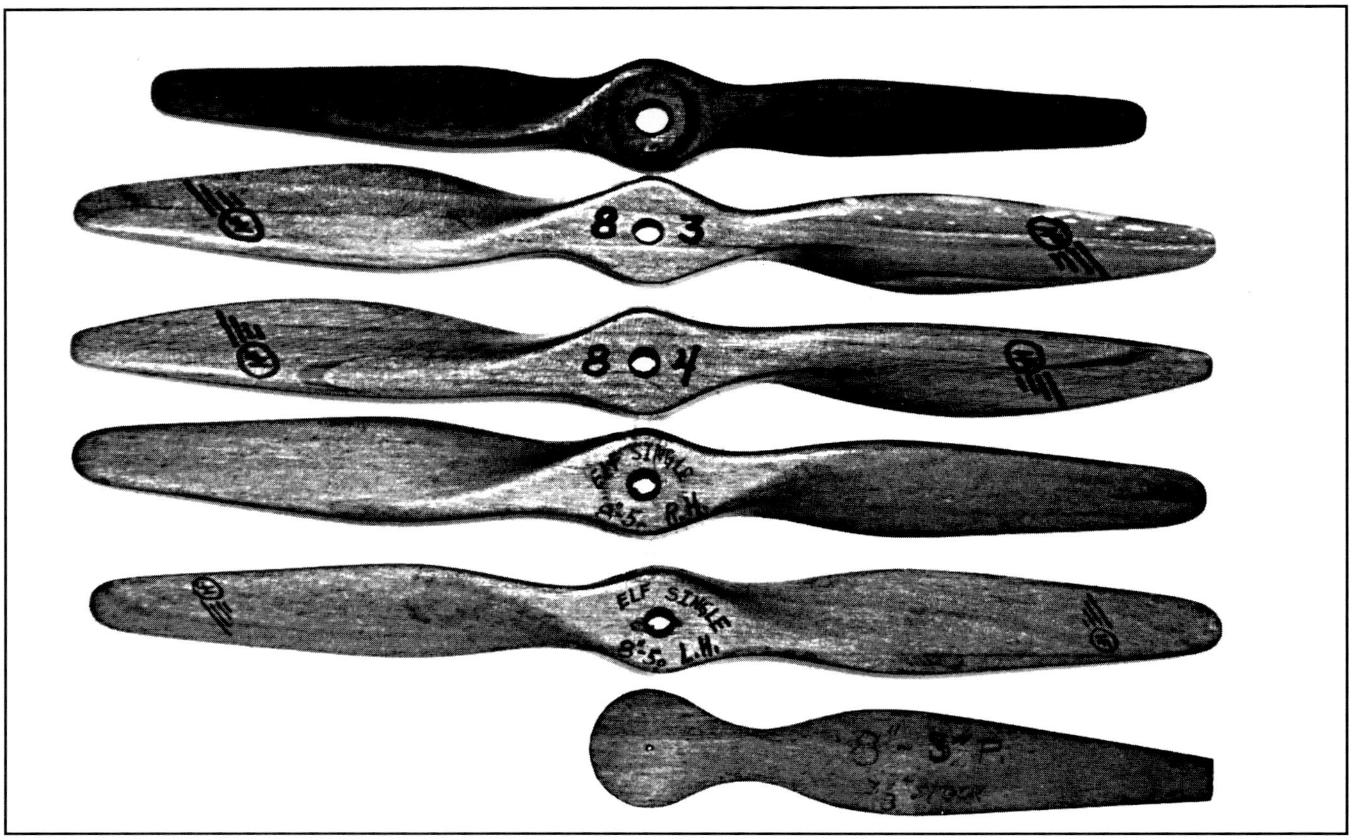

Figure 13-5 An assortment of props made by Harry Fosbury for ELF engines. The plywood template at the bottom was made for Harry by Dan Calkin. (Dick Tichenor photo)

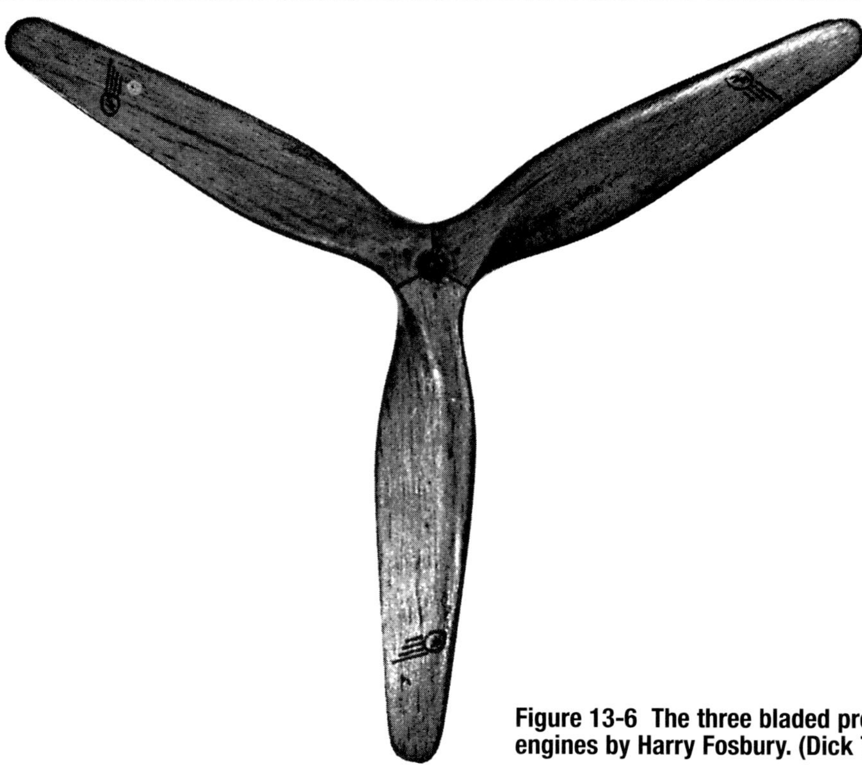

Figure 13-6 The three bladed prop for the Model 40 and 60 made for ELF engines by Harry Fosbury. (Dick Tichenor photo)

rising demand for propellers for all sizes of engines.

One of these new entrepreneurs was another Portland resident and associate of Dan's, Harry Fosbury. Harry was well known and highly respected in the Pacific Northwest for his high quality, beautifully smooth balsa that he supplied to most dealers and modelers in the area. Harry designed and manufactured a series of propellers with his HF logo, some of which were made especially for ELF engines. Dan designed the ELF series of props and provided templates to Harry, who manufactured and distributed them. Dan Calkin and Harry Fosbury are shown in Figure 13-9 discussing the performance of a twin engined ELF-powered sea plane.

Later, Harry designed and produced the props for other companies, most notably the highly regarded Y & O (Yates and Orwick) props. The name Harry Fosbury may be familiar to some old time modelers because he manufactured the "Pacificoaster" free flight, an Elbert Weathers design.

Model 10, 20, 40 and 60

The Harry Fosbury props may be identified by the stylized HF logo shown on Figures 13-5 and 13-6. The HF props designed for ELF engines included an 8 x 4 and an 8 x 6 for the Single, 9 x 6 for the Twin and a 12 x 6 for the Four. The Six could use the same 12 x 6 prop as the Four, but the engine speed would increase by about 25%. The Six could also use a 13 x 5 or the three bladed 12 x 6 inch propeller.

Figure 13-5 shows an assortment of props for a Model 10 ELF. The top prop is a 7 x10, next an 8 x 3, an 8 x 4, an 8 x 5, then a left handed (pusher) 8 x 5 prop. The bottom item is the plywood prop template for the 8 inch props made by Dan Calkin.

The Fosbury three-bladed 12 x 6 prop for the Model 40 or 60 is shown in Figure 13-6.

Figure 13-7 Dick Rothrock with his ELF Twin-powered free flight. (Dick Rothrock photo)

Figure 13-8 (L-R) Rufus Rothrock, Phil Foster, Unknown, and Leonard "Lucky" Lucas.

Figure 13-9 Dan Calkin (Left) and Harry Fosbury examining a float plane powered by two ELFs (Phil Foster photo)

Chapter 14 ELF ENGINE BOXES

ALL THE BOXES USED FOR SHIPPING ENGINES

If you read this chapter it proves that you are a true ELF engine enthusiast. A less interested person would just roll his eyes if the subject of ELF boxes came up. Boxes are important because a serious collector will assign a greater value to an engine complete with the original box and papers.

Dan Calkin used, as his standard box, a kraft-paper-covered box made of tagboard. These were ordinary, off-the-shelf, boxes. Any ornamentation, such as pasted-on labels or rubber stamping, was done by Dan or Josephine.

Box sizes are for the bottom of the box. The top was merely a cover for the box. Measurements of the length and width are made at the base of the bottom of the box with the length always being the greater dimension of the two. Dimensions are length, width, and height and are given in that sequence.

MODEL 1 BOXES

The Model 1 engines were sold in a large compartmented box $12\frac{3}{4}$ x $5\frac{1}{2}$ x $4\frac{3}{4}$ inches. The box had to be large enough to hold the twelve inch propeller and an engine mounted on a piece of $\frac{1}{2}$ inch, five ply, douglas fir plywood. The engine came mounted on the plywood stand, all wired and ready to run. There were compartments in the box for the engine with stand, tank, coil and propeller. See Figures 14-1 and 14-2.

MODEL 10, 20, 40 and 60 BOXES

The boxes used for the slant plug engines were similar to that of the Model 1, both in materials and construction, but they were of a much smaller size.

Three sizes of boxes were used corresponding roughly, but not always, to the Singles,Twins and Fours. Dan used whatever boxes he had on hand, and after October 12, 1949, he bought and used only the large box. The three box sizes were:

Small

$4\frac{1}{2}$ x $3\frac{1}{2}$ x $2\frac{1}{2}$, standard for Model 10

Medium

$5\frac{1}{2}$ x $4\frac{1}{2}$ x $2\frac{1}{2}$, standard for Model 20

Large

6 x $4\frac{1}{2}$ x 3, standard for Model 40, 60

The only other box used was the wooden box for the Model 40, Series 2 and the first Model 40, Series 3 engines.

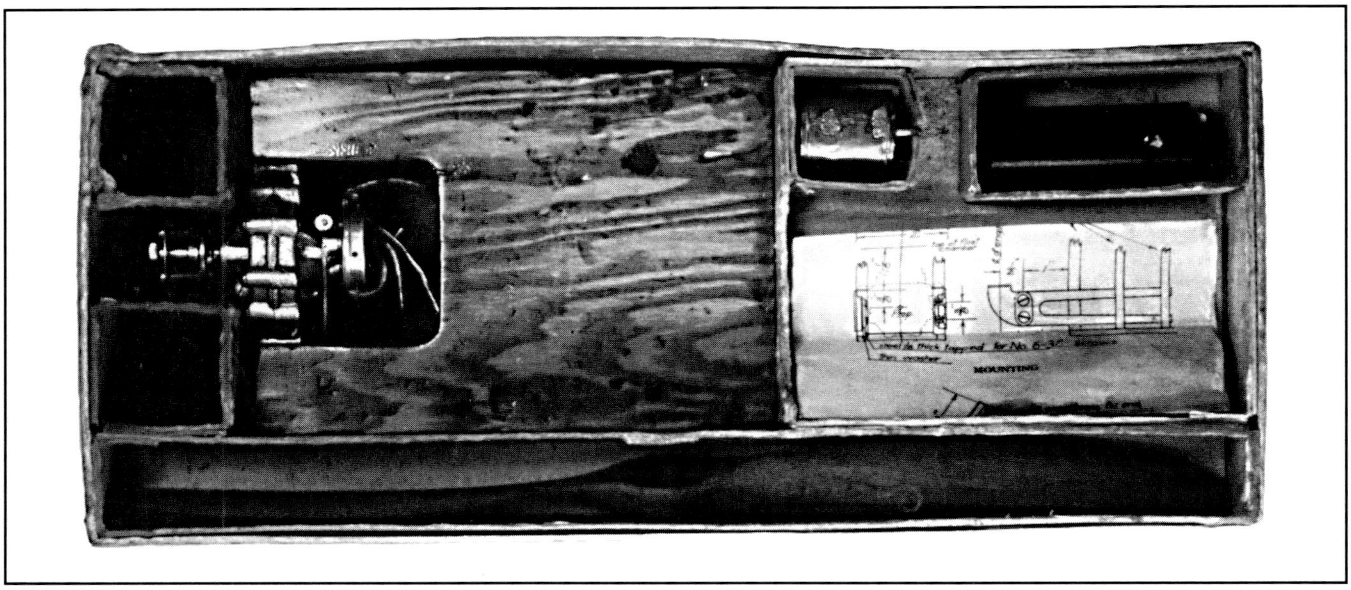

Figure 14-1 The Model 1 ELFs were shipped, ready to run, in a box like this.

Figure 14-2 The outside of a Model 1 box.

The records of box purchases exist back to 1945 and show that Calkin bought 502 of the small size boxes on March 19, 1946. No purchases of the medium size boxes were made between 1946 and 1956, so there must have been an existing stock of this size. Several purchases of the large size box were made: 105 were purchased on October 10, 1949, another 102 on November 11, 1951, and 102 on January 26, 1954. A final purchase of 2,750 boxes was made on February 3, 1955. On January 1, 1956, there were still 2,575 large boxes still on hand.

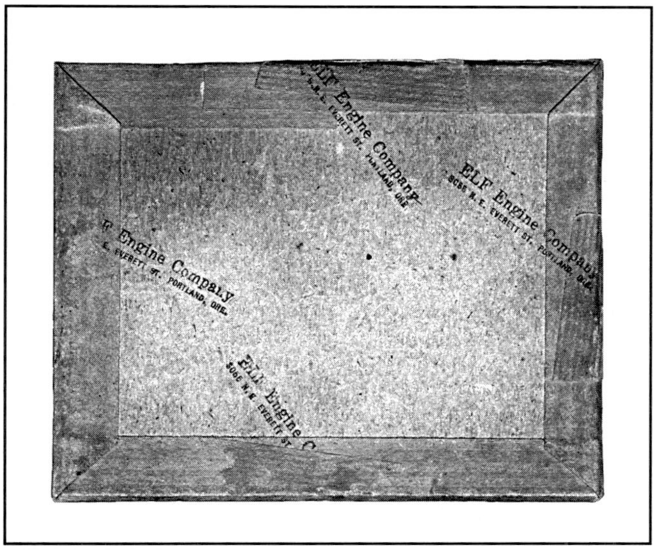

Figure 14-3 The bottom stamping used on some Model 10 and 20 engines from the Everett Street location.

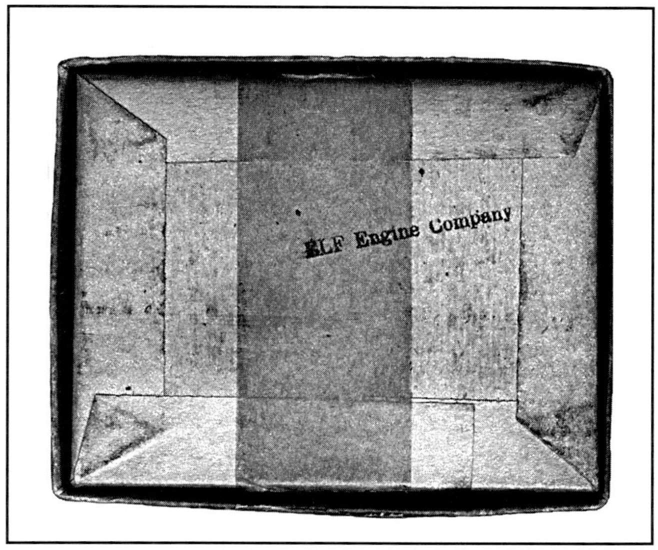

Figure 14-4 The bottom stamping used during the first few years on some of the Washington Street boxes.

Keep in mind that not only were new engines shipped in the boxes, but also engines returned to customers after factory repairs as well as orders for parts.

Markings

The boxes are characterized by markings in three different places: top, end and bottom. Some had no markings, some had one, some two, none had all three.

Bottom Markings

Two slightly different bottom markings were used. Both were made with a rubber stamp, probably the same rubber stamp.

The earliest bottom markings were used on some Model 10 and 20 boxes at Everett Street and consisted of the company name and address. See Figure 14-3.

The other bottom marking was probably made using the same rubber stamp as in Figure 14-3, but with the address removed. This was used on some of the Model 10 and 20 boxes shipped from Washington Street, especially during the first few years there (1942-1945). See Figure 14-4.

Model 10 Boxes

The early Model 10, Series 1 engines, first shipped in late October 1939, used the medium, rather than the small size box. A label was glued on the box cover showing the Series 1 Single and Twin. (Figure 14-5, also Figure 14-9.) This box was also used for the first of the Series 2 engines until early 1942.

After the introduction of the Series 2 Singles, a new box was used for shipping the Model 10 engines. This was the small size box with a label, taken from the ELF 1941 three-color brochure, glued on the top. These were used until the wartime shutdown and may have been used for a short time after production resumed in 1944. (Figure 14-6)

The next Singles, from 1944-1949, were shipped in the same small box as previously, but were marked using a rubber stamp on the end of the box instead of the glued-on label. See Figure 14-7.

ELF Single No...........

The same small size box was also the next box used, starting in late1949. It had a rubber stamp

Figure 14-5 The first Model 10 and 20 Box. (1939-1942)

Figure 14-6 The Model 10 box used in 1942.

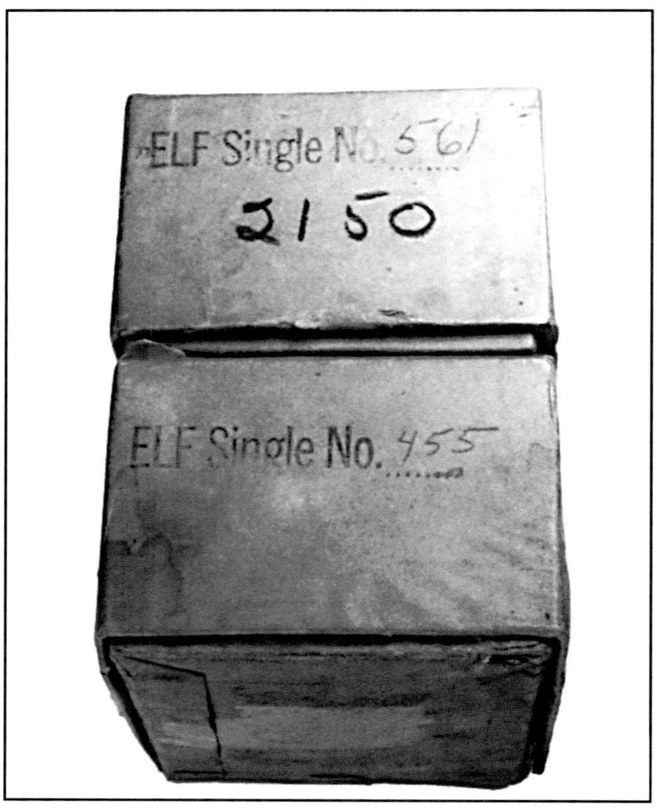

Figure 14-7 Two of the Model 10 boxes used from 1944 to 1949.

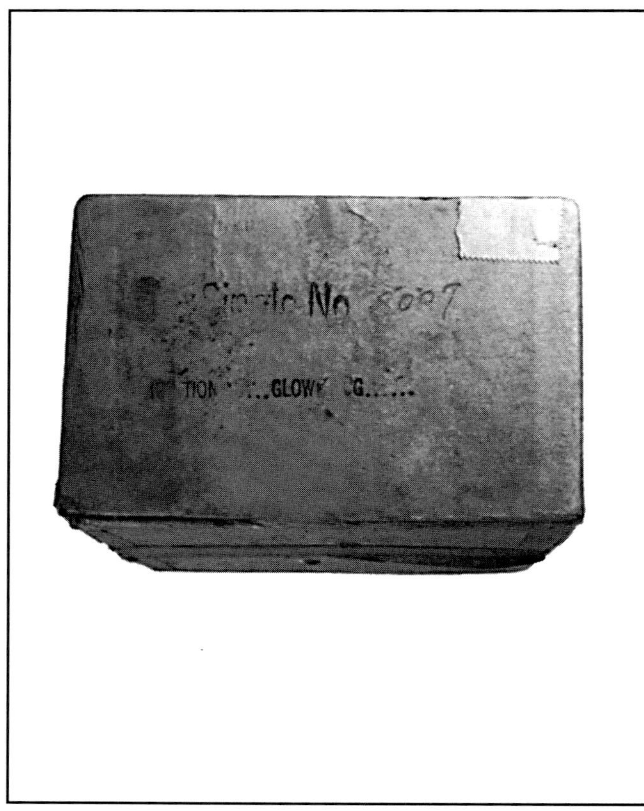

Figure 14-8 The ELF Single box from late 1949 to the last Single sold in 1951.

Figure 14-9 The Model 20, Series 2 box used from 1940 to late 1941.

Figure 14-10 The ELF Twin box from late 1941, and perhaps into 1944, when production resumed.

Figure 14-11 The ELF Twin box from late 1944-1950.

Figure 14-12 The ELF Twin box used from 1950-1952.

Figure 14-13 The large size box that was used for most engines shipped from 1950-1956.

Figure 14-14 The Model 40 wooden box, 1941-1948.

similar to the above, but a second line was added to indicate either ignition or glow model. Figure 14-8.

ELF Single No...........

IGNITION..... GLOW PLUG.....

By the summer of 1950, when the Series 4 engines were introduced, Dan ran out of the small boxes and started shipping the Singles in the large box, the same one used for the four cylinder engines. Sales of the Single were so small that additional purchases of the small boxes were not justified. The same rubber stamp was used as in Figure 14-8. No Model 10 engines were shipped after 1951.

Model 20 Boxes

The first Twins were shipped in the same medium size box as the early Model l0 engines. There was a place on the label to check if the engine inside was a Single or a Twin. These boxes were used until late 1941. (Figure 14-9) The box is similar to the one shown in Figure 14-5.

The second box for the Model 20 engines was the same medium size box as previously used, but it had a picture (from the 1941 brochure) glued on the box cover showing a Model 20, Series 2 engine. This was used starting in late 1941, until production ceased in late 1942. A few more may have been used when production resumed in 1944. (Figure 14-10)

The same medium size box was used from 1944 to 1949 using a rubber stamp. (Figure 14-11)

ELF Twin No...........

After that, by late 1949, a new rubber stamp was used with another line added to allow the selection of either ignition or glow model. (Figure 14-12)

ELF Twin No...........

IGNITION..... GLOW PLUG.....

The last box used for the Twins from 1950 on was the same large box used previously, but the rubber stamp was modified to remove the word 'Twin'. Calkin continued using the large box with

Figure-14-15 The large size box used for the Model 40 from 1949 -1950.

the rubber stamp until production ceased in 1956. (Figure 14-13)

ELF NO..........

Model 40 Boxes

The first four cylinder engines were shipped in a wood box made of alder. They were used on all Model 40, Series 2 engines and many of the Series 3 engines until September 30, 1949. There were no markings of any kind on the box, which measured 6$\frac{3}{8}$ x 5 x 3$\frac{7}{8}$ inches. The engines were mounted on the standard ELF metal motor mounts. Some engine mounts were attached to the box with small wood screws while other engines were loose in the box. (Figure 14-14)

By October 5, 1949, with the stock of wooden boxes exhausted, Dan started using the large size tagboard boxes. These were used initially with a rubber stamp as shown in Figure 14-15 and reading:

ELF Four No..........

IGNITION..... GLOW PLUG.....

Later, a similar stamp was used with the word 'Twin' removed, as was the second line with the ignition or glow plug designation. This box was the same as the last Twin box and was used until the end of ELF production in 1956. (Same as Figure 14-13)

ELF NO..........

MODEL 60 BOXES

Some sixes sold locally were sold without a box. The mail order engines went out in the large size box that was used for everything after 1949. See Figure 14-15.

Chapter 15 ELF TOOLING

THE MAKING OF THE ENGINES

MODEL 1 TOOLING

Dan Calkin made nearly everything himself, purchasing only raw materials. He also made all his own tooling, some of which are quite unique.

Photos of the tooling are included here to give you an appreciation for the methods Dan Calkin used to make his ELF engines.

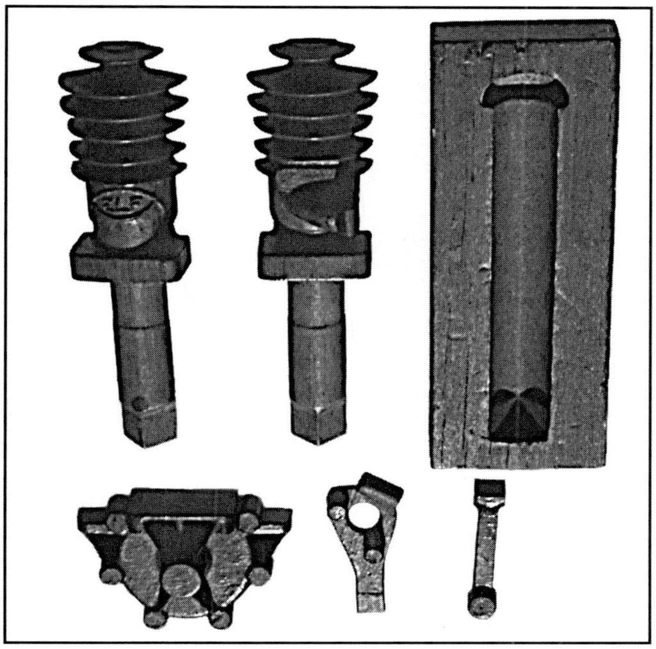

Figure 15-1 The original master wood patterns for the Model 1 ELF.

SLANT PLUG ENGINE TOOLING

Figure 15-2 Match plate for cylinder castings.

Figure 15-3 Match plate for cylinder heads.

Figure 15-4 The match plate for Model 10 reed valve engines. The rear case pattern has been modified, from the Series 3 version, with wax to add the rear intake tube.

Figure 15-5 The crankcase match plate for all multi-cylinder ELF engines from Series 2 on, except the Series 4 Twins.

Figure 15-6 The modified match plate for the Model 20 reed valve engines. As with the Model 10, the rear case pattern was built up with wax to add the rear intake tube.

Figure 15-7 The crankcase extender for the Model 40 and 60, Series 4 engines were made from this pattern.

Figure 15-8 The match plate for timer arms for the round timers. It was used for Series 3B and 4B engines.

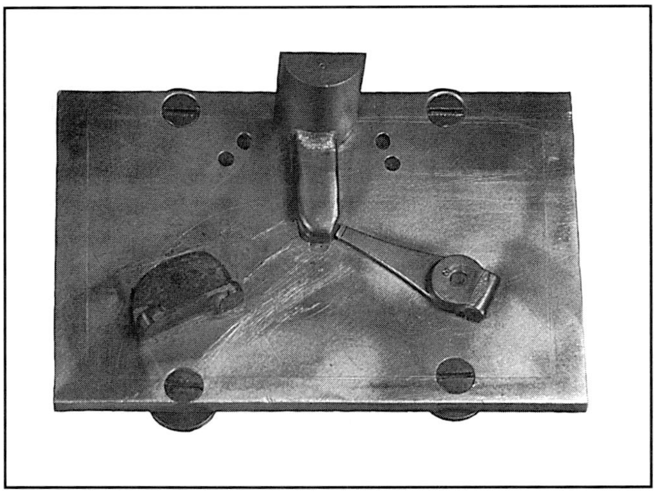

Figure 15-9 The match plate for casting the open cast timer for the Model 20, Series 4C engines.

Figure 15-10 The match plate for Model 40, Series 4C open cast timer.

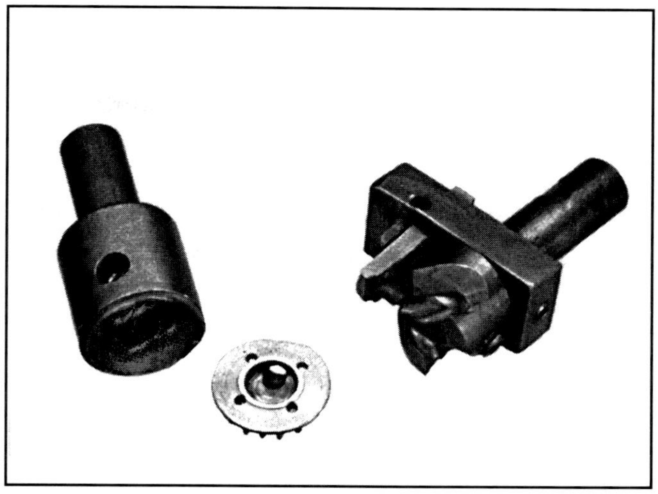

Figure 15-11 The tool on the left holds the cylinder head, while the cutter on the right machines the bottom surfaces and the outside diameter.

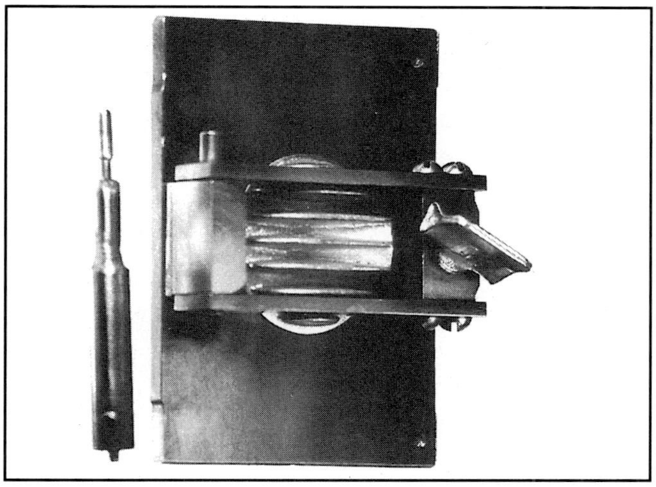

Figure 15-12 Tool for holding a cylinder head while the four holes are drilled. The tool for spot facing the holes is also shown.

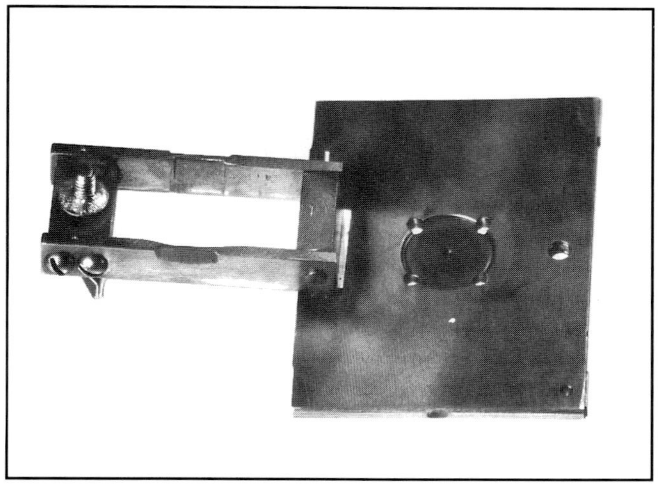

Figure 15-13 The same tool as shown in Figure 15-12, with the head removed and the clamp open.

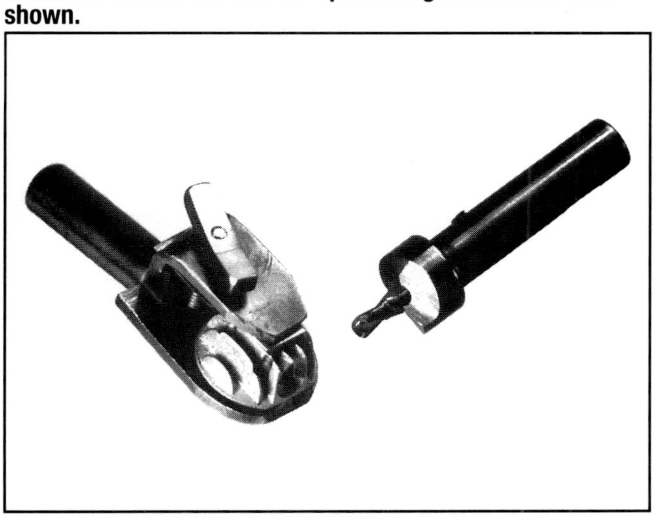

Figure 15-14 Calkin used the tool on the left to hold the head at a 15 degree angle while drilling and spot facing with the other tool.

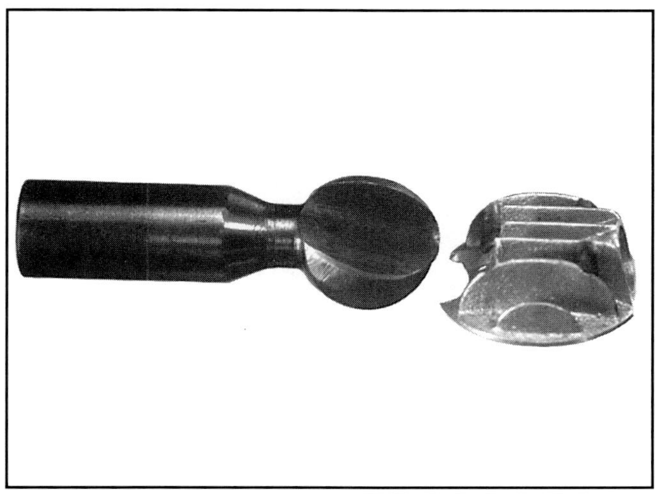

Figure 15-15 This ball end-mill was used to relieve the front of the rear head on some Model 40, Series 3 and 4 engines and some Model 60 engines.

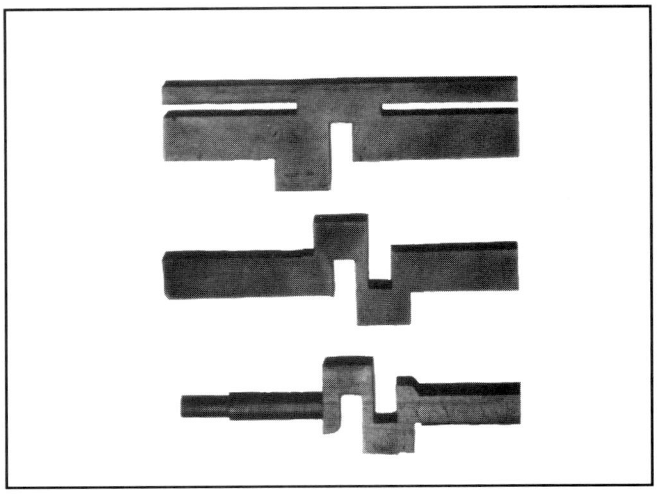

Figure 15-16 Examples of the way the crankshafts were rough cut to size on the power hack saw or band saw. The bottom blank has been partly machined.

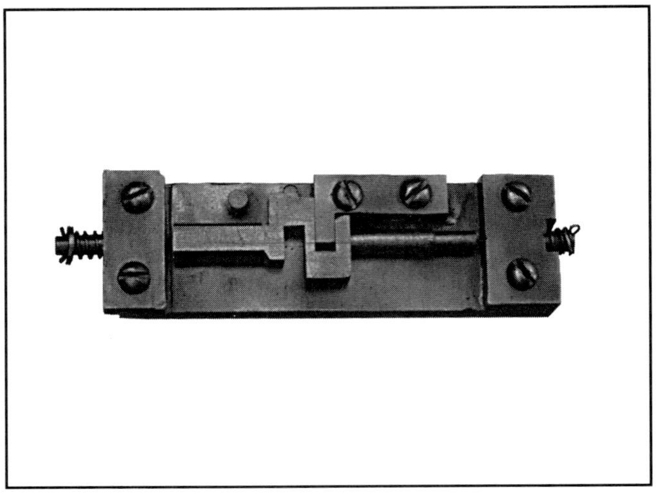

Figure 15-17 The fixture Dan Calkin used to center punch both ends of the crankshaft at one time.

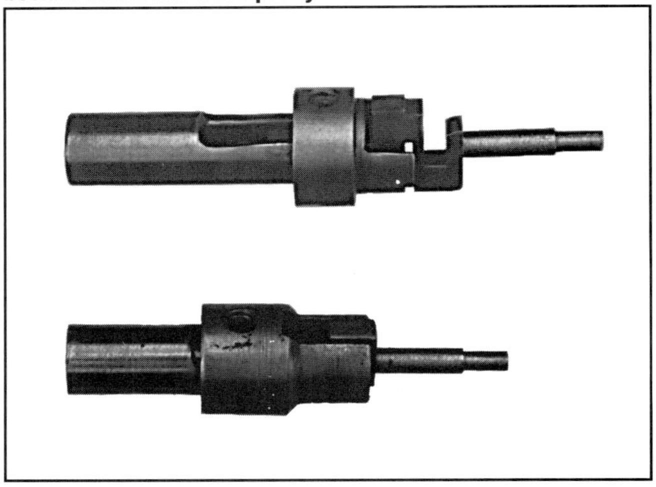

Figure 15-18 Two of the tools used to hold the crankshafts for machining the front end.

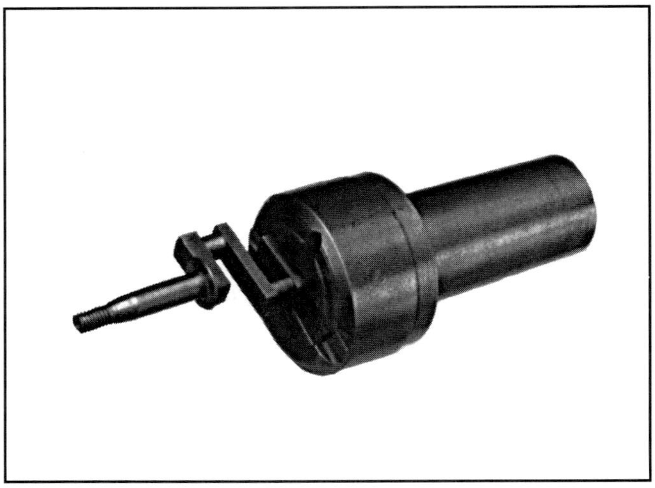

Figure 15-19 A tool for holding the crankshafts off-center for machining the throws.

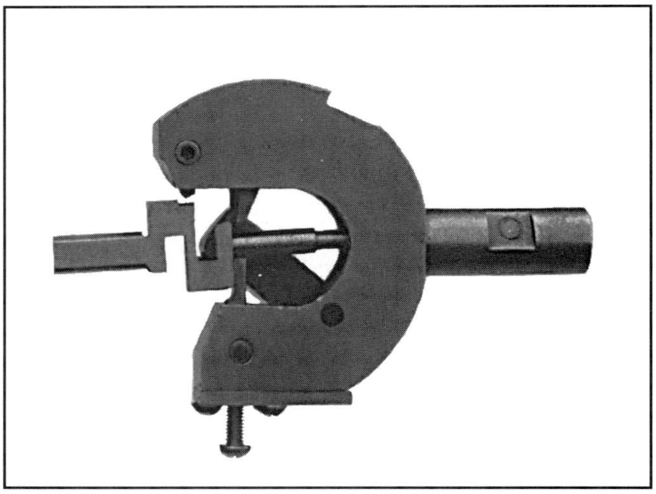

Figure 15-20 The multi-cutter tool for turning the end of the shaft.

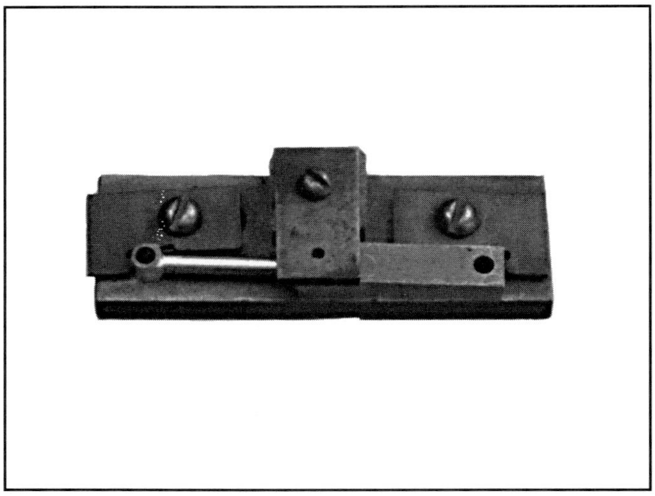

Figure 15-21 Fixture for holding a blank for two connecting rods while drilling the center hole.

Figure 15-22 The fixture for drilling the two holes for the small end of the connecting rod.

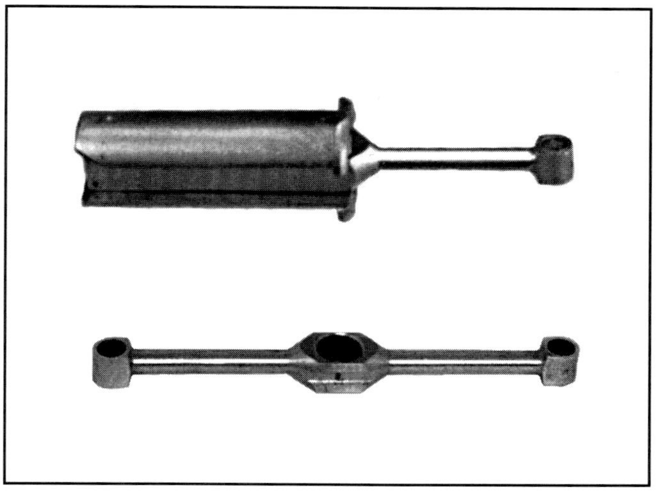

Figure 15-23 A split collet for turning the connecting rods and a connecting rod before being cut in two.

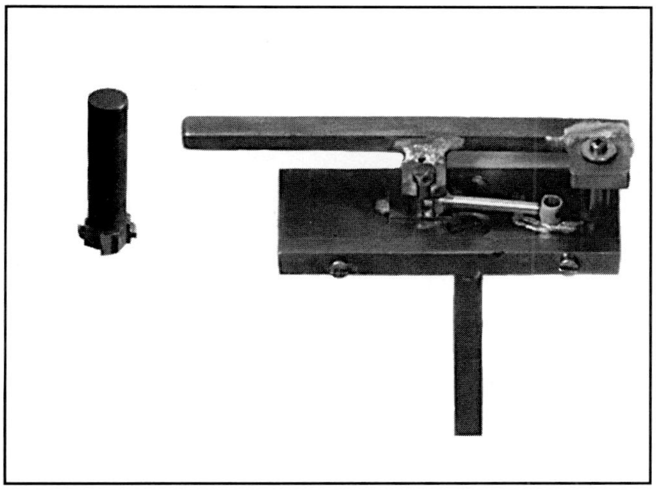

Figure 15-24 The tool and holding fixture for cutting the side of the rod where the strap attaches.

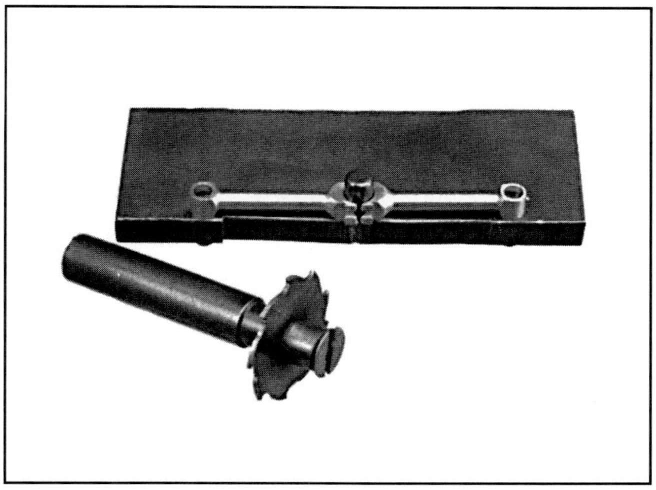

Figure 15-25 The tool and holding fixture for cutting the rod blank into two rods.

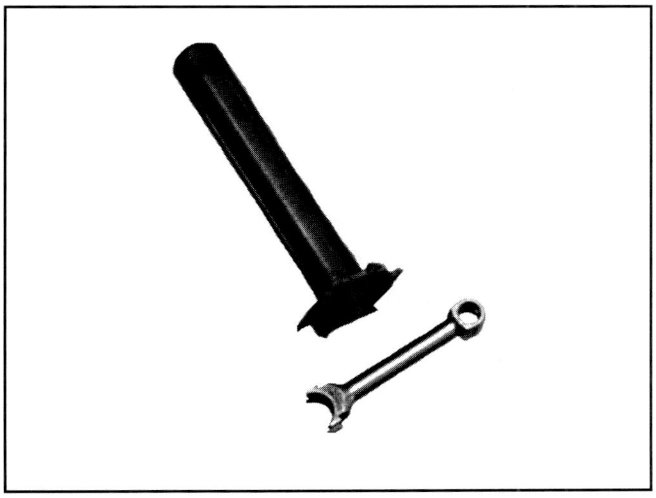

Figure 15-26 The tool for cutting the sides of the connecting rod where the strap attaches. It was used with the holding fixture shown in Figure 15-24.

Figure 15-27 The two-piece fixture for holding the crankcase in a lathe while the shaft hole and interior surfaces were machined.

Figure 15-28 Dan Calkin's drill jig for drilling the eight holes in the crankcase.

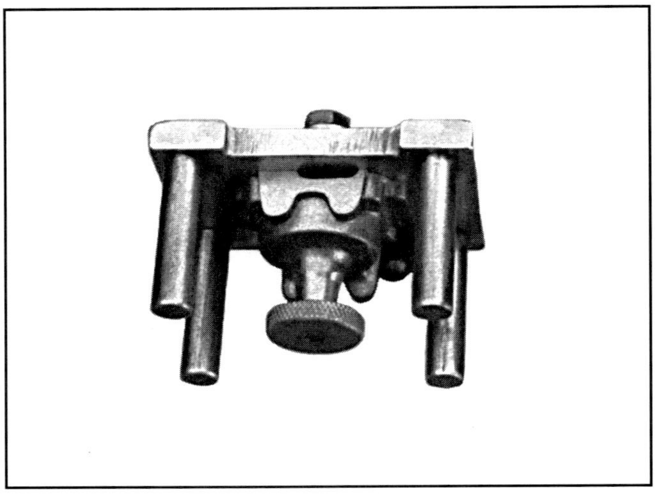

Figure 15-29 Another look at the drill fixture in Figure 15-28.

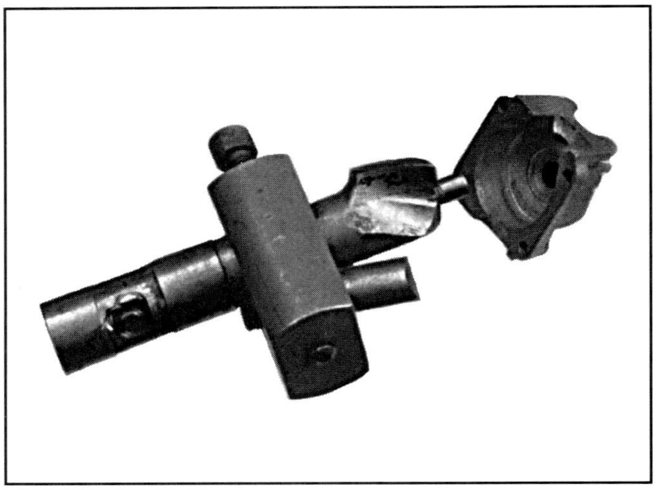

Figure 15-30 The cutting tool for boring the inside of the crankcase and facing the mating surface.

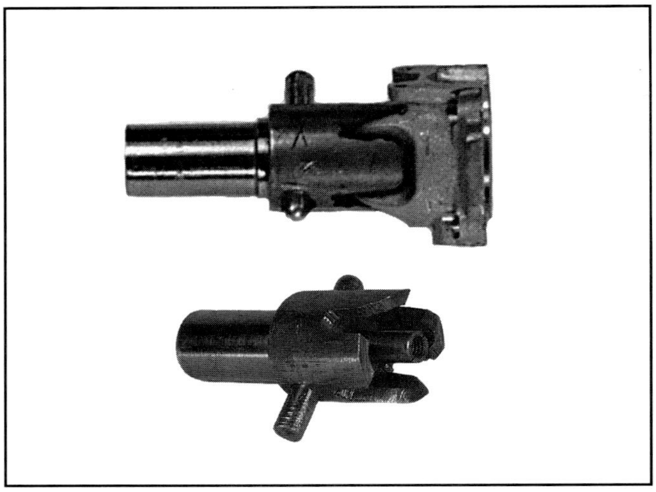

Figure 15-31 Two views of a tool for holding the front crankcase by the front of the casting.

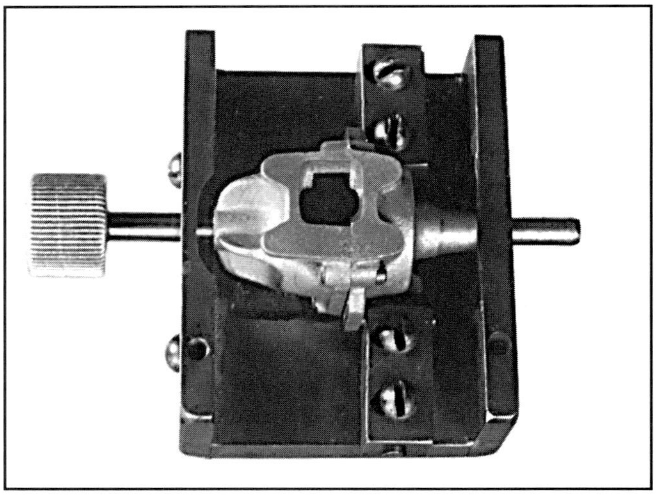

Figure 15-32 Holding fixture for milling the crankcase-to-cylinder surfaces.

Figure 15-33 The milling fixture in Figure 15-32 without the crankcase installed.

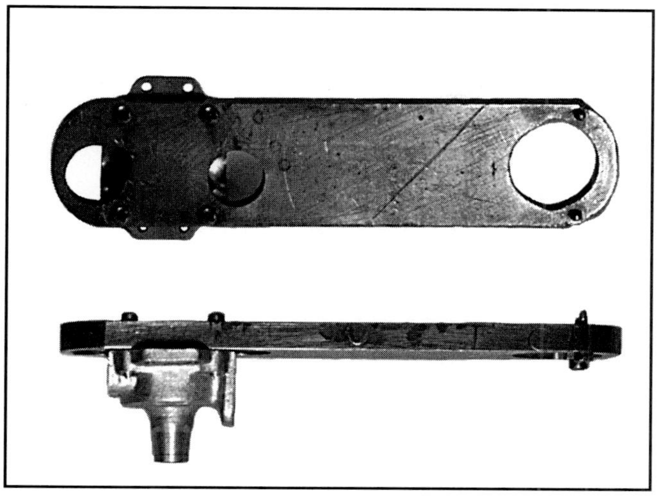

Figure 15-34 Two views of a tool for locating the cutter for cutting relief holes in the crankcase.

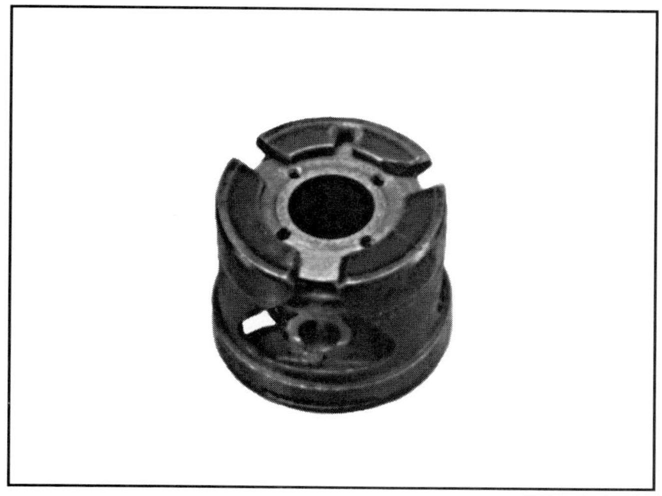

Figure 15-35 The fixture for holding a cylinder for machining.

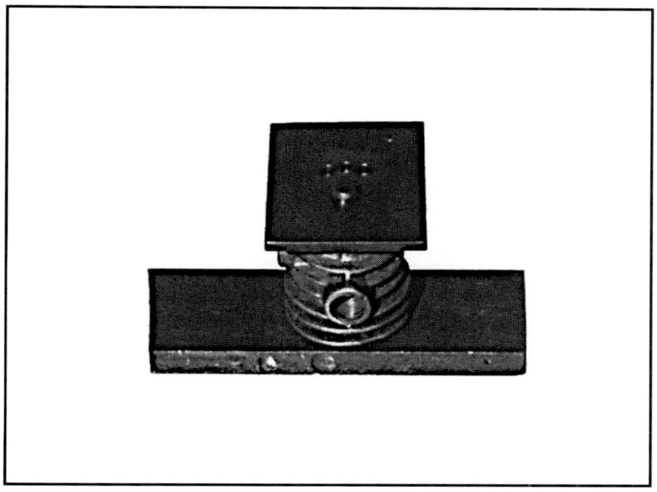

Figure 15-36 A jig for drilling the three holes for porting in the cylinder casting.

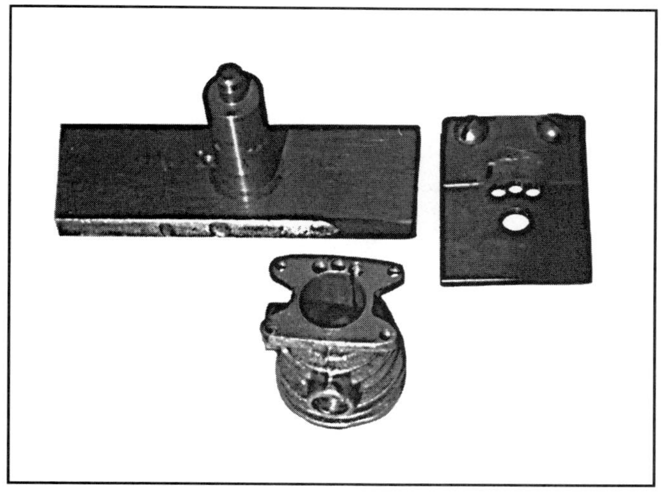

Figure 15-37 The tool from Figure 15-36 disassembled.

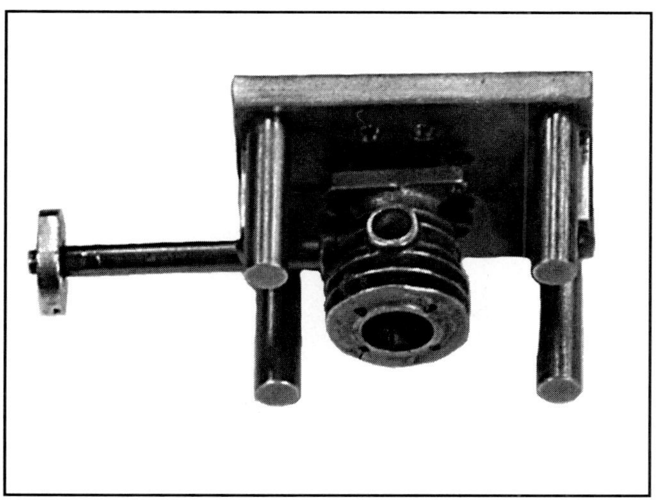

Figure 15-38 A drill jig for drilling holes in the cylinder for the crankcase attachment.

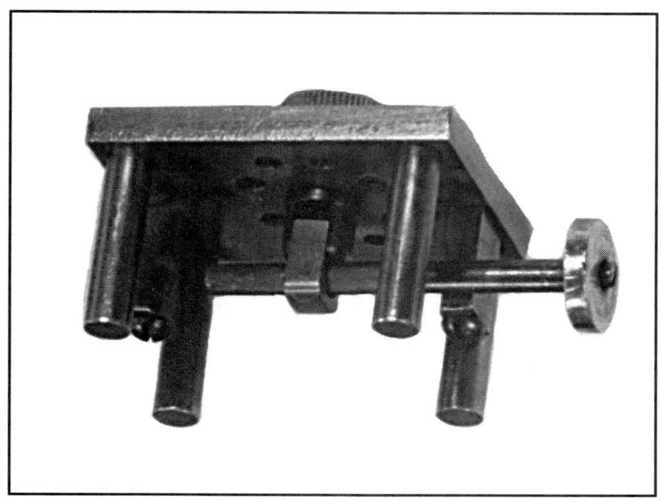

Figure 15-39 Another look at the drill jig in Figure 15-38.`

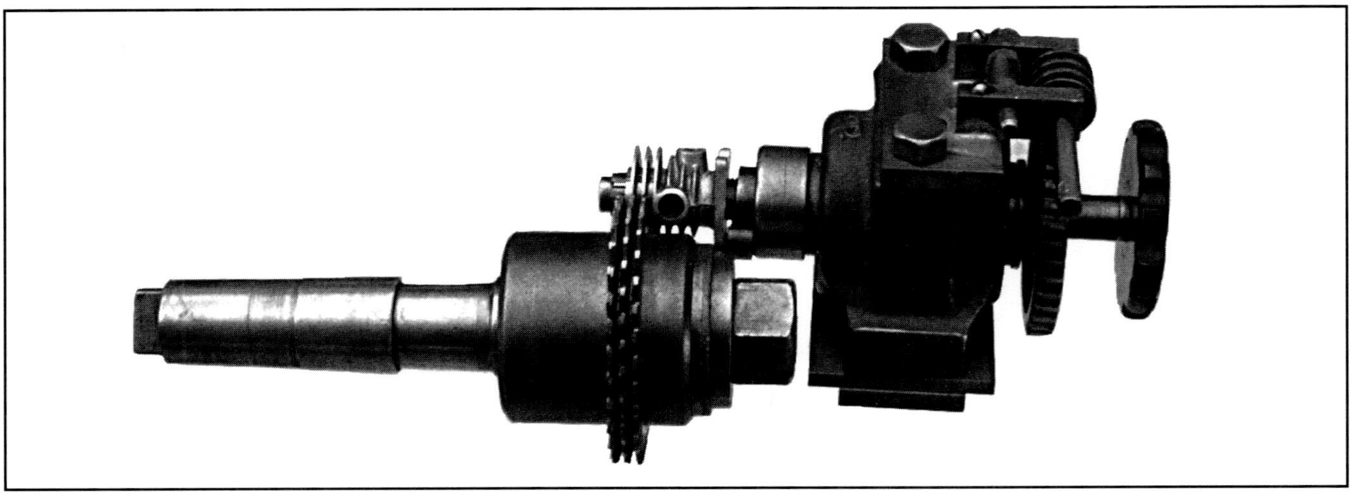

Figure 15-40 The mechanism Calkin used for milling the cylinder fins. The gear box on the right had a draw screw tightened by the knob. The gear box was driven by a small electric motor to rotate the cylinder against the mill cutter.

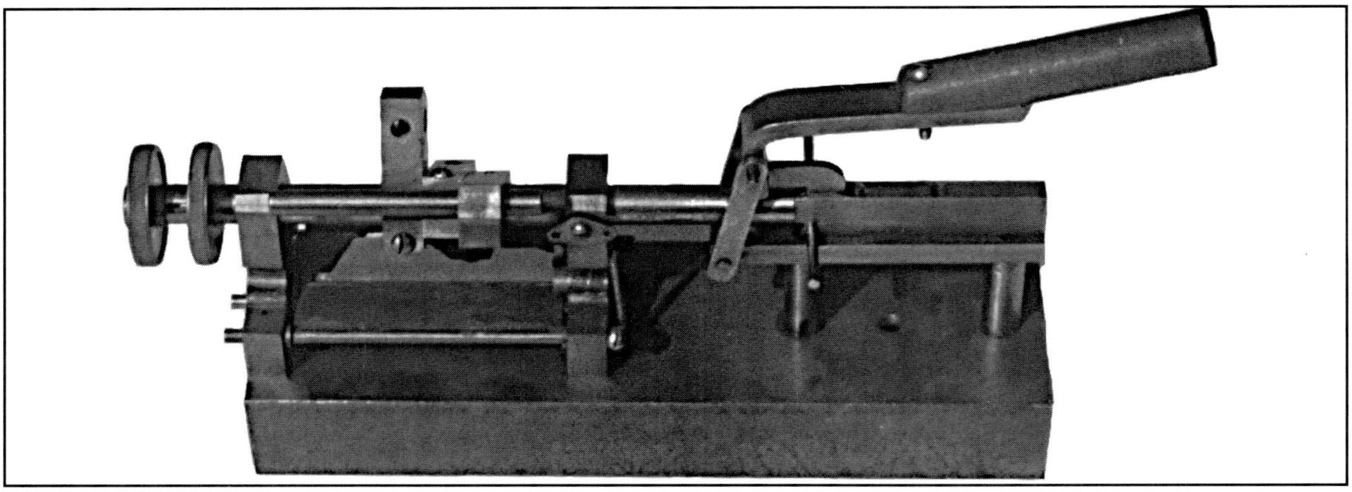

Figure 15-41 A clever tool for punching the ports in the cylinder liner. A system of stops was used to assure that the holes were punched in the correct position. The knob on the left was for the draw screw that held the liner securely. Two sets of punches were used, one for the intake ports and one for the exhaust. One punch and die set is installed in the punch.

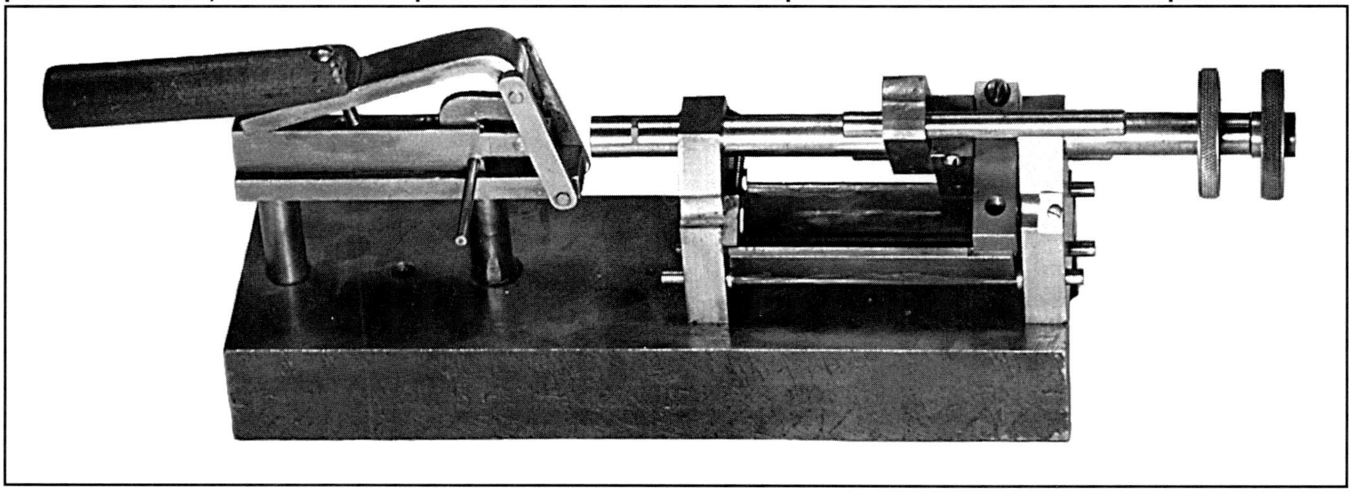

Figure 15-42 The opposite side of the above tool. This tool was also used for punching tabs on the round timer body.

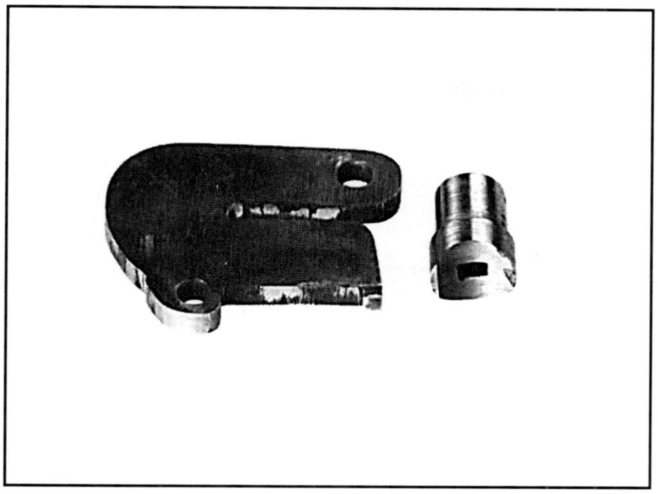

Figure 15-43 One of the two punches used with the tool in Figures 15-41 and 15-42 for punching the ports in the cylinder liner.

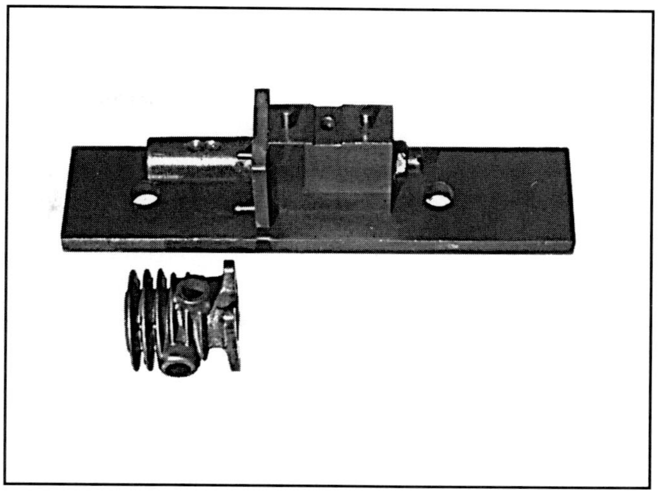

Figure 15-44 The drill jig for drilling the intake and exhaust ports on the cylinder casting.

Figure 15-45 The fixture to install the liner in a cylinder.

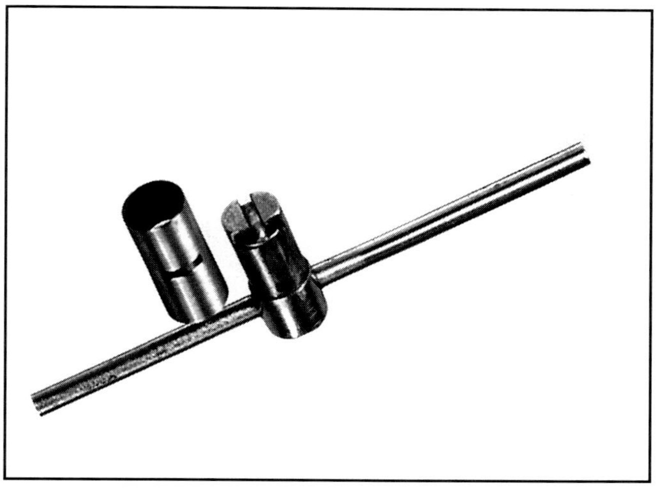

Figure 15-46 The special tool for holding the cylinder liner while installing it in the cylinder. A spring loaded brass lip fitted into the exhaust port. (Dick Tichenor photo)

Figure 15-47 The holding fixture and mill cutter for cutting the intake side of the piston baffle.

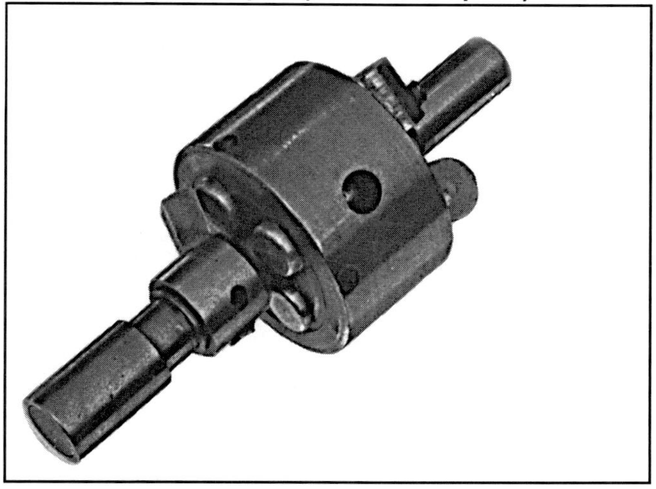

Figure 15-48 The mill cutter used with the piston-holding fixture for milling the exhaust side of the baffle.

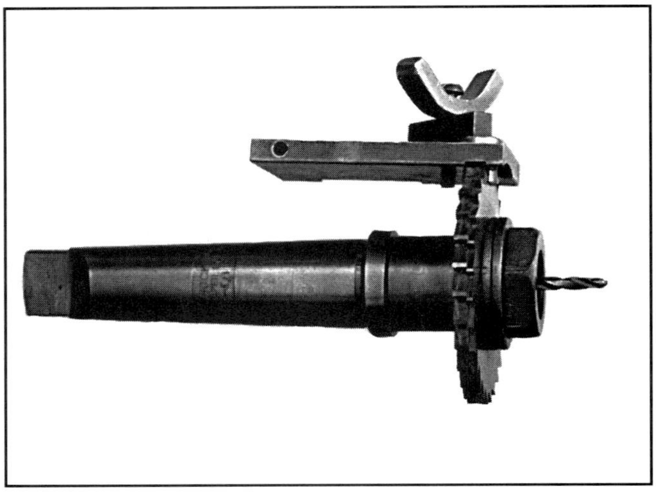

Figure 15-49 The cutter used to cut the slot for the connecting rod in the top of the two-piece piston.

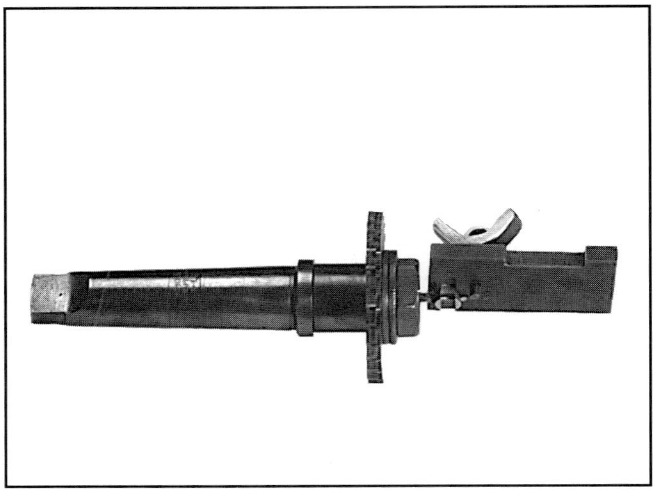

Figure 15-50 The drill was used to make the hole for the wrist pin in the piston top. These are the same two items shown in Figure 15-49.

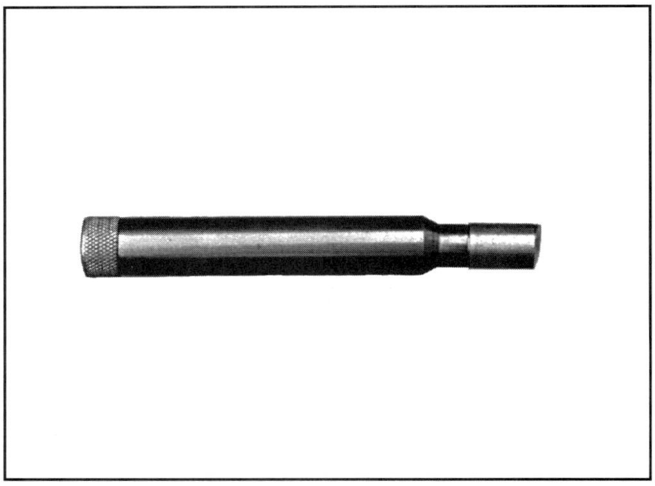

Figure 15-51 The unique tool to hold a piston, with wrist pin and connecting rod installed.

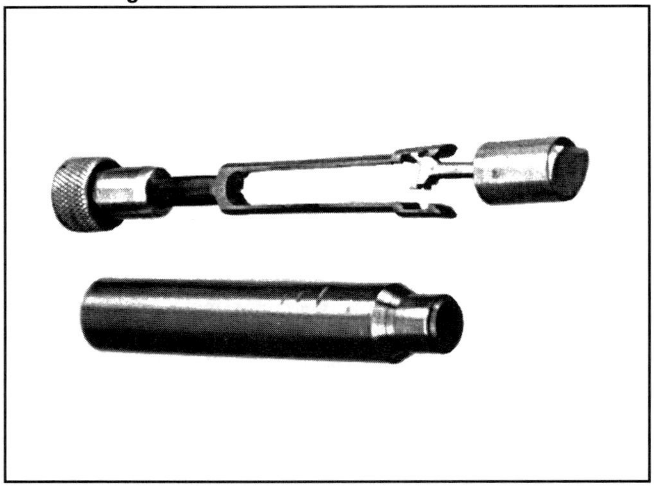

Figure 15-52 The tool from Figure 15-51 is shown dismantled to show how it caught the lips on the rod.

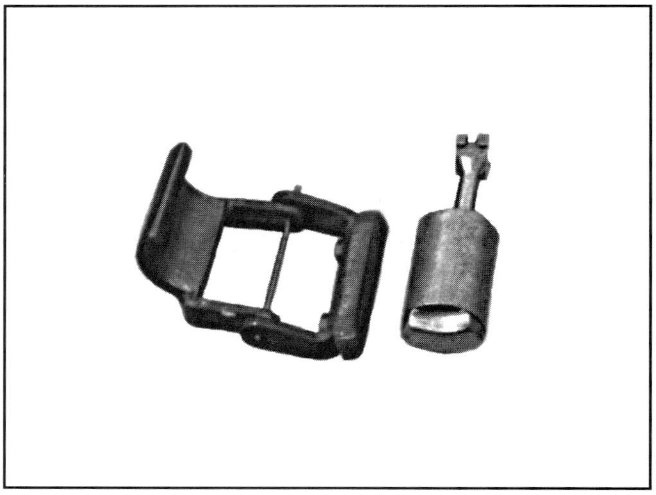

Figure 15-53 The small hand-held lap for pistons. It used a stone from the Sunnen hone.

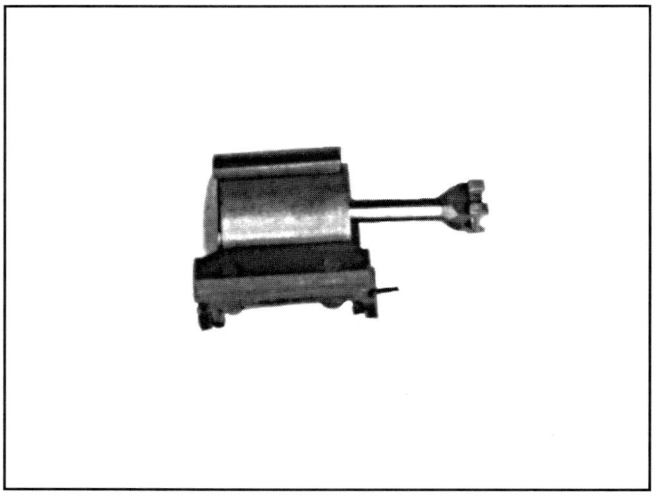

Figure 15-54 The lap for the pistons with the piston in place.

Figure 15-55 Punch for the crankcase-to-cylinder paper gasket. This punch made all the holes in the interior.

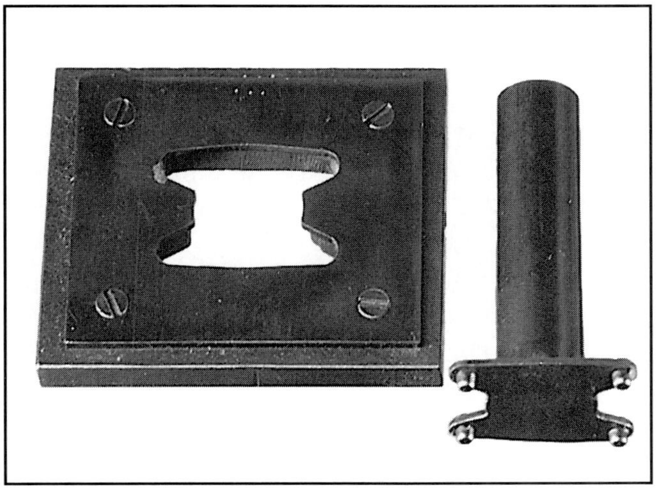

Figure 15-56 This punch was used with the punch in Figure 15-55 for the paper gaskets. This punch cut the outside edge of the gasket.

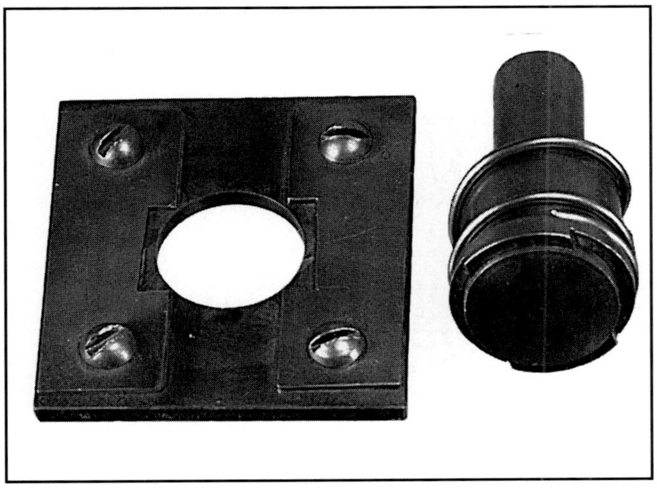

Figure 15-57 The punch and die used to make the reed valves for the Series 4 engines. This one cuts the outside shape (rounded with flat sides).

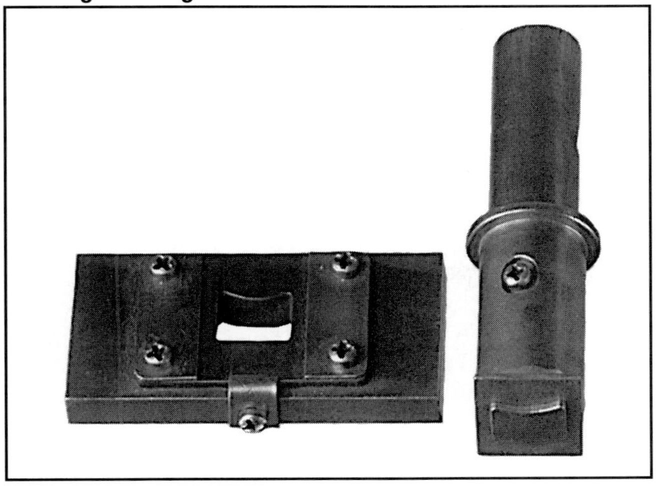

Figure 15-58 This punch and die was used to punch out the hole in the reed valve.

Figure 15-59 The punch and die used for making the straps which held the connecting rod to the crankshaft.

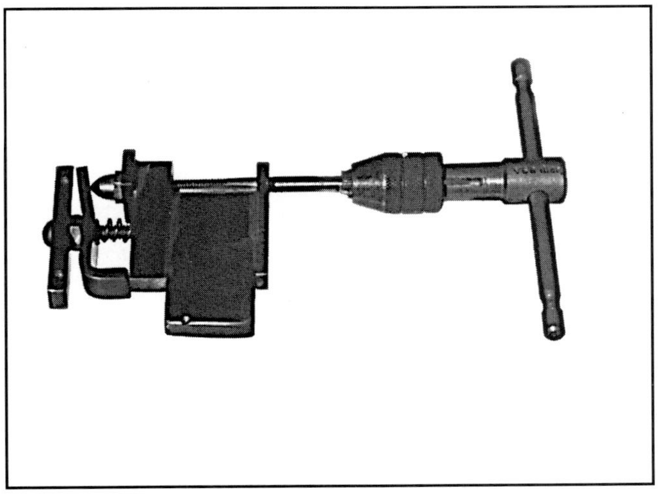

Figure 15-60 The fixture used to tap the 8-32 threads in the prop nut.

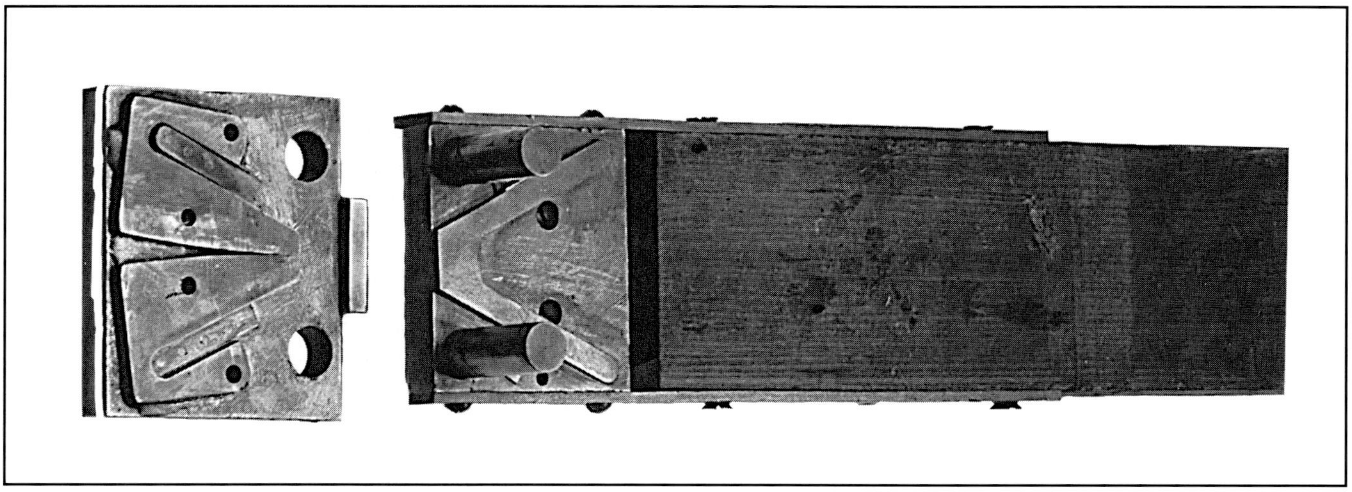

Figure 15-61 The punch and die Calkin used to stamp out the ELF motor mounts.

Figure 15-62 The bending fixtures for forming the motor mounts.

Figure 15-63 The drill jig for drilling the two holes to attach the motor mount to the engines. The holes were tapped 3-48.

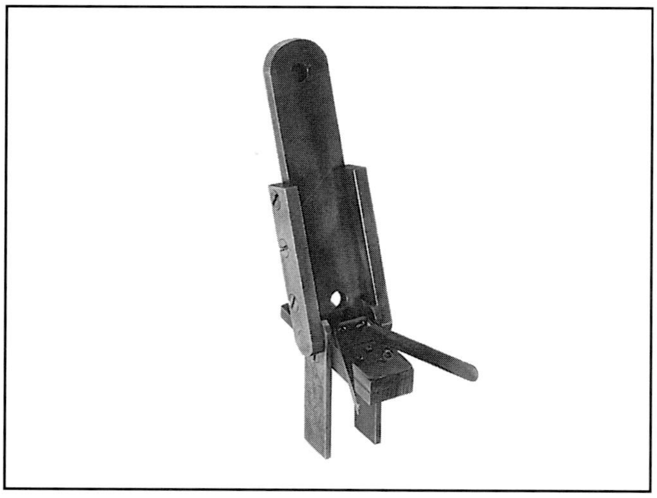

Figure 15-64 The bending jig used for bending the front motor mount tab over on itself to form a double thickness

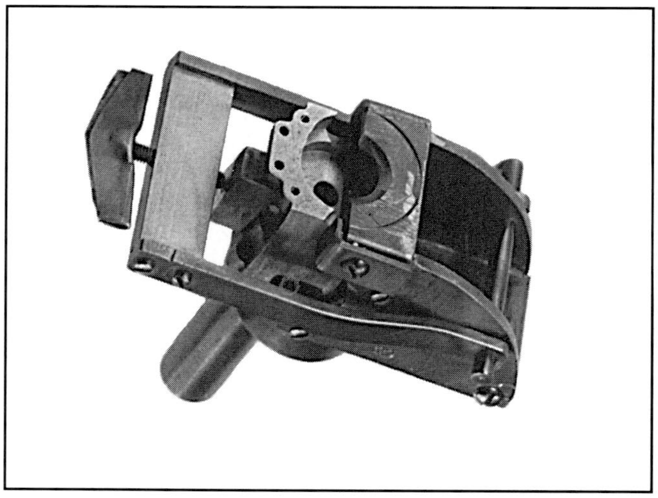

Figure 15-65 The holding fixture for the center crankcase extender for the Model 40 and 60 engines.

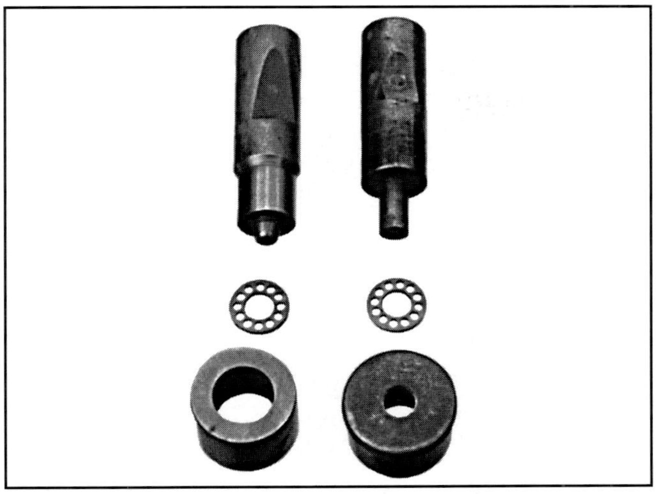

Figure 15-66 The punches for the ID and OD for the thrust bearings.

Figure 15-67 The small rotary table for punching the 12 (6 for the Model 10) holes in the thrust bearing. The piece on the right is a spring loaded ejector that fits over the punch.

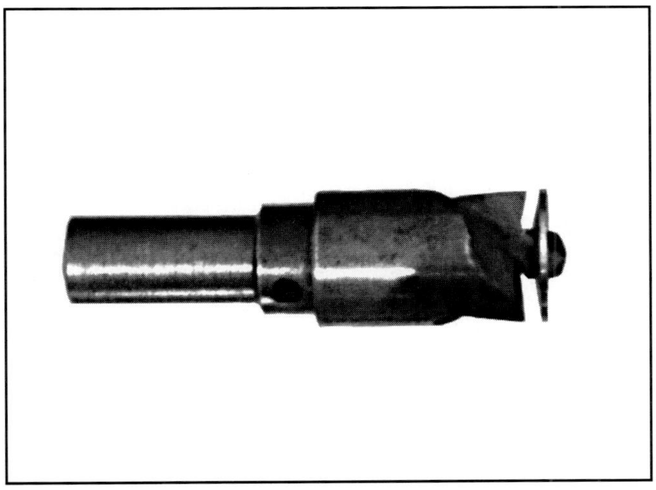

Figure 15-68 The tool for cutting the beveled face on the front prop washer. A prop washer is sitting on the pilot of the tool.

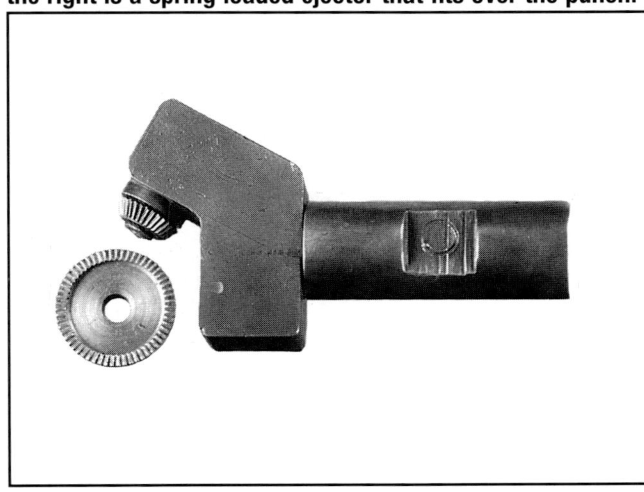

Figure 15-69 The tool used to knurl the prop drive washers.

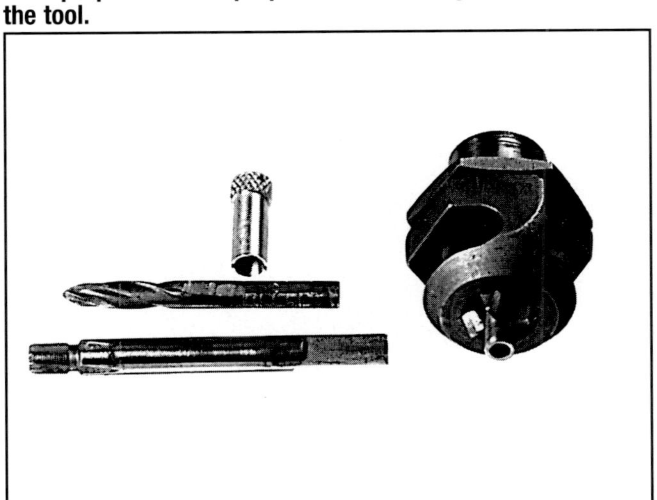

Figure 15-70 The tap drill and tap and a die used for the needle valve (5/16-72 thread).

Figure 15-71 Dan Calkin used this Starret counter with this stop watch to determine engine RPM. This is the same counter used on his engine test stand in Figure 2-13.

Figure 15-72 The turret Calkin built and automated to replace the tailstock on one of his South Bend lathes.

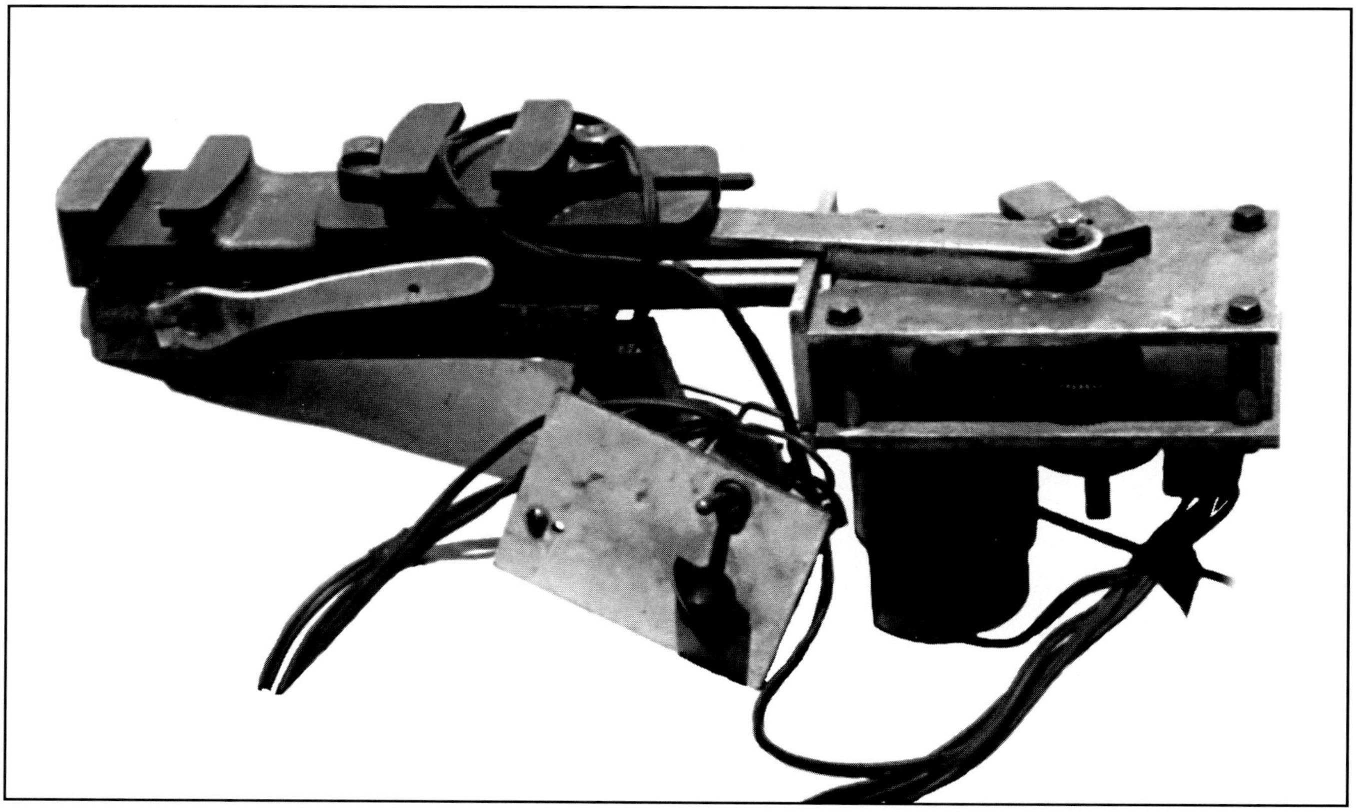

Figure 15-73 The automated tool post for a Calkin South Bend lathe. The 3-position toggle switch controlled the motor that drove the tool post in or out and was off in the center position. The rheostat knob changed the speed of the tool post motor.

Chapter 16 FOUR STROKE PROJECT

THE LOST ELF

Near the end of WW II, Carl VanCourt and Dan Calkin had many discussions about making four stroke cycle engines for model airplanes. At that time a few four stroke engines had been built, by Feeney and others, but none had been nearly as small as the engine Dan proposed.

The plan was for Dan to produce the "top end" consisting of the cylinder, head, piston, connecting rod, push rods and valves. Carl would produce the crankcase in a variety of configurations. Possible configurations would be singles and twins of the usual sorts, as well as several versions of radial engines.

The patterns for the cylinder and heads were made and a few parts of each were cast.

Carl never made the crankcases. Changing market conditions made it imperative that Carl go on to other things. A plan had been considered to incorporate ELF and make Carl a key part of it, but that plan was never implemented due to changing economic necessities.

Instead, Carl started a dental laboratory in Los Angeles. The workmanship in his lab was of such high quality that he soon had so much work that it was not possible to return to the four stroke project.

Figure 16-1 The beautifully made match plate for the cylinders.

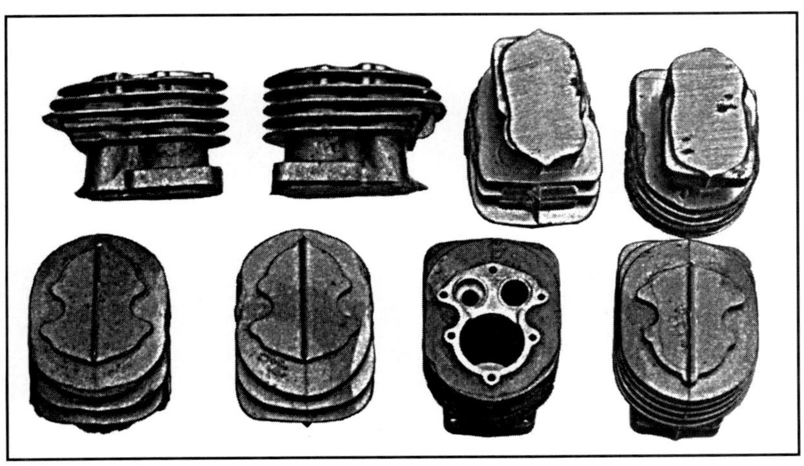

Figure 16-2 A group of the existing cylinders, including one that was machined.

Figure 16-3 The match plate for making the heads for the four stroke engine.

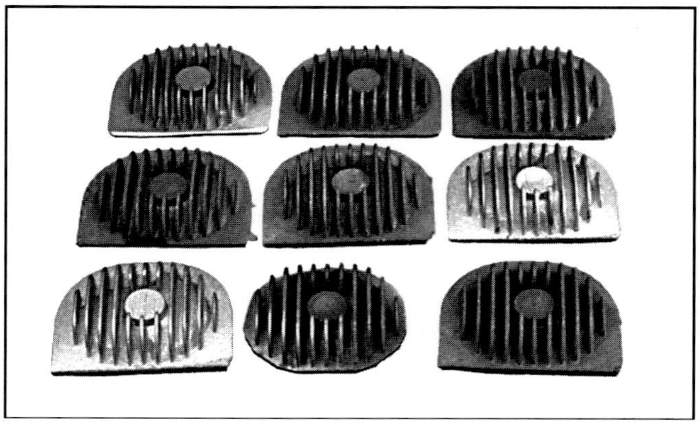

Figure 16-4 Some of the heads made from the match plate.

Evidently Dan Calkin, in his usually methodical way, did spend a lot of time developing a four stroke engine. Plans exist for a twin with a displacement of 0.1995 cubic inch, stroke 0.588 inch, and a bore of 0.656 inch. The engine may have been only a platform on which to test the "top end" that he was to have been responsible for developing.

The camshaft was to be located below the crankshaft to operate the pushrods to each cylinder. A gear type oil pump would be located in a sump below the cam shaft.

There is no certainty that Dan actually built a four stroke engine, but what we have is intriguing: a small supply of cylinder and head castings, including one machined cylinder casting, as well as the match plate patterns for making these. The patterns are beautifully made, being built of layers of machined pieces of brass.

There is considerable tooling for machining the crankshaft, cylinder, heads and other items.

There are fairly extensive notes about the project. Included are notes about testing that suggests that at least one engine was made, listing changes that should be incorporated. A portion of the drawings are included here to give an insight into Dan's thinking.

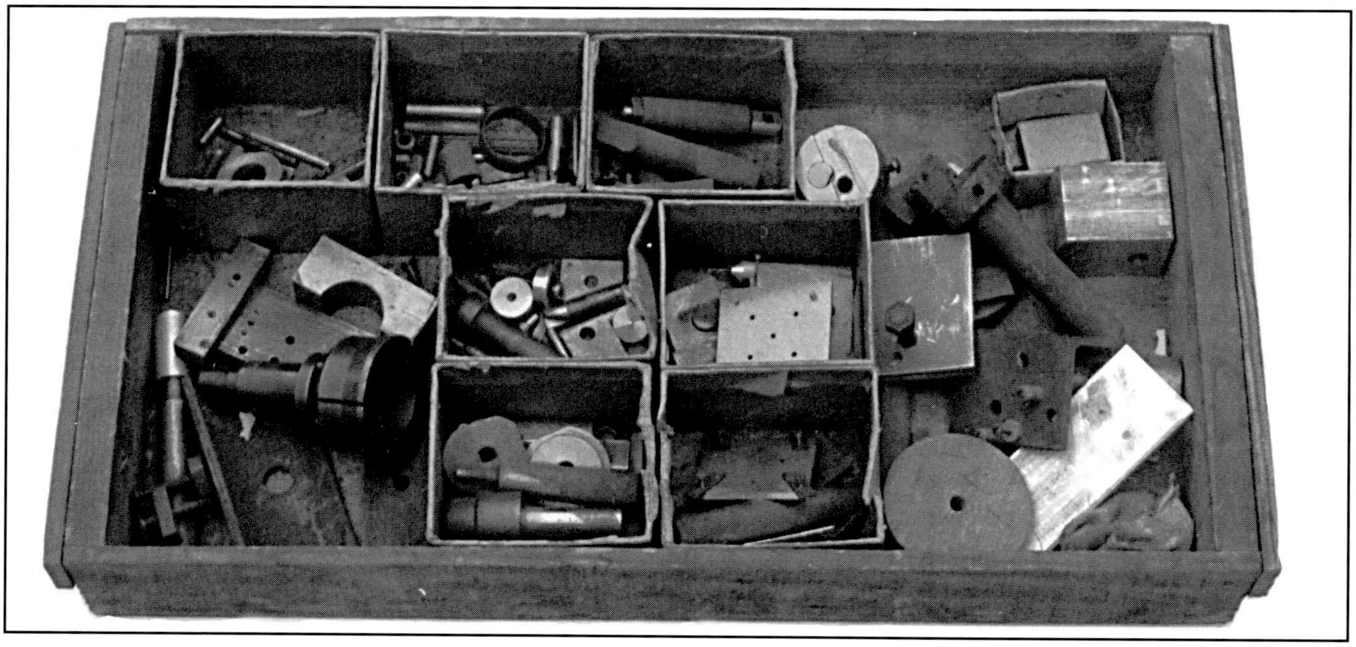

Figure 16-5 Dan Calkin's box of tooling for building the four stroke ELF.

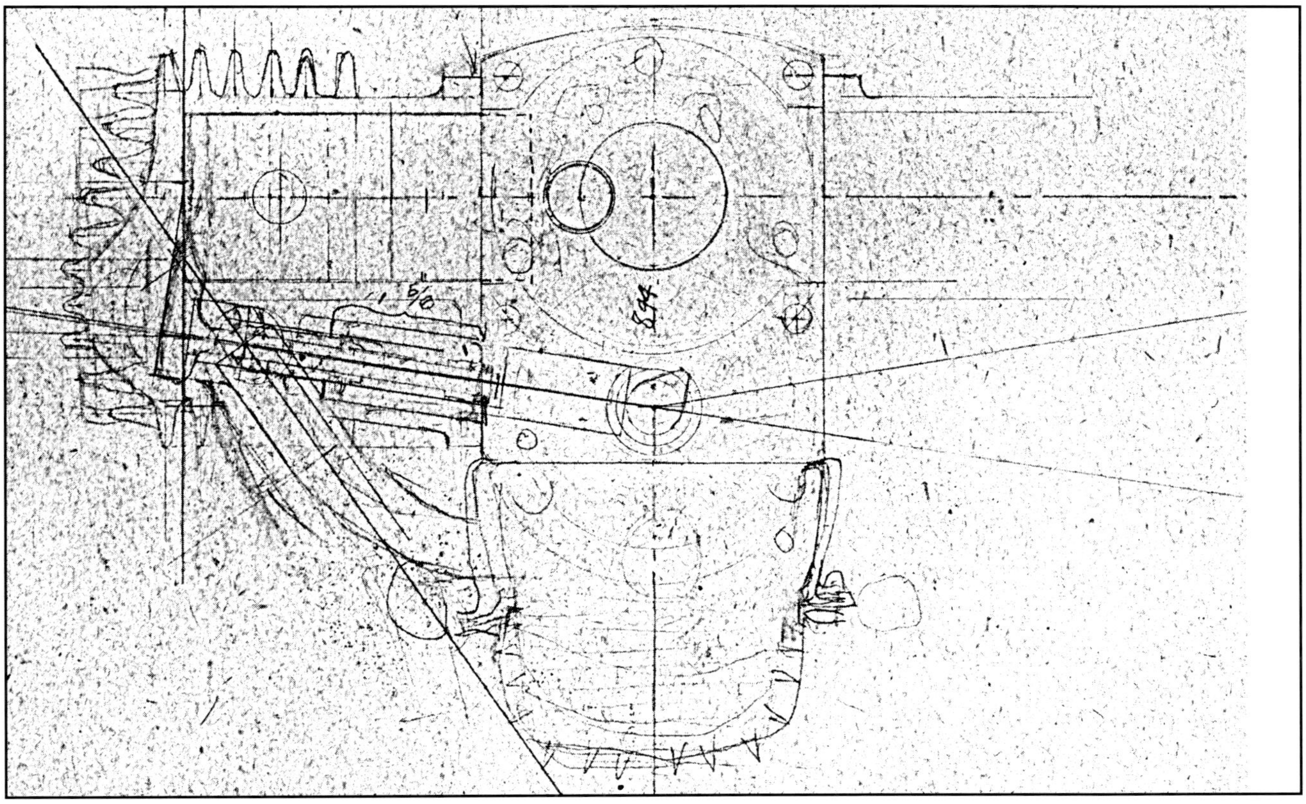

Figure 16-6 The front view, showing details of one cylinder and the crankcase. Notice the removable section of the crankcase pan.

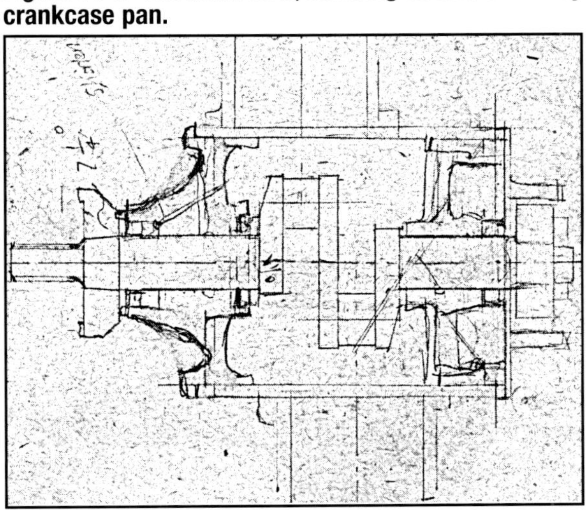

Figure 16-7 Details of the crankcase and crankshaft.

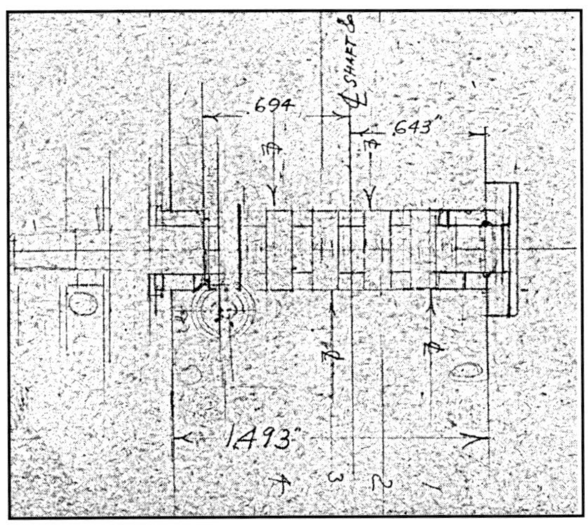

Figure 16-8 Camshaft details.

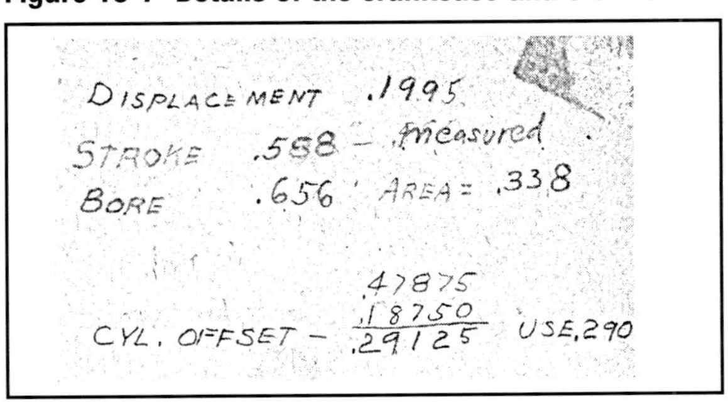

Figure 16-9 Engine specifications.

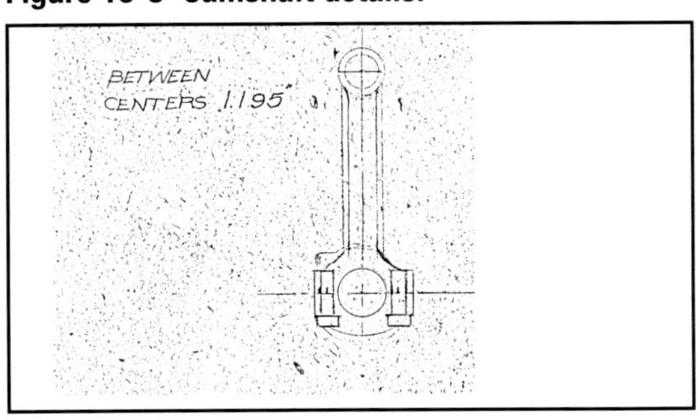

Figure 16-10 Some connecting rod details.

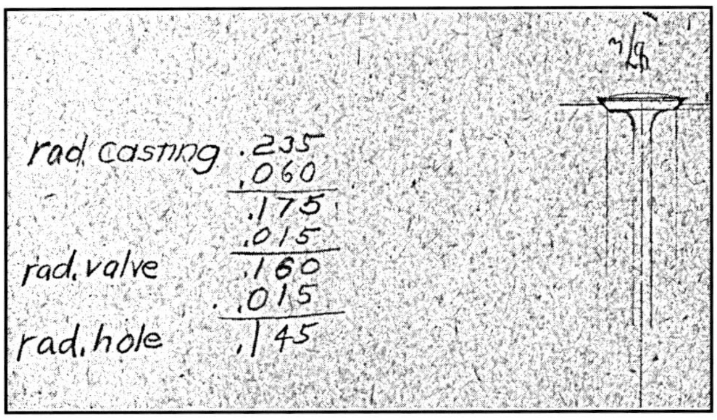

Figure 16-11 Some sketchy information for the valves.

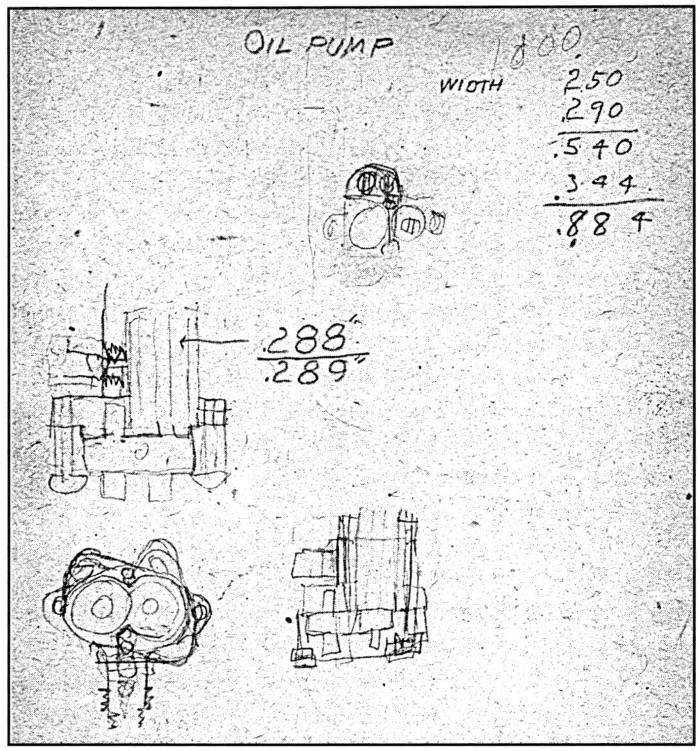

Figure 16-12 Details of the gear type oil pump.

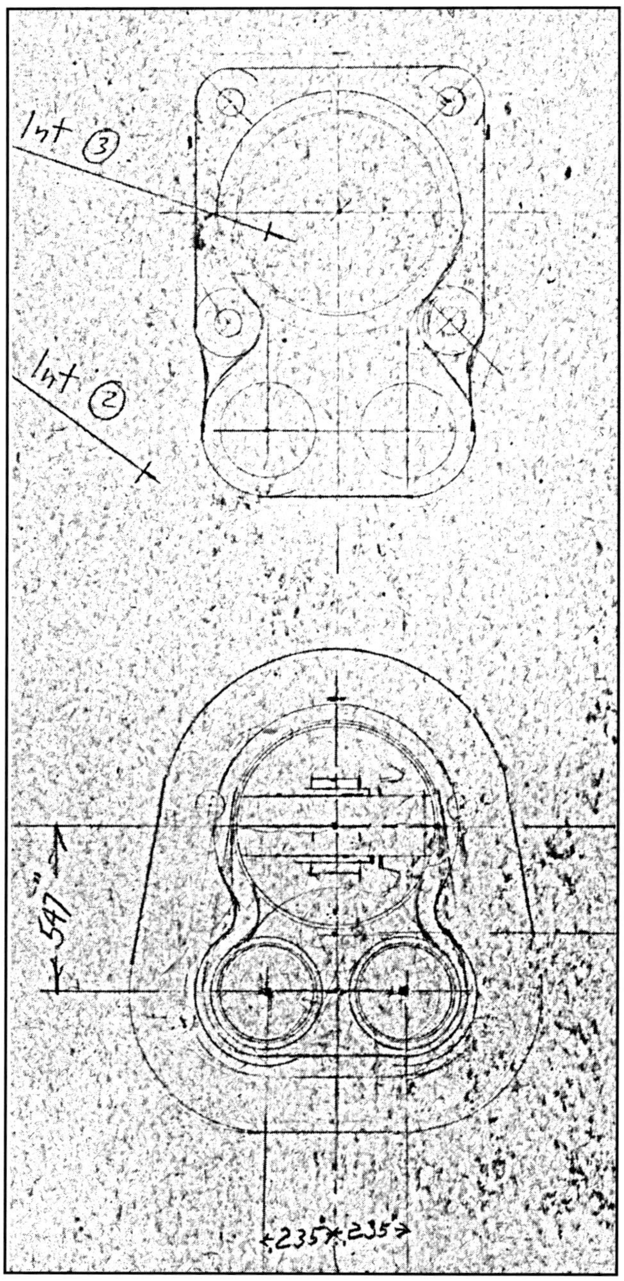

Figure 16-13 A drawing of the machining of the cylinder.

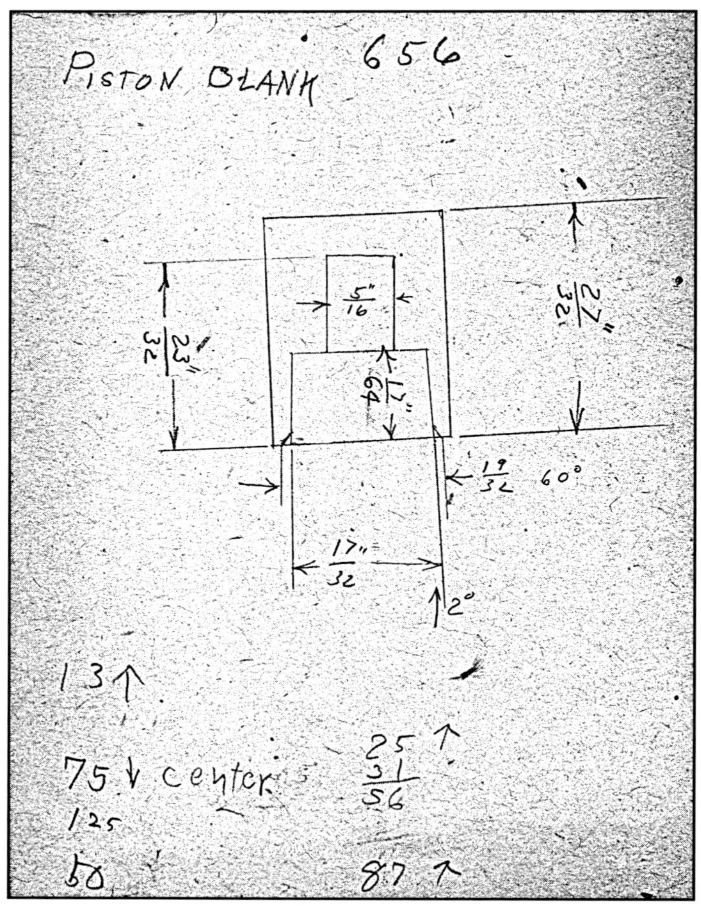

Figure 16-14 Details for the piston blank.

Chapter 17 R/C EXPERIMENTS

THE CHALLENGE

The late 1930s and early 1940s was a time when both the electronics and the mechanics for controlling an airplane via radio were still in a primitive state. A standard system consisted of a single channel transmitter and receiver, both of which used vacuum tubes, and an escapement relay. This operated a single control surface, usually the rudder, in a sequence. The pressing of a button on the transmitter would cause the rudder to move to the next position in a sequence of: left, center, right, center. The operation of a second control surface required a second complete system, of transmitter, receiver and escapement relay on a different radio frequency. Only a handful of modelers in the country even attempted radio control. The most notable were Jim Walker, Clint DeSoto, and the Good brothers. To control a plane via radio, the modeler was required to have an extensive knowledge of radio and electronics fundamentals and to obtain an amateur radio license.

Soon after moving to Portland, Dan Calkin met Jim Walker. Jim was one of the leading radio control modelers of the day, as well as the owner of the American Junior Company, a manufacturer of a variety of model airplane kits and other items. Jim occasionally hired Dan to help with mechanical problems. In the course of their relationship, Dan could see the challenges and opportunities awaiting in radio control of model airplanes. Little by little, Dan's interest in radio control was building, but, while at Everett Street, he was so preoccupied with getting his business established that he had little time left to devote to RC.

A few months after Dan moved to Washington Street, at the beginning of WWII, things became less hectic because the government would not let him build engines. With some time on his hands, Dan started investigating ways to solve the problems associated with RC. After a careful analysis, he devised a way to get multiple control channels at a time when almost everyone else had but one channel. Moreover, he achieved true proportional control. This was a great accomplishment.

Dan never obtained an amateur radio license. When he was testing and adjusting his radio equipment he worried about the FCC tuning in on his signal and coming to the shop and finding him operating illegally. He would try to do the adjustment quickly and then would hide all the radio apparatus.

Dan Calkin designed and built a number of experimental planes for use with his radio control systems. One of these airplanes is shown in Figures 17-1 and 17-2. This may have been the airplane he sold to Jim Walker in 1946 for $100.00.

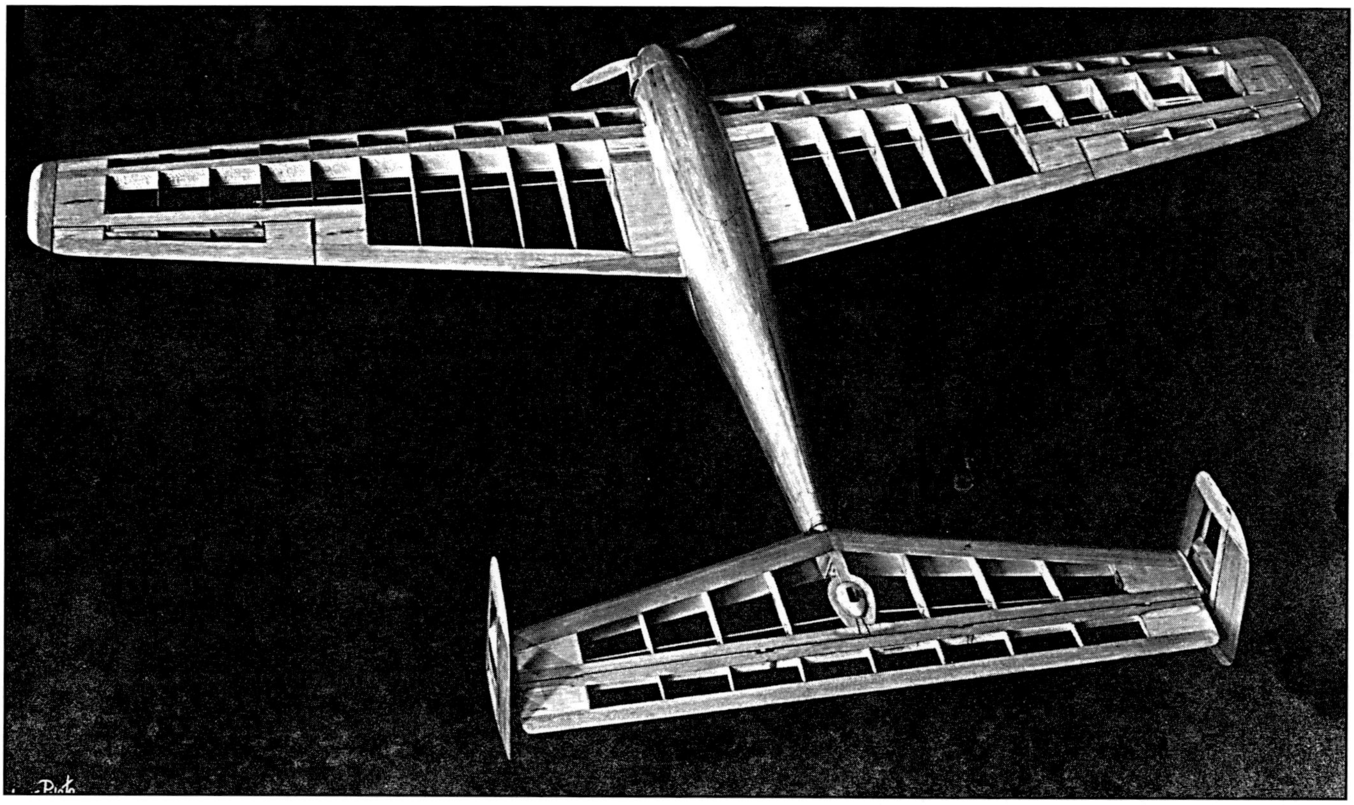

Figure 17-1 The high performance radio control airplane Dan Calkin built in the 1940s when other modelers were using very simple, stable airplanes; planes that would fly themselves if the radio failed. (Calkin family photo)

Figure 17-2 Dan, with the same airplane as shown in Figure 17-1. An ELF Twin provided the necessary power. (Calkin family photo)

Figure 17-3 The top view of the relay bank that Calkin devised to provide multiple controls. Proportional control was achieved by having a small motor drive a potentiometer.

Figure 17-4 The bottom side of the decoding and control relays Calkin used for his airplanes.

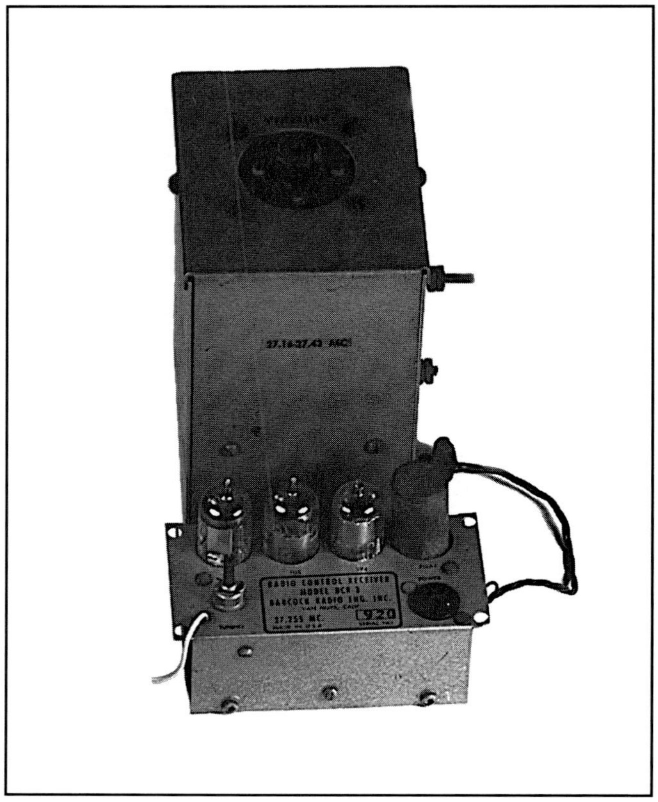

Figure 17-5 The Good brothers' transmitter and receiver from the late 1930s and early 1940s.

Figure 17-6 A Calkin-built transmitter and Babcock receiver from the late 1940s.

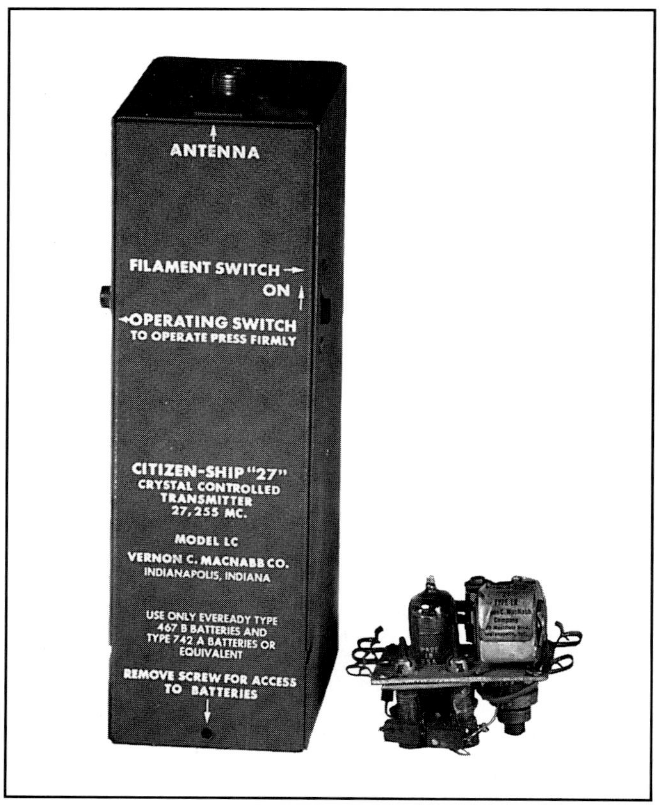

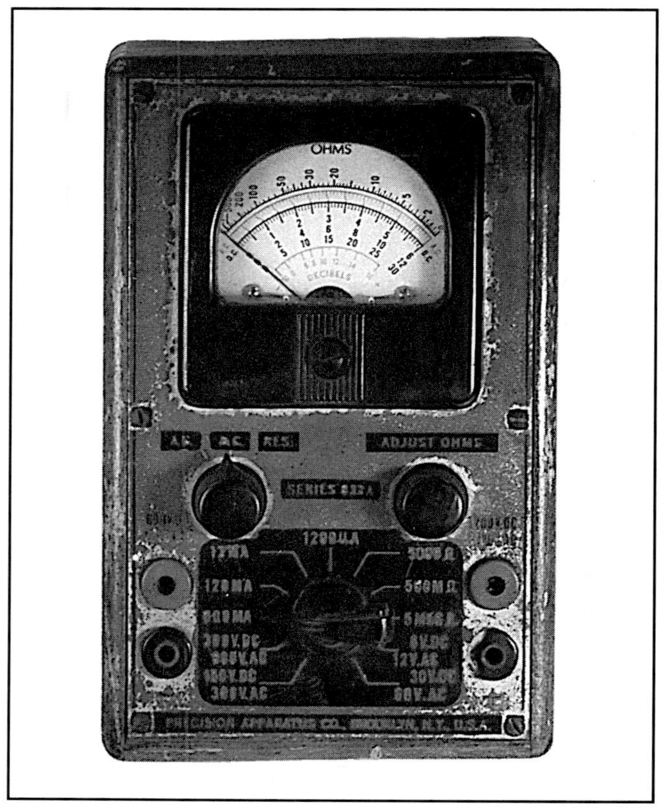

Figure 17-7 The 27 Megahertz transmitter and receiver made by Vernon McNabb from the early 1950s.

Figure 17-8 Dan Calkin's personal volt-ohmmeter that he used for his radio and electrical work.

Figure 17-9 The high performance airplane Dan built to test his radio control systems. This was his first plane based on the Convair 240, but with high wing. (Calkin family photo)

Figure 17-10 Another look at the big RC airplane Dan built. It was powered by two ELF Four engines. (Calkin family photo)

Figure 17-11 A look at Josephine holding Dan Calkin's radio control plane showing the interior. (Calkin family photo)

Figure 17-12 A peek at the inner workings of the radio controlled plane. (Calkin family photo)

Dan designed his system to obtain multiple proportional controls using a single channel radio. He used a bank of coded relays to provide the multiple controls. These are shown in Figures 17-3 and 17-4. In 1953, he wrote a construction article and submitted it to *Model Airplane News*. The article was rejected, possibly because it would have been too complex for most modelers. But who else in those days had multiple controls from a single channel radio and proportional controls as well?

To make his control system work efficiently, weight had to be kept to the minimum, so it was necessary to use very small electric motors and relays. This task occupied Dan for some time. He experimented with many kinds of commercially available units. Finally, finding that nothing meeting his requirements was available, he designed his own electric motors and relays. By about 1945, after much experimentation, he had the motors and relays he needed. Over a period of years, a complete system of relays was put together to enable the control of multiple surfaces on a model airplane. Tooling was made to produce the motors and relays and a limited production ensued. Despite doing all he could to minimize weight, the system was still very heavy due to the weight of the relays, electric motors and batteries.

Proportional control was obtained by having a small electric motor drive a potentiometer. The position of the potentiometer was proportional to the motor position, thus providing feed-back. This was also a "fail-safe" system, in that if there was a radio failure, the control surfaces would return to the neutral settings.

Dan became more and more interested in solving the problems of RC. As time went on, it became his main interest, perhaps surpassing that of his engines. He dreamed about having the very best

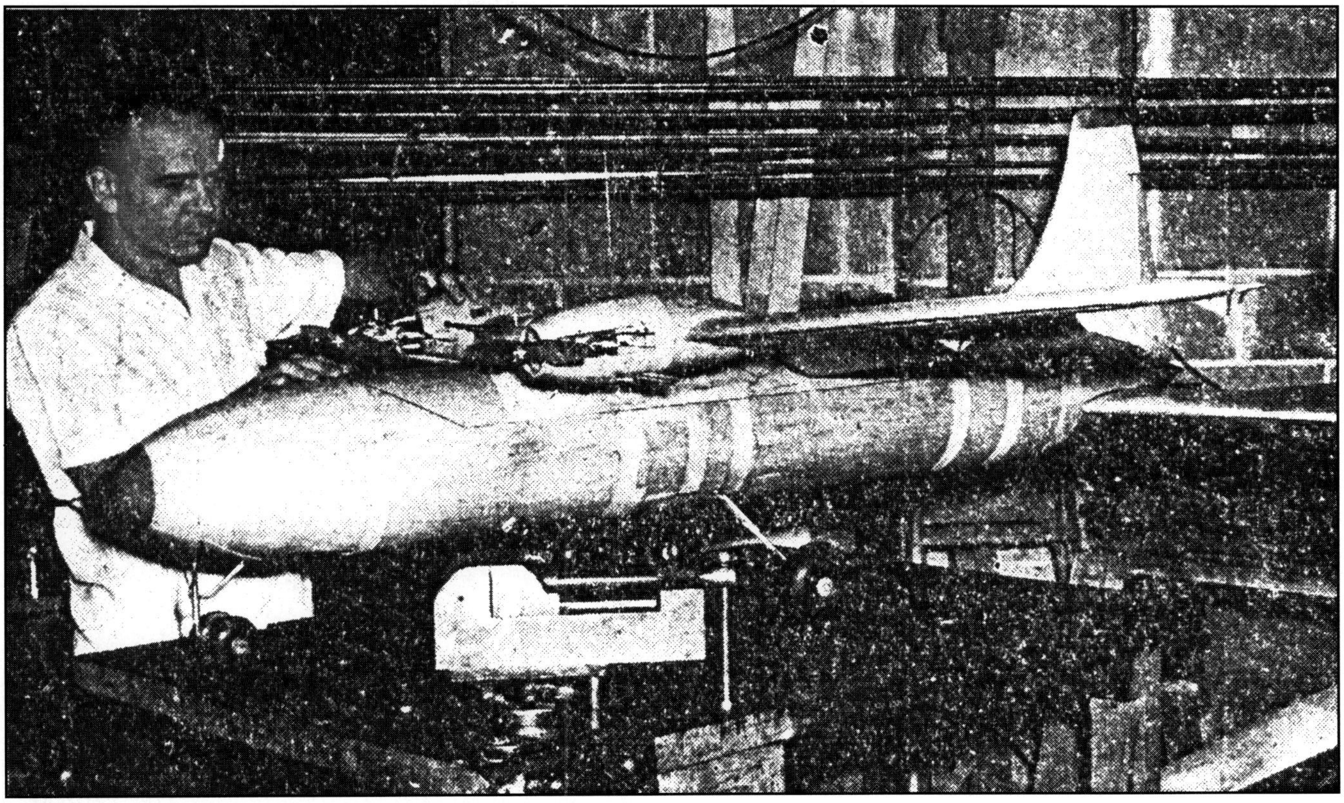

Figure 17-13 Dan Calkin at the 21st Street Shop working on his large radio control airplane. (Calkin family scrapbook)

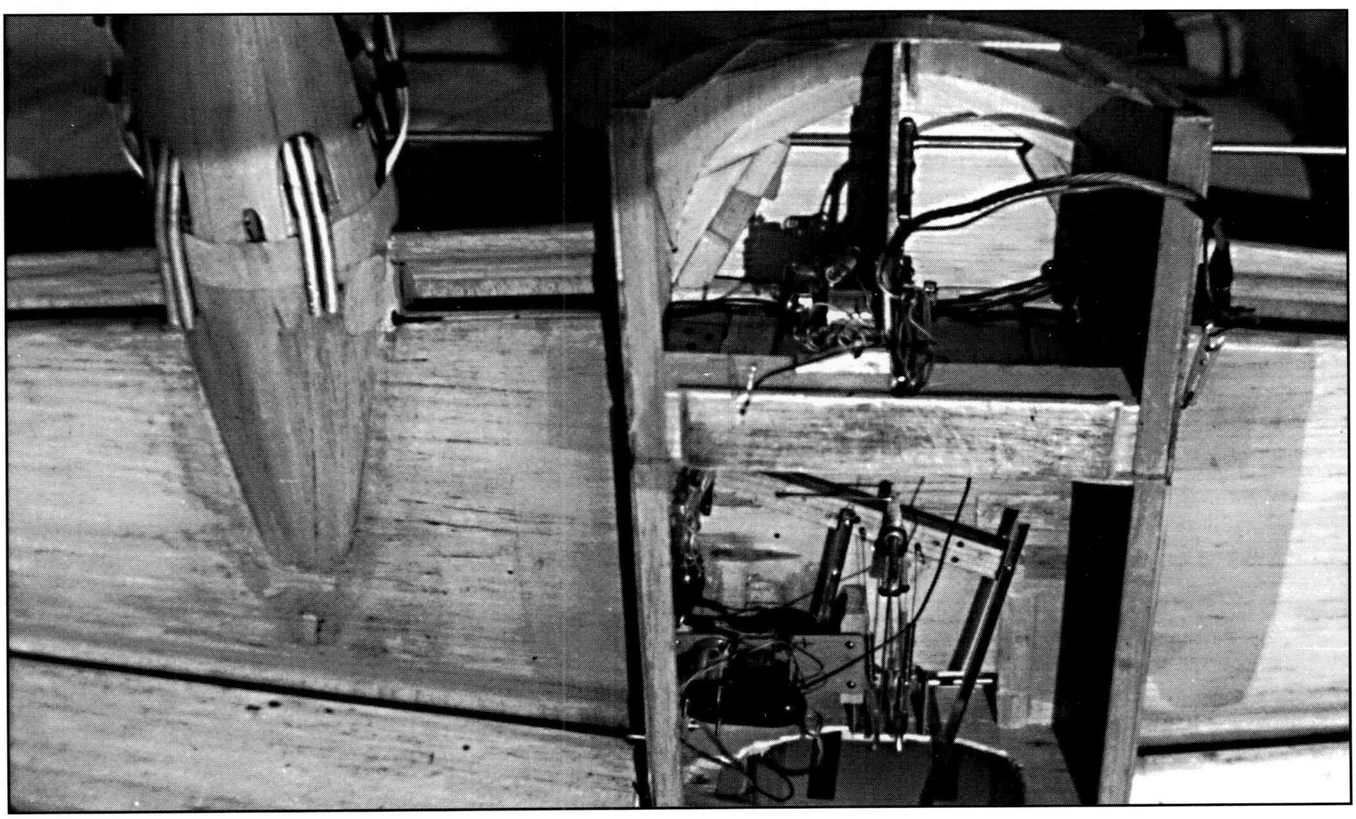

Figure 17-14 The same airplane viewed from under the wing. Note the elaborate mechanism. (Calkin family photo)

Figure 17-15 The Calkin hydraulic control for his radio plane.

Figure 17-16 The hydraulic controls from another angle. The scale in the rear is for testing the aileron mechanism. Dan spent thousands of hours in design, development and machining time on this system.

in a radio controllable airplane, and set about building just that. He designed an airplane, roughly based on the Convair 240 airliner of the day. This was to be the ultimate in airplanes, and would be controlled by the ultimate in control systems. The building of the plane was started while he was at the Washington Street shop, and continued while he was at 21st Street. The first version of the airplane was made with a wing much above the scale position, as shown in Figures 17-9 and 17-10.

The airplane was large (over 6 foot wingspan) and powered by a pair of ELF Four engines. The needle valves and carburetor chokes of the two engines were coupled together to maintain speed and synchronization of the engines. The plane had a number of technically advanced features, such as wing slots. A 12 foot long catapult was employed to assure perfect launches. In his usual methodical way, everything was designed mathematically first, then the plan was executed.

Dan seemed to be more concerned with the technical problems of developing the guidance system and controls than he was in actually flying his plane, although he did fly it a number of times through the years. The airplane had a very deep fuselage to house the heavy, bulky equipment and batteries and he just kept developing better and better systems for it.

The first few times he flew his radio guided airplane he had a string attached to a wing tip so it would fly in a circle and he could occupy himself with just the elevator and throttle controls. Each flight brought a new source of design information to improve the flight characteristics. What he wanted was no less than an airplane that would fly as smoothly and perfectly as a real airplane. He wanted perfection. That became his goal for the remainder of his life.

The Radio Equipment

Dan tried many different radio systems on his airplanes. Some have been lost through the years, but a few still exist that he used in the earlier years. The first radio system he used was a Good brothers unit manufactured by Beacon Electronics. See Figure 17-5. For a time, Dan also used a Babcock radio receiver system with a transmitter that Dan made himself. (Figure 17-6) In the 1950s he obtained a single channel, 27 Megahertz McNabb Citizen-ship transmitter, receiver and relay which he used for many years. (Figure 17-7)

Figure 17-17 The rate gyro Calkin built to stabilize the yaw axis.

The Hydraulic System

After moving to California, Dan began developing a hydraulic/mechanical system of control. A small hydraulic pump driven by an electric motor provided the power for the system. The associated accumulator, actuators, valves and fittings completed the system.

Dan worked on his hydraulic control system for many years, constantly refining it. He was still working on it at the time of his death in 1978. The system provided precise proportional control of the ailerons and other flight controls.

A complete rate gyro was included in the control system to enable the rudder to damp oscillation in the yaw axis. This was a very sophisticated system for its time.

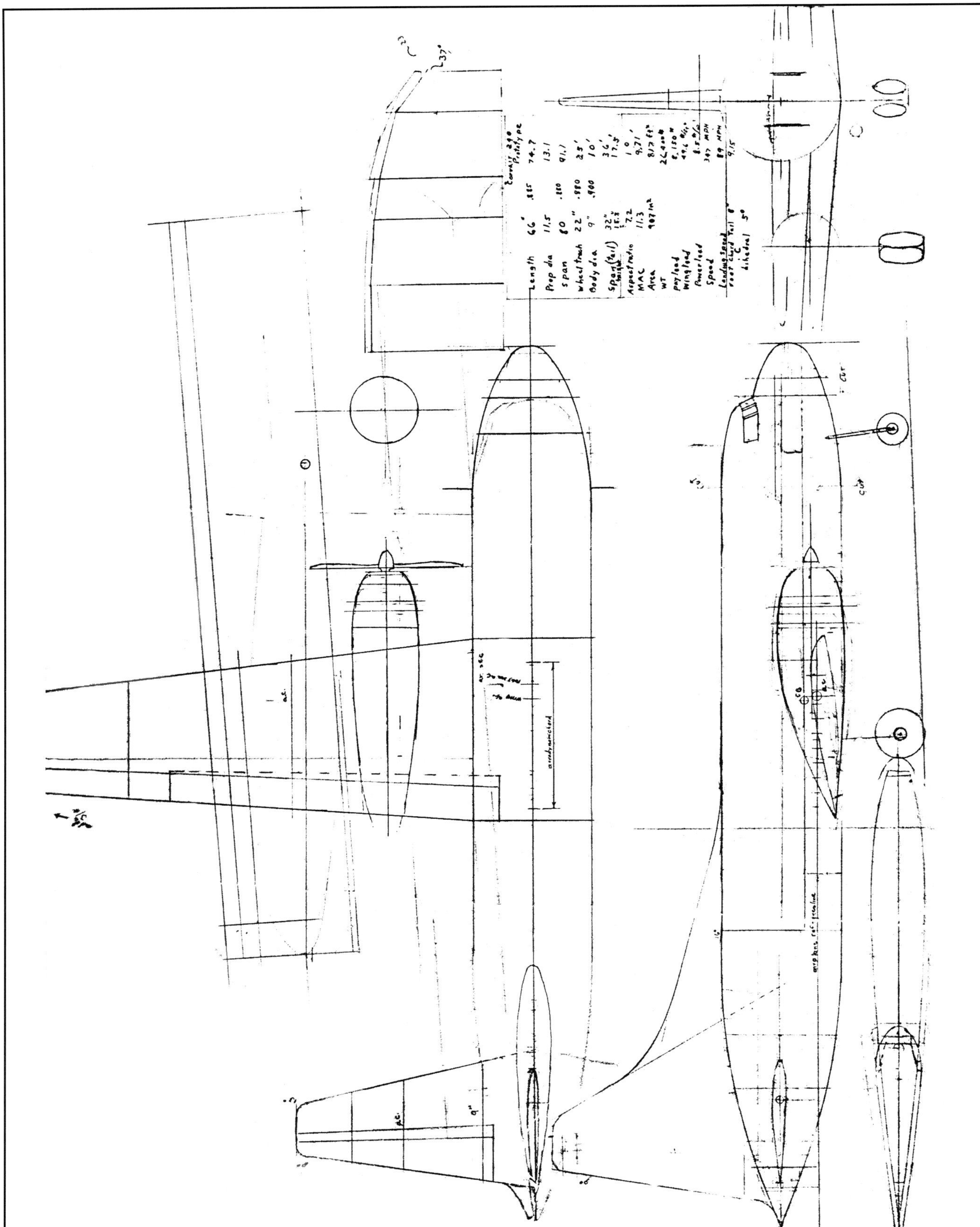

Figure 17-18 This was the original plan for the radio planes that were based on the Convair 240. (Calkin family files)

Chapter 18 IN PERSPECTIVE

A LOOK BACK

THE MAN

The story of Dan Calkin is a complex one. He was greatly admired by those who knew him, although few really did. He was not an outgoing person. He kept to himself. He didn't talk a lot, but when he talked, those who knew him listened, because he didn't engage in idle chatter and he had a keen insight. Dan Calkin was a caring husband and father.

He was highly educated, and had a master's degree from MIT to prove it, but he was highly educated in other ways. He taught himself whatever he needed to know to accomplish the task at hand, learning to be a machinist, tool maker, a pattern maker, foundry man, aircraft designer, boat designer, boat builder, electronics expert, hydraulics engineer, and dynamicist. He could read a complicated mathematical formula and easily understand how changing any term would affect the result.

Dan accomplished an amazing amount of work, all the more remarkable because of his health.

He had a highly inquisitive mind. He had a lot of ideas and he brought them to fruition. He wasn't one to copy the work of others. He had his own vision of how things should be done and he did things his way.

The Formative Years

In Dan Calkin's early years he is characterized as a serious, studious, brilliant young man, somewhat a loner, who delved into science, chemistry, mathematics, and mechanics and excelled in his studies. He was somewhat brash, with the confidence to try unusual things. He diligently pursued his studies while maintaining his interest in small engines. He also pursued and caught his bride.

The Early ELF Years

The years at Everett Street seemed to be a period of great enthusiasm and productivity. In his zeal to establish his engines, Dan not only put together a small manufacturing plant, but he developed and improved many engines and products for model airplanes. In his spare time he somehow even managed to build a cabin cruiser.

Dan wrote numerous letters to various newspapers and magazines to obtain free publicity. He used paid advertising in 1938 for the Model 1 engines, as well as in 1941 to establish his Series 2 Model 10, 20 and 40. Through these years he also had several instruction sheets and flyers printed.

Dan Calkin's engines were unique enough that other engine manufacturers bought them and were influenced by them.

The maker of the *Husky* engines visited Dan's factory and bought an early Model 1 ELF, serial number 10 and adopted its split crankcase, thin steel liner and float carburetor.

Bob Chunn, maker of the *Chunn* engines also bought a Model 1 and studied it carefully.

Two of the earliest sales of the smooth nose Model 10 were to the French engine maker, R. Stab.

The Middle ELF Years

The middle years at Washington Street were a time of less intense engine effort, although there was continuous development and improvement of the engines and related products. The most notable engine development was the conversion of the side port ELF into a reed valve engine. Dan developed glow plug versions of his engines, including the ELF Six. This period marked the real beginnings of his interest in radio controlled model planes, which were to consume more and more of his time.

There was no paid advertising during this era, and only instruction sheets and flyers announcing new products were printed. This period established his main business as building multi-cylinder engines.

The Last ELF Years

When the ELF Engine Company was at 21st Street in Milwaukie, Oregon, there was little development of engines. The only new product was the open cast timer which was part of his effort to reduce machining time and costs. More machines were automated and his sand casting machine was improved. Most of his creative genius went into solving the problems associated with the control of model planes by radio.

Again, there was no paid advertising and no brochures were printed. Only instruction sheets were printed, and these were printed on a mimeograph machine instead of a printing press, as had been done earlier. For advertising, he was dependent on a few loyal dealers, especially the Miniature Aircraft Company of New York. They specialized in scale models and the ELF engines they offered for these models were always a good seller, especially the twin and four cylinder models. Miniature featured ELF engines in their annual catalog, although they always showed photos of the pre-war ELF engines. The second most important ELF dealer was Colonel Bob's Hobby Shop in Los Angeles.

After ELF

After closing up the ELF Engine Company, Dan spent his spare time developing his RC machine. He hoped it would make possible the flying of a model airplane with performance similar to a full sized plane. He was able to bring a hydraulic-mechanical system to a point of near perfection at a time when electronic systems were showing their superiority for control purposes and surpassing other means. He continued development of this system until his death.

THE ROCKET SCIENTIST

Dan was a highly regarded senior dynamics expert at Douglas, where he was among the very best of this elite engineering group. He was often asked to conduct classes and assist the less experienced scientists there. Even some of the most experienced scientists would ask Dan for advice in solving complex problems.

THE ENGINES

Dan built the kind of hand-crafted engines that you couldn't help but admire. These were not always the engines that competition model builders bought to win contests, they were bought by ordinary modelers to power the airplane of their dreams; the scale model they always hoped to build. Who couldn't picture a real four cylinder engine that resembled a Lycoming engine powering a replica of a real airplane?

Every part needed for the ELF engines was made by Dan Calkin. All he bought was raw materials and spark/glow plugs and, in the beginning, he even made the spark plugs. Dan made the tooling, including cutting tools, to make his own gaskets, reed valves, connecting rod straps, etc. He did heat treating, plating, and made celluloid tanks. He was inventive in the things he did; no one could accuse Dan Calkin of being a copy-cat.

It must be kept in mind that Dan's development of the ELF occurred at a time when there were no other comparable small engines being manufactured. So the challenges he faced were much greater than those of the engine manufacturers who came along years later. Dan carried on his engine building virtually alone for over 20 years, long after many larger companies had disappeared.

PRIZED BY COLLECTORS

The ELF engines are among the most collectable of all vintage model airplane engines. Anyone owning an ELF has a treasure. None of the ELF engines can be considered common. In general, the spark ignition engines are more desirable than the glow plug engines. The reed valve engines are slightly less desirable than the earlier side port engines, except for the Singles and Sixes. The rarest and most desirable ELF engines are:

Model 10, Series 1, smooth nose Single

Model 20, Series 1, smooth nose Twin

Model 40, Series 2, Four with distributor

Model 40, Series 3, post-war Fours

Model 10, Series 4, reed valve Single

Model 60, The Six

CONTRIBUTION TO THE MODEL HOBBY

Dan's contributions to the hobby were not limited to just engines. He designed several free flight and radio controlled models, as well as several boats. He designed and manufactured propellers, the first 1.5 Volt spark coils and the first $\frac{1}{4}$ inch spark plugs. He was keenly interested in radio control and is partly to be credited for the success of Jim Walker in that area. Before the days of proportional radio control, he made relays and electric motors smaller and lighter than anything else available.

He published an excellent small booklet on the design of models and propellers at a time when virtually nothing of any value was available. (See Appendix C for a reprint of his booklet.) Finally, he did ground-breaking research into a hydraulic flight control system.

His work with model airplanes was a success in many ways, but not financially. He was forced to depend on his parents for financial aid on many occasions. The ELF Engine Company barely provided him with a subsistence but, after his retirement, he confided to a friend that building the ELF engines gave him more satisfaction than his work at Douglas (now a part of the Boeing Co.).

To quote the final paragraph of an article called "The ELF Engine for Gas Models," published in the January 1937, *Popular Aviation* magazine:

> The only wish of the writer is that this article may be the nucleus for future well-deserved recognition of this excellently constructed and engineered engine in the Miniature-Gas-Engine Hall of Fame.

That was the acclaim for the very first ELF production engine. There were many more, and far better, engines to come.

Dan was a brilliant, quiet man. He was a true pioneer in manufacturing model airplane engines and devoted his considerable engineering skill to their development. He literally went from designing and manufacturing model airplane engines to being a rocket scientist.

Dan Calkin's wife, Josephine, passed away just as this book was going to the printer. He is survived by his daughter, Cecelia. He has three grandchildren: Allanna Lea Jackson, born in 1963, Ian Allan Jackson, born in 1966, and Joy Janene Jackson, born in 1981. All are the children of Cecelia and Martin Jackson.

Dan's reputation as a master craftsman and engineer survives along with his legacy of the ELF engines; "built like a watch," as Irwin Ohlsson once said. The legacy of Edward Daniel Calkin and his ELFs will remain long after many other engines are forgotten. What else would you expect from a man whose guarantee was, "Your ELF engine is guaranteed to give you satisfactory performance in every way." No one else gave such a guarantee, no one else produced engines like the ELF.

The Favorite Photograph

Josephine Calkin kept this February, 1976, photograph nearby after Dan died. (Figure 18-1). It was her favorite photograph of the man she had known for 50 years and been married to for 44 years.

Figure 18-1 Dan Calkin, 1911-1978 (Calkin family photo)

Epilogue:

The search for answers about Dan Calkin has resulted in a startling number of occasions when I was fortunate enough to be just in time to get information and other materials before it was too late. Three of the important participants in this story have died this year over a period of three months, and all were severely impaired for a year or so earlier. If I had started on the story just a little later it would have been too late.

I am still collecting information about ELF engines and would welcome any additional information, and I also may be able to answer specific questions.

If you have information especially about the few items I was unable to fully document, such as the flywheel for the Model 1, the stamped metal tanks for the Model 40 and the small tank for the Model 10, as well as the double timer (two speed), it would like to hear from you. Also of interest would be the instruction sheets for the Model 10, Series 4 ELF and for the fuel pump, throttle for the carburetor and the double timer. Copies of any ELF correspondence would also be appreciated.

Thanks again to all who contributed to this book.

Appendix A ELF SERIAL NUMBERS

Many questions related to ELF serial numbers have arisen. Some of the most often asked questions are:

Why Weren't The Engines Shipped in Consecutive Order?

If we could have seen how Dan operated, we might have noticed that he would machine a batch of crankcases and completely assemble the bottom end of the engine with the crankshaft and thrust bearings. Only then, did he hand-stamp the serial number on the back of the crankcase. At this point the crankcase assemblies were stacked in a box until needed. As orders were received, Dan would reach in the box and get an assembly, and it didn't really matter to him what number it was. Then he matched pistons and cylinders to the crankcase and finished it up. Every engine was test run by Dan. If it didn't come up to his expectations he would rework it. Sometimes it would be reworked immediately, other times it would be put aside to be reworked later. If numbers had been assigned to the engines as they were sold they would have been in order, but that wasn't the way he worked.

By the end of ELF production in 1956, Dan was stamping serial numbers on the crankcases as soon as the case was machined internally. When he changed to this method is unknown.

Why Were Some Numbers Skipped?

It is believed that when he stamped the serial number on the crankcase he didn't knowingly skip any numbers. Some crankcases probably had to be scrapped for various reasons, and that number was not replaced on another crankcase.

Why Were Some Numbers Duplicated?

One reason is that some engines were shipped, then returned and resold so that the serial number appears as sold twice in the ELF log book. Dan didn't always record the return of the engines. In these cases the number wasn't actually duplicated, but the log book of engines sold indicates that two engines had been sold with the same number. These cases exist but they are not common.

Valid duplicate-numbered engines do exist. Dan probably mistakenly stamped the same number more than once. There is one instance where he shipped four Twins to Comet Model Company in California; two engines were #478 and two were #488. Dan noted the numbers as such. Evidently that didn't bother him.

Why Were Some Engines Not Recorded In The Log?

"Lucky" Lucas tells that, in his lifetime, he has owned between 40 and 50 ELF engines. If he crashed one, or had problems with one, or needed another one, Dan would just replace it with a new one. Certainly that is what happened to the first Model 10, Series 1 ELF that was delivered to him. Lucky knew that Dan was working on strengthening the front crankcase and making other improvements and, when the new engines were ready, his old engine was exchanged for a new model. Lucky is only listed in the log book as receiving three engines. The same thing must have happened with other active modelers, contest flyers, friends and employees. Carl VanCourt certainly received engines that were not recorded. Surely, other engines were given to friends or flyers for promotional or good-will purposes or as replacement for returned engines.

Next, quite a few engines, especially pre-war Fours, were shipped by Dan without him listing the serial number. There are also some Singles and Twins in this category.

In some cases, Dan apparently just forgot or didn't bother to write the transaction down. He probably forgot to check the serial number until the engine was wrapped and ready for shipment. Finally, remember that a former employee assembled engines from defective parts and was selling them before WWII. This is the period when most of the questionable engines appeared.

Possibly a few of the duplicate serial numbers may be the result of reproduction engines, although reproduction engines usually don't have serial numbers.

Hard-to-Read Numbers

By the way, the task of stamping the serial numbers on the engines was one that Dan always did himself. The steel number stamps he used were quite distinctive, and it is believed that he used the same set throughout all the ELF years. Some of the numbers are very difficult to read and the numbers may have been recorded incorrectly. The 1's and 7's look similar and the 3's, 5's, 8's and 9's are difficult to tell apart. Both Dan and some owners may have read the the number incorrectly.

Dan didn't keep close track of the engine serial numbers. It probably mattered much less to him at the time than it does to ELF enthusiasts today. In his log book he admits to oversights when he wrote, "Twins 621-660 Twice, 721-760 none." There were also some Twins in the 600s and 700s on which the numbers were stamped upside down. He also repeated the numbers in the 500s used for the Model 10, Series 1 engines, probably believing they had all been recalled and replaced by Series 2 engines.

In the ELF log book of engine repairs, the serial number of the repaired engine only appears three times out of the nearly 200 engines repaired

Also, remember that he was plagued by serious health problems throughout the ELF years. Surely there were days when he really felt ill, but still managed to ship a few engines, anyway. Why would he have cared if he skipped some numbers, or used some twice, or had a batch of crankcases, or even complete engines with lower numbers, sitting around for a while?

MODEL 1 ENGINE SERIAL NUMBERS

The First Model ELF Single engines were numbered consecutively starting with 1 and going through nearly 500. No log book for these early engines exists. The highest known serial number is 432. No multi-cylinder First Model ELF engines were ever built. The serial number location was on the left side of the cylinder flange, just above the rear-most mounting lug. (Figure A-1)

MODEL 10 ENGINE SERIAL NUMBERS

The earliest slant plug Single serial numbers started as a continuation of the First Model engine numbers. These engines, numbered in the 500s, were the beam mount, smooth nose, Series 1 variety. Their numbers range from 502 through 578. (Not all numbers were used.) Known survivors are numbers 549, 555, 567 and Dan's factory cutaway engine 52_, the third number having been cut off. (Figure A-2)

Most Series 1 engines were recalled by Dan and replaced by Series 2 engines.

Serial numbers for the newer, radial mount ELF Singles (Series 2 and 3) range from 01 through 561 and from 1500 to 1578. The engines with numbers in the 1500s really were engines with a serial number in the 500s with a "1" added before the numbers. These crankcases had been built for some time and changing them to the 1500s brought them more in line with the numbers being used on the other, more popular multi-cylinder engines.

At least twenty-two slant plug Singles were shipped without the serial numbers being recorded in the log book.

The serial numbers were located on a flat on the rear of the crankcase, opposite the ELF logo. (Figure A-3)

In 1950, the only batch of reed valve, Series 4, slant plug Singles was built. Sales were recorded for numbers 5000 through 5009, except 5006,

Figure A-1 Model 1 ELF Serial number 339

Figure A-2 Model 10, Series 1; #549 on rear crankcase

Figure A-3 Model 10, Series 2 and 3; serial number 402. Some, like this engine are very difficult to read.

Figure A-4 Model 10, Series 4; #5006

Figure A-5 Model 20, Series 1; #T 34

Figure A-6 A rear crankcase for the Model 20, Series 2 and 3 or the Model 40, Series 2 and 3. This is #1176.

Figure A-7 Model 20, Series 4 engine; Engine # 1566.

Figure A-8 Model 40 and 60 numbers. This is #1389.

which Dan kept for himself. These serial numbers were divided on the rear crankcase with the 50 being to the left of the intake tube. The 50 referred to the year of manufacture.The two digits placed to the right of the intake tube represent the actual serial number. (Figure A-4)

ELF MODEL 20 SERIAL NUMBERS

The slant plug Twins (Series 1) started with the smooth nose variety. The prototype Twin was serialized TX, which was scratched in by hand. T-7 was also scratched by hand. No others known to still exist were numbered by hand. The first production engines were numbered from T1 through T75, but not all numbers were used. Known existing smooth nose Twins, in addition to TX and T7, are T18 and T24 and the factory cutaway engine T34. (Figure A-5)

Dan changed his numbering system a couple of times in the first few months of the Twin production. The T numbers continued until 51, which was listed as T2-51. After that, the T was dropped, and numbers like 2-51 and 2-68 are listed. On the engines themselves, there was no hyphen. On engine number 260 there is a space between the 2 and the 6; thus 2 60. With 2-78, Dan dropped the hyphen in his records, and the serial number was listed as 278. Because of this unusual numbering system, there are no Twins with numbers under 200, except the few with T numbers. (Figure A-6)

Like the smooth nose Singles, most of the smooth nose Twins were recalled and replaced with Series 2 engines. The Series 2 engines had numbers in the high 200s and low 300s range. Serial numbers continued for the Series 3 and 4 engines to the 1500s. The highest known number listed in the ELF log book for a Twin is 1573. (Figure A-7)

Four Twins were shipped with only a partial serial number being recorded; that is, he listed a number such as 9-8. A few serial numbers have a question mark next to them; 487? for example. Eight Twins were shipped with no serial number being recorded.

One batch of engines, in the 600s and 700s, has the serial numbers stamped upside down. That is, the engine has to be held in the inverted position to correctly read the number.

MODEL 40 SERIAL NUMBERS

The ELF Fours are in the range of #299, the prototype, through #1561. The first production four cylinder engine delivered was #375, which went to Dan's friend, Carl VanCourt. The second Four, #397, was purchased by VanCourt for his flying buddy, Bill Sweet.

For some unknown reason, Dan shipped quite a few Series 2 Fours without recording the serial numbers (39 of the 55 built). One reed valve Four was shipped without the number being recorded.

MODEL 60 SERIAL NUMBERS

The Sixes were the crowning achievement of the ELF line. They were made only on special order. According to the log book, only 10 were shipped. One six was sold to Earl Vivell with the number unrecorded, though it is now known to be #1018. Two additional Sixes were kept by Dan for his own use. They are #1369 and #1373. Known examples of the Six have serial numbers in the range from #986 to #1378. (Figure A-8)

A few replica Sixes have been made by hobbyists. One or more were made from all replica parts. Others have been made by taking an original ELF Four and a Twin or two Fours and fitting a new crankshaft to it with a few new small parts. The known replica Sixes all have 8-32 prop nuts instead of the 10-32 size that Calkin used.

GENERAL

All engines below serial number 200 were Single cylinder models. The numbers for the Singles and Twins were numbered in parallel from #201 through #561, (except #299, the ELF Four, Series 2 prototype). That is, there are Singles and Twins with the same serial number in that range, but certainly not all numbers were used for both. Very often, the date that a Single and a Twin with the same number was shipped were months or years apart. Some numbers were skipped. A few numbers were used three times for the same series engine.

From serial number #362 through #561, the four cylinder engines were numbered in parallel with the Singles and Twins. Twenty-one duplicate numbers were used for Singles and another twenty-one duplicate numbers were used for Twins. Three numbers were used three times for the same kind of engine; two for Singles, one for Twins.

From #572 through #897, all engines were Twins except #583, 800, 813 and 840 which were Fours. There were thirty-five duplicate numbers in this range, all Twins. One batch of numbers (#722-#760) was not used.

From #800 to 1578 all engines were numbered in one sequence with seven instances of duplicate numbering being recorded.

Dan kept some engines for his own use, mostly reed valve engines, without recording the numbers.

A few minutes spent looking through Appendix B should make it clear to you how the engines were serialized. Also, refer to page 87 for a description of the various models and the production period of each.

DUPLICATE-NUMBERED SINGLE CYLINDER ENGINES

001	018	087	089	106	138	158
268	275	277	282	284	286	291
292	294	295	296	298	300	301
302	304	306	307	309	311	312
315	317	319	320	321	322	323
324	325	326	327	334	337	378
396	455	460	502	507	510	522
526	527	528	529	530	534	535
536	540	542	544	549	561	

TRIPLICATE-NUMBERED SINGLE CYLINDER ENGINES

283	290	297	314	316	523	554

DUPLICATE-NUMBERED TWIN CYLINDER ENGINES

254	268	324	459	461	462	465
467	472	473	476	477	478	480
481	482	485	486	491	492	497
584	598	623	624	625	627	628
629	630	631	633	634	635	636
638	639	640	641	642	643	644
645	647	649	650	652	653	659
660	668	686	788	808	847	870
1107	1238	1497				

TRIPLICATE-NUMBERED TWIN CYLINDER ENGINES

488 657

OMITTED BLOCK OF TWIN CYLINDER NUMBERS

The serial numbers from 722 through 760 were not assigned.

DUPLICATE-NUMBERED FOUR CYLINDER ENGINES

1259

Appendix B ELF SERIAL NUMBER LOG

ELF RECORD OF ENGINE SALES

The following pages represent a computerized copy of the original ELF log book of engines shipped, from the first Model 10 in 1939 to the last ELF in April 1956. The original record was in chronological order. This record has been sorted by the serial number of the engines in ascending order. This change was for the convenience of persons wanting information on a certain engine.

The first column is the date of shipment. The second is the number of cylinders of the engine shipped. The third column is the serial number of the engine. The fourth is the designation as to whether the engine was a spark ignition or a glow plug type. The last column in the log book was the name of the customer the engine was shipped to. Omitted are Dan Calkin's notes about changes made to the engines and the price of the engine.

You will notice a few notations of an engine being returned by the dealer. In many cases the same engine was resold, thus causing a duplicate number in the log. You will also find other duplicate numbers, some for the same kind of engine. More commonly, especially before WWII, the same numbers were used for Singles, Twins and Fours, as was Dan's practice at that time.

If you own an ELF engine, the first thing you may want to do is to look up the serial number to see when it was sold and to whom, and if the engine originally was spark or glow plug ignition. All ELF engines were spark ignition until December 22, 1949, when an ELF Four with glow plugs was sold.

Date	Cyl	SN	Ign/gp	Buyer
6/16/41	1	01	ign	Meyers & Ballman
8/7/41	1	01	ign	Berkeley
4/16/41	1	02	ign	Comet-Oakland
3/29/41	1	03	ign	Lumsden
4/16/41	1	04	ign	Comet-Oakland
5/24/41	1	06	ign	Skyway
6/16/41	1	07	ign	Good's
4/16/41	1	08	ign	Comet-Oakland
4/16/41	1	09	ign	Comet-Los Angeles
4/16/41	1	010	ign	Comet-Los Angeles
4/16/41	1	011	ign	Comet-Los Angeles
4/16/41	1	013	ign	Comet-Oakland
5/24/41	1	014	ign	Skyway
4/16/41	1	016	ign	Comet-Los Angeles
5/24/41	1	017	ign	Skyway
5/24/41	1	018	ign	Skyway
8/28/41	1	018	ign	Dishman
4/16/41	1	019	ign	Dr Burke
5/29/40	1	022	ign	Al Dippert 549 022
4/16/41	1	023	ign	Comet-Oakland
4/16/41	1	024	ign	International
4/16/41	1	026	ign	Comet-Los Angeles
8/7/41	1	029	ign	Berkeley
7/2/41	1	031	ign	Lau (Hobby Supply House)
6/14/41	1	032	ign	Carr
5/24/41	1	035	ign	Skyway
6/28/41	1	036	ign	Wyoming Cycle
5/24/41	1	037	ign	Skyway
5/24/41	1	039	ign	Skyway
5/24/41	1	042	ign	Skyway
5/24/41	1	043	ign	Skyway
6/16/41	1	044	ign	Good's
8/1/41	1	046	ign	Comet Western
8/7/41	1	047	ign	Berkeley
5/24/41	1	048	ign	Skyway
6/14/41	1	049	ign	Richard Phillips
5/24/41	1	050	ign	Skyway
7/2/41	1	051	ign	Lau (Hobby Supply House)
10/3/46	1	052	ign	Church (Phil's)
5/24/41	1	054	ign	Skyway
12/23/40	1	55	ign	Amer Jr
5/24/41	1	055	ign	Skyway
7/2/41	1	056	ign	Roth
8/7/41	1	057	ign	Berkeley
8/1/41	1	058	ign	Comet Western
5/24/41	1	059	ign	Skyway
5/24/41	1	061	ign	Skyway
6/16/41	1	062	ign	Good's
6/16/41	1	063	ign	Allen Bros
8/1/41	1	064	ign	Contest prize
5/24/41	1	065	ign	Skyway
6/16/41	1	066	ign	Allen Bros
6/27/41	1	067	ign	Lumsden
5/24/41	1	069	ign	Skyway
7/2/41	1	070	ign	Roth
8/1/41	1	071	ign	Comet Western
8/20/41	1	072	ign	Sherman (Hillboro)
8/7/41	1	073	ign	Berkeley
6/27/41	1	074	ign	Lumsden
7/2/41	1	074	ign	Arrow
7/8/41	1	075	ign	Comet Western (Oakland)
8/1/41	1	076	ign	Comet Western
8/1/41	1	077	ign	Comet Western
7/10/41	1	078	ign	Western
7/8/41	1	079	ign	Comet Western (Oakland)
8/1/41	1	080	ign	Comet Western
8/1/41	1	081	ign	Comet Western
8/1/41	1	082	ign	Comet Western
5/24/41	1	084	ign	Skyway
8/1/41	1	085	ign	Comet Western
8/7/41	1	086	ign	Berkeley
7/8/41	1	087	ign	Comet Western (Oakland)
8/20/41	1	087	ign	Sherman (Hillboro)
8/7/41	1	088	ign	Berkeley
7/8/41	1	089	ign	Comet Western (Oakland)
7/8/41	1	089	ign	Comet Western (Oakland)
8/7/41	1	090	ign	Berkeley
7/8/41	1	092	ign	Comet Western (Oakland)
8/1/41	1	093	ign	Comet Western

7/8/41	1	094	ign	Comet Western (Oakland)
7/2/41	1	095	ign	Roth
8/20/41	1	096	ign	Sherman (Hillboro)
7/8/41	1	098	ign	Comet Western (Oakland)
7/8/41	1	100	ign	Comet Western (Oakland)
8/7/41	1	101	ign	Berkeley
7/2/41	1	104	ign	Lau (Hobby Supply House)
7/8/41	1	105	ign	Comet Western (Oakland)
7/2/41	1	106	ign	Lau (Hobby Supply House)
8/7/41	1	106	ign	Berkeley
8/1/41	1	107	ign	Comet Western
8/1/41	1	109	ign	Comet Western
8/7/41	1	110	ign	Berkeley
7/2/41	1	111	ign	Lau (Hobby Supply House)
8/1/41	1	112	ign	Comet Western
7/2/41	1	113	ign	Roth
7/8/41	1	114	ign	Comet Western (Oakland)
8/1/41	1	115	ign	Comet Western
8/7/41	1	116	ign	Berkeley
8/1/41	1	117	ign	Comet Western
8/7/41	1	119	ign	Berkeley
8/1/41	1	120	ign	Comet Western
8/1/41	1	121	ign	Comet Western
8/1/41	1	123	ign	Comet Western
8/7/41	1	124	ign	Berkeley
8/7/41	1	125	ign	Berkeley
7/8/41	1	126	ign	Comet Western (Oakland)
7/8/41	1	127	ign	Comet Western (Oakland)
7/8/41	1	128	ign	Comet Western (Oakland)
8/20/41	1	129	ign	Herd (for Contest)
7/8/41	1	131	ign	Comet Western (Oakland)
7/8/41	1	132	ign	Comet Western (Oakland)
7/8/41	1	133	ign	Comet Western (Oakland)
7/8/41	1	134	ign	Comet Western (Oakland)
12/27/41	1	135	ign	Barr
1/7/42	1	136	ign	Comet (Oakland)
1/7/42	1	137	ign	Comet (Oakland)
12/31/41	1	138	ign	Baxter
1/7/42	1	138	ign	Comet (Oakland)
1/29/42	1	140	ign	Berkeley
12/24/41	1	141	ign	Western Model
12/27/41	1	142	ign	Skyway
1/29/42	1	143	ign	Berkeley
1/7/42	1	144	ign	Western Model
12/27/41	1	146	ign	Skyway
1/7/42	1	147	ign	Comet (Oakland)
1/7/42	1	148	ign	Comet (Oakland)
1/7/42	1	149	ign	Comet (Oakland)
12/24/41	1	150	ign	Western Model
1/7/42	1	151	ign	Weeks (Bon Marche)
1/7/42	1	152	ign	Weeks (Bon Marche)
1/7/42	1	153	ign	Comet (Oakland)
12/27/41	1	155	ign	Skyway
12/27/41	1	156	ign	Skyway
12/27/41	1	157	ign	Skyway
12/27/41	1	158	ign	Skyway
1/29/42	1	158	ign	Berkeley
1/7/42	1	160	ign	Simpson (engine only)
1/29/42	1	161	ign	Berkeley
1/7/42	1	163	ign	Comet (Oakland)
1/29/42	1	164	ign	H & H
1/29/42	1	165	ign	Berkeley
1/7/42	1	166	ign	Comet (Oakland)
1/7/42	1	167	ign	Comet (Oakland)
1/7/42	1	168	ign	Western Model
1/29/42	1	169	ign	Berkeley
1/7/42	1	170	ign	Comet (Oakland)
1/7/42	1	171	ign	Comet (Oakland)
12/24/41	1	172	ign	Western Model
3/17/42	1	173	ign	Milford S/N ?
3/9/42	1	174	ign	Walther
6/2/42	1	175	ign	Comet
3/9/42	1	176	ign	Plecan
3/14/42	1	177	ign	International
3/9/42	1	178	ign	International
3/14/42	1	179	ign	Dallaire
6/2/42	1	180	ign	International
6/2/42	1	181	ign	International
3/16/42	1	182	ign	Comet Western (L.A.)
3/14/42	1	184	ign	International
3/9/42	1	185	ign	International
6/2/42	1	186	ign	Comet
3/9/42	1	187	ign	Dallaire
3/7/42	1	189	ign	Comet

6/2/42	1	190	ign	International	
3/9/42	1	191	ign	International	
5/27/42	1	192	ign	Dallaire	
3/19/42	1	193	ign	Ortiz	
3/14/42	1	194	ign	International	
3/14/42	1	195	ign	Auler	
3/9/42	1	196	ign	Roth	
6/2/42	1	197	ign	International	
3/9/42	1	199	ign	Pohl	
3/16/42	1	200	ign	Comet Western (L.A.)	
4/22/40	2	201	ign	T1 Phil Foster	T1
6/2/42	1	201	ign	International	
5/27/42	1	202	ign	Dallaire	
3/28/40	2	203	ign	C Van Court	T3
3/9/42	1	204	ign	Weeks	
4/20/40	2	205	ign	Aerodiddler (Harold Carter)	T5
3/9/42	1	205	ign	Weeks	
4/20/40	2	206	ign	American Jr.	T6
3/14/42	1	206	ign	Betters	
5/29/40	2	207	ign	Bill Crandell (Portland M.S. Co)	T7
3/9/42	1	207	ign	Dallaire	
5/9/40	2	208	ign	Roth	T8
6/2/42	1	208	ign	International	
5/21/40	2	209	ign	Eagle Wing (Morihiro)	T9
5/27/40	2	210	ign	Monmouth M S Co	T10
3/14/42	1	210	ign	International	
5/27/42	1	211	ign	Dallaire	
7/8/40	2	212	ign	Fosbury	T12
5/27/42	1	212	ign	Auler	
3/16/42	1	213	ign	Comet Western (L.A.)	
3/14/42	1	214	ign	Auler	
5/21/42	1	215	ign	Skyway	
7/17/40	2	216	ign	Newman	T16
3/7/42	1	216	ign	Comet	
7/1/40	2	217	ign	Rosenthal	T17
8/3/40	2	218	ign	Amer Jr (Vivell)	T18
3/9/42	1	218	ign	Dallaire	
5/31/40	2	219	ign	Amer Jr for Nationals	T19
6/2/42	1	219	ign	Comet	
5/27/42	1	220	ign	Auler	
3/16/42	1	221	ign	Post Exchange	
7/1/40	2	222	ign	Rothrock (Malkson)	T22
3/16/42	1	222	ign	Post Exchange	
6/28/40	2	223	ign	Burkhardt	T23
5/27/42	1	223	ign	Auler	
3/9/42	1	224	ign	H & H	
3/14/42	1	225	ign	Dallaire	
7/17/40	2	226	ign	Newman	T26
3/9/42	1	226	ign	H & H	
3/14/42	1	227	ign	Auler	
7/6/40	2	228	ign	Laubenthal	T28
3/14/42	1	228	ign	Auler	
7/6/40	2	229	ign	Laubenthal	T29
7/12/40	2	230	ign	Club	T30
3/7/42	1	230	ign	Comet	
7/13/40	2	231	ign	Polks	T31
3/14/42	1	231	ign	Dallaire	
7/13/40	2	232	ign	Wolther	T32
7/13/40	2	233	ign	Tropical	T33
3/14/42	1	234	ign	International	
5/31/40	2	235	ign	Polks	T35
5/27/42	1	235	ign	Dallaire	
3/9/42	1	237	ign	Dallaire	
8/24/40	2	238	ign	Gibson	T38
3/14/42	1	238	ign	Rothrock	
5/27/42	1	239	ign	Dallaire	
5/21/42	1	240	ign	Milford	
8/3/40	2	241	ign	Handicraft Hobbies	T41
5/27/42	1	242	ign	Dallaire	
3/14/42	1	243	ign	Rothrock	
3/9/42	1	244	ign	International	
3/16/42	1	246	ign	Comet Western (L.A.)	
3/14/42	1	247	ign	Betters	
5/27/42	1	248	ign	Dallaire	
5/27/42	1	249	ign	Dallaire	
3/14/42	1	250	ign	Auler	
10/5/40	2	251	ign	Scientific	T2-51
3/16/42	1	251	ign	Comet Western (L.A.)	
3/14/42	1	252	ign	Dallaire	
7/27/40	2	253	ign	Stanley Jones	T53
3/9/42	1	253	ign	International	
8/3/40	2	254	ign	Cullen	T54
11/13/40	2	254	ign	Good's	2-54
3/14/42	1	254	ign	International	

1/14/41	2	255	ign	Burke & Co	
6/30/42	1	255	ign	Dallaire	
11/16/40	2	256	ign	Pouch	2-56
3/9/42	1	256	ign	Dallaire	
10/31/40	2	257	ign	Allen Bros	2-57
3/7/42	1	257	ign	Comet	
11/30/40	2	258	ign	Polks	2-58
6/2/42	1	258	ign	International	
3/14/42	1	259	ign	Dallaire	
8/31/40	2	260	ign	Polks	T60
11/13/40	2	261	ign	Good's	2-61
3/7/42	1	261	ign	Comet	
10/31/40	2	262	ign	C Van Court (Wm Sweet)	2-62
5/21/42	1	262	ign	Skyway	
6/2/42	1	263	ign	Comet	
3/7/42	1	264	ign	Comet	
11/20/47	2	264	ign	Sim Bros	
6/2/42	1	265	ign	Comet	
5/27/42	1	266	ign	Auler	
3/14/42	1	267	ign	Kos	
11/5/40	2	268	ign	Pouch	2-68
11/13/40	2	268	ign	C Van Court (Vincent Robert	2-68
3/14/42	1	268	ign	Auler	
11/13/40	2	269	ign	Good's	2-69
11/5/40	2	270	ign	Ballentine	2-70
8/31/40	2	271	ign	Polks	T71
3/9/42	1	271	ign	International	
6/2/42	1	272	ign	International	
11/12/40	2	273	ign	Allen Bros	2-73
3/14/42	1	273	ign	Dallaire	
7/6/42	1	274	ign	Miami	
11/29/40	2	275	ign	Allen Bros	2-75
6/30/42	1	275	ign	Comet	
6/30/42	1	276	ign	International	
6/30/42	1	277	ign	Dallaire	
9/28/44	1	277	ign	Don Fulton	
12/17/40	2	278	ign	Grose's Bike Shop	278
7/1/42	1	278	ign	Berkeley	
12/20/40	2	279	ign	Sonneborn's	
6/30/42	1	279	ign	Comet	
1/10/41	2	280	ign	Good's	
8/15/44	1	280	ign	Laveielle	
6/30/42	1	281	ign	Dallaire	
1/10/41	2	282	ign	Good's	
6/30/42	1	282	ign	Dallaire	
7/17/44	1	282	ign	Harry Allison	
1/10/41	2	283	ign	Good's	
7/1/42	1	283	ign	Willson Co Ltd	
9/12/44	1	283	ign	Roseway	
5/25/45	1	283	ign	Roth	
1/8/41	2	284	ign	Miami Novelty	
7/2/42	1	284	ign	Wolther	
8/14/44	1	284	ign	Bari Fletcher	
7/17/44	1	285	ign	Dick Wonderly	
12/20/40	2	286	ign	Lul	
6/30/42	1	286	ign	Comet	
9/12/44	1	286	ign	Roseway	
12/30/40	2	287	ign	Savage	
9/15/44	1	287	ign	Davenport	
9/12/44	1	288	ign	Roseway	
6/30/42	1	289	ign	International	
6/30/42	1	290	ign	Dallaire	
7/17/42	1	290	ign	Weeks	
9/13/44	1	290	ign	Vic's Hobby Supply	
1/14/41	2	291	ign	Schafer	
6/30/42	1	291	ign	Dallaire	
2/8/46	1	291	ign	Western Hobbycraft	
1/14/41	2	292	ign	Kipp Bros	
6/30/42	1	292	ign	Dallaire	
12/20/44	1	292	ign	Johnson	
1/14/41	2	293	ign	Polks	
6/30/42	1	293	ign	Comet	
1/14/41	2	294	ign	Eagle Wing	
7/1/42	1	294	ign	Berkeley	
12/27/44	1	294	ign	Bob's	
6/30/42	1	295	ign	Comet	
2/28/45	1	295	ign	Habe's	
1/14/41	2	296	ign	Laurence	
6/30/42	1	296	ign	Dallaire	
2/28/45	1	296	ign	Keep	
1/2/41	2	297	ign	Austin-Craft	
6/30/42	1	297	ign	Comet	
6/30/42	1	297	ign	Comet	
12/27/44	1	297	ign	Bob's	

6/30/42	1	298	ign	Dallaire	
5/25/45	1	298	ign	Roth	
3/9/48	2	298	ign	Simmons	
1/23/46	1	299	ign	Bob's	
1/21/41	2	300	ign	Lau	
6/30/42	1	300	ign	Comet	
12/29/45	1	300	ign	Whitman	
1/17/41	2	301	ign	Model Aircraft Institute	
6/30/42	1	301	ign	Comet	
12/23/44	1	301	ign	Hedgepath	
1/17/41	2	302	ign	Toledo	
6/30/42	1	302	ign	Dallaire	
8/25/45	1	302	ign	Chamberlain	
1/17/41	2	303	ign	Paradise	
3/7/45	1	303	ign	Huffman	
1/21/41	2	304	ign	Lau	
6/30/42	1	304	ign	Auler	
5/25/45	1	304	ign	Sims	
1/31/46	1	305	ign	Lt. Regrutto	
1/21/41	2	306	ign	Enr-Roe	
6/30/42	1	306	ign	Auler	
8/24/45	1	306	ign	May	
7/1/42	1	307	ign	Berkeley	
12/23/44	1	307	ign	Glen's	
1/14/41	2	308	ign	Millers Inc	
2/28/45	1	308	ign	Habe's	
1/8/41	2	309	ign	Miami Novelty	
6/30/42	1	309	ign	Dallaire	
2/28/45	1	309	ign	Habe's	
2/28/45	1	310	ign	Habe's	
7/17/42	1	311	ign	Weeks	
12/23/44	1	311	ign	Glen's	
10/5/40	2	312	ign	Austin Craft	~~2-50~~ 312
6/30/42	1	312	ign	Comet	
10/9/45	1	312	ign	Wherry	
11/8/40	2	313	ign	Berkeley	
6/30/42	1	313	ign	Auler	
1/24/41	2	314	ign	Modelcrafters	
6/30/42	1	314	ign	Dallaire	
3/21/45	1	314	ign	Carl VanCourt	
9/13/45	1	314	ign	Powell	
6/30/42	1	315	ign	Comet	
12/23/44	1	315	ign	Vic's	
6/30/42	1	316	ign	International	
12/23/44	1	316	ign	Vic's	
5/25/45	1	316	ign	Sims	
8/10/42	1	317	ign	Merritt	
12/23/44	1	317	ign	Western Model	
6/30/42	1	318	ign	Auler	
2/5/41	2	319	ign	Bay Ridge	
6/30/42	1	319	ign	International	
12/27/44	1	319	ign	Bob's	
7/1/42	1	320	ign	Berkeley	
12/23/44	1	320	ign	Glen's	
2/6/41	2	321	ign	O H Harms 1416 NE Euclid	
6/30/42	1	321	ign	International	
12/23/44	1	321	ign	Hedgepath	
7/1/42	1	322	ign	Willson Co Ltd	
12/23/44	1	322	ign	Glen's	
6/30/42	1	323	ign	Dallaire	
5/25/45	1	323	ign	Paterra	
2/6/41	2	324	ign	United	
3/29/41	2	324	ign	Roth	
7/1/42	1	324	ign	Berkeley	
12/23/44	1	324	ign	Hedgepath	
2/21/41	2	325	ign	Good's	
6/30/42	1	325	ign	Comet	
12/23/44	1	325	ign	Hedgepath	
6/30/42	1	326	ign	Dallaire	
3/7/45	1	326	ign	Huffman	
6/30/42	1	327	ign	International	
8/24/45	1	327	ign	Milburn	
3/12/41	2	328	ign	International	
5/25/45	1	328	ign	Cinquanta	
12/23/44	1	329	ign	Glen's	
12/23/44	1	330	ign	Vic's	
12/13/45	1	331	ign	Vic's	
2/6/41	2	332	ign	Weeks	
2/11/41	2	333	ign	O H Harms 1416 NE Euclid	
12/23/44	1	334	ign	Glen's R Backus	
12/13/45	1	334	ign	Vic's	
1/17/46	1	336	ign	Rauch	
1/3/46	1	337	ign	Robert E. Holmes	
1/23/46	1	337	ign	Bob's	

12/28/45	1	338	ign	Roy's Hobbyland
12/21/45	1	339	ign	Huffman
5/27/46	1	340	ign	Glen's
1/23/46	1	341	ign	Glen's
2/11/41	2	342	ign	O H Harms 1416 NE Euclid
2/4/46	1	342	ign	Beth
12/29/45	1	343	ign	Wolther
12/28/45	1	345	ign	Roy's Hobbyland
2/21/41	2	346	ign	Polks
12/28/45	1	346	ign	Roy's Hobbyland
12/13/45	1	347	ign	Haywards
12/28/45	1	348	ign	Al Dippert
1/23/46	1	349	ign	Glen's
2/21/41	2	350	ign	Good's
12/13/45	1	350	ign	Air King
12/13/45	1	351	ign	Air King
3/18/42	2	353	ign	Rothrock
12/13/45	1	353	ign	Glen's
3/1/41	2	355	ign	Don's Model Shop
10/9/45	1	355	ign	Sims
1/23/46	1	357	ign	Air King
1/23/46	1	358	ign	Chouinard
2/26/41	2	359	ign	Burke & Co
12/29/45	1	359	ign	Wolther
12/29/45	1	360	ign	Wolther
2/26/41	2	361	ign	Bill Sweet
2/8/46	1	361	ign	Vic's
3/12/41	2	362	ign	Skyway
7/11/41	4	362	ign	Williams
10/9/45	1	363	ign	Sims
2/26/41	2	364	ign	Burke & Co
1/23/46	1	364	ign	Air King
3/12/41	2	365	ign	Burk & Co
12/21/45	1	365	ign	Huffman
10/9/45	1	366	ign	Sims
12/13/45	1	367	ign	Glen's
3/12/41	2	368	ign	Burk & Co
12/13/45	1	370	ign	Bob's
3/18/41	2	371	ign	Cartwright
10/9/45	1	371	ign	Sims
12/29/45	1	372	ign	Jack Ross
3/15/41	2	373	ign	Webb
10/9/45	1	373	ign	Sims
3/15/41	2	374	ign	Webb
12/13/45	1	374	ign	Bob's
5/30/41	4	375	ign	C Van Court
3/15/41	2	376	ign	Webb
10/12/45	1	376	ign	Ferbel S/N ?
11/9/45	1	377	ign	Cinquauta
10/9/45	1	378	ign	Sims
12/13/45	1	378	ign	Haywards
6/7/41	4	379	ign	Amer. Jr.
6/6/46	1	380	ign	Moscow
2/13/46	1	381	ign	Rouge
2/12/46	1	382	ign	Gresham
7/11/41	4	383	ign	Morton's
5/28/46	1	383	ign	Western
5/1/46	1	384	ign	Forster's
3/28/41	2	385	ign	Balaban pd in 5/13/order
2/13/46	1	385	ign	Sims
3/22/41	2	386	ign	S O Fisher
2/12/46	1	386	ign	Gresham
3/25/41	2	387	ign	Miami Novelty Co
2/8/46	1	387	ign	Western Hobbycraft
5/28/46	1	388	ign	Meccan
5/28/46	1	389	ign	Meccan
5/28/46	1	390	ign	Hobbyland
6/12/41	4	392	ign	Comet-Los Angeles
5/29/46	1	392	ign	Jr Aircraft, Walltree
4/27/46	1	393	ign	Jim Walker
6/12/41	4	394	ign	Comet- Oakland
5/27/46	1	394	ign	Air King
2/8/46	1	395	ign	Roth
5/28/46	1	396	ign	Walker's Hobby Center
6/5/46	1	396	ign	Huffman
6/12/41	4	397	ign	Sweet
6/3/46	1	397	ign	Harvey
5/28/46	1	398	ign	Everett
6/5/46	1	399	ign	Huffman
2/14/46	1	401	ign	Fandel
5/24/41	2	402	ign	Skyway
2/8/46	1	402	ign	Vic's
5/31/41	2	403	ign	Portland
5/27/46	1	403	ign	Bob's
5/13/41	2	404	ign	Comet-Oakland
2/12/46	1	404	ign	Gresham
6/5/41	2	405	ign	Berkeley
2/13/46	1	405	ign	Rouge
2/13/46	1	406	ign	Huffman
5/31/41	2	407	ign	E E Wilson (Polk's)
5/28/46	1	407	ign	Hobbyland
8/9/41	2	408	ign	Comet-Los Angeles
6/3/46	1	408	ign	Harvey
6/4/41	2	409	ign	Polk's
6/5/46	1	409	ign	Huffman
6/5/41	2	410	ign	Berkeley
5/28/46	1	410	ign	Everett
6/4/41	2	411	ign	Polk's
5/29/46	1	411	ign	Jr Aircraft, Walltree
8/9/41	2	412	ign	Comet-Los Angeles
2/14/46	1	412	ign	Bon Marche
6/5/41	2	413	ign	Berkeley
4/6/46	1	413	ign	Marrahi
8/9/41	2	414	ign	Comet-Los Angeles
2/8/46	1	414	ign	Roth
5/1/46	1	415	ign	Forster's
8/9/41	2	416	ign	Comet-Los Angeles
5/13/41	2	417	ign	Comet-Oakland
6/5/46	1	417	ign	Huffman
6/3/41	2	418	ign	Myers & Ballman
6/6/46	1	418	ign	Moscow
5/24/41	2	419	ign	Skyway
6/5/46	1	419	ign	Spokane
5/28/46	1	420	ign	Walker's Hobby Center
5/13/41	2	421	ign	Comet-Oakland
2/13/46	1	421	ign	Sims
5/13/41	2	422	ign	Comet-Los Angeles
3/26/46	1	422	ign	Neal Augus
6/5/41	2	423	ign	Berkeley
2/13/46	1	423	ign	Huffman
5/13/41	2	424	ign	Comet-Los Angeles
2/14/46	1	424	ign	Bon Marche
5/1/46	1	425	ign	Forster's
5/24/41	2	426	ign	Skyway
6/5/46	1	426	ign	Dick's
8/9/41	2	427	ign	Comet-Los Angeles
6/5/46	1	427	ign	Dick's
5/24/41	2	428	ign	Skyway
6/5/46	1	428	ign	Spokane
6/5/46	1	429	ign	Dick's
8/9/41	2	430	ign	Comet-Los Angeles
2/13/46	1	430	ign	Murf's
5/24/41	2	431	ign	Skyway
6/16/41	2	432	ign	Good's
6/5/46	1	432	ign	Wan's
5/13/41	2	433	ign	Comet-Los Angeles
5/27/46	1	433	ign	Ace (Amer Jr.)
6/4/41	2	434	ign	Polk's
4/6/46	1	434	ign	Fliedner
6/4/41	2	435	ign	Polk's
2/14/46	1	435	ign	Bon Marche
5/31/41	2	436	ign	Karwelkow
6/5/46	1	436	ign	Wan's
5/26/41	2	437	ign	Model Aircraft Institute
5/29/46	1	437	ign	Jr Aircraft, Walltree
5/13/41	2	438	ign	Comet-Los Angeles
5/27/46	1	438	ign	Vic's
6/5/41	2	439	ign	Berkeley
4/27/46	1	439	ign	Leonard Lucas
6/3/41	2	440	ign	Hawkeye
5/1/46	1	440	ign	Stanley-Winthrop
8/9/41	2	441	ign	Comet-Los Angeles
2/13/46	1	441	ign	Sims
8/9/41	2	442	ign	Comet-Los Angeles
2/14/46	1	442	ign	Bert's
6/5/41	2	443	ign	Berkeley
2/14/46	1	443	ign	Bert's
5/13/41	2	444	ign	Comet-Los Angeles
6/6/46	1	444	ign	Colonel Bob's
5/26/41	2	445	ign	Demus
3/27/46	1	445	ign	Kopp
8/9/41	2	446	ign	Comet-Los Angeles
8/26/46	1	446	ign	Bryan
8/9/41	2	447	ign	Comet-Los Angeles
2/20/47	1	447	ign	Bentley's
5/31/41	2	448	ign	Karwelkow
7/23/46	1	448	ign	Dick Brown
5/13/41	2	449	ign	Comet-Oakland

9/30/46	1	449	ign	B. Paul
5/24/41	2	450	ign	Skyway
6/3/41	2	451	ign	Hobby Supply (Law)
2/26/47	1	451	ign	Almeida & Veiga
5/24/41	2	452	ign	Skyway
8/26/46	1	452	ign	Douglas
5/13/41	2	453	ign	Comet-Oakland
9/11/46	1	453	ign	Fosbury
6/16/41	2	454	ign	Wiley Clayton
9/5/46	1	454	ign	Rouge
12/20/48	1	454	ign	Versage
8/9/41	2	455	ign	Comet-Los Angeles
11/19/46	1	455	ign	Cincinatti
7/18/47	1	455	ign	Stewart's
8/9/41	2	456	ign	Comet-Los Angeles
2/26/47	1	456	ign	Almeida & Veiga
5/13/41	2	457	ign	Comet-Oakland
2/26/47	1	457	ign	Almeida & Veiga
12/5/41	4	458	ign	Miami Novelty
8/26/46	1	458	ign	Douglas
9/9/41	2	459	ign	Roth
2/26/42	2	459	ign	Dallaire
8/26/46	1	459	ign	Douglas
8/25/41	2	460	ign	Comet (Oakland)
6/12/46	1	460	ign	Haywards
2/18/47	1	460	ign	Jay Simmons
9/6/41	2	461	ign	Polk's
12/5/41	2	461	ign	Comet (Oakland)
9/9/41	2	462	ign	People's Drug
1/12/42	2	462	ign	Kos
9/5/46	1	462	ign	Wolther
12/5/41	4	463	ign	Hobby Supply House
1/12/42	2	463	ign	Ruth
2/26/47	1	463	ign	Almeida & Veiga
9/6/41	2	464	ign	Polk's
11/12/42	4	464	ign	Cpl Charles H Gilligan
9/9/41	2	465	ign	People's Drug
1/30/42	2	465	ign	Skyway
2/20/47	1	465	ign	Bentley's
12/5/41	2	466	ign	Comet (Oakland)
12/5/41	4	466	ign	Gord's
9/12/41	2	467	ign	Huffman
2/26/42	2	467	ign	Week's
4/5/48	1	467	ign	Smith-Watkins
9/6/41	2	468	ign	Polk's
9/10/46	1	468	ign	Colonel Bob's
9/6/41	2	469	ign	Polk's
9/5/46	1	469	ign	Rouge
8/9/41	2	470	ign	Gibson
8/25/41	2	471	ign	Comet (Oakland)
9/5/46	1	471	ign	Wolther
9/6/41	2	472	ign	Polk's
2/26/42	2	472	ign	Good's
8/29/46	1	472	ign	Huffman
8/25/41	2	473	ign	Comet (Oakland)
2/26/42	2	473	ign	Good's
2/20/47	1	473	ign	Girard
1/30/42	2	474	ign	Skyway
9/28/46	1	474	ign	Huffman
10/7/41	4	475	ign	C M Adams
7/30/47	1	475	ign	Burk & Co
9/6/41	2	476	ign	Good's
1/12/42	2	476	ign	Polk's
9/6/41	2	477	ign	Good's
2/26/42	2	477	ign	Good's
9/5/46	1	477	ign	Hall, Inc
8/25/41	2	478	ign	Comet (Oakland)
8/25/41	2	478	ign	Comet (Oakland)
2/26/47	1	478	ign	Almeida & Veiga
1/24/42	4	479	ign	Rothrock
10/1/46	1	479	ign	Bon Marche
5/24/41	2	480	ign	Skyway
8/25/41	2	480	ign	Comet (Oakland)
1/30/42	4	480	ign	Skyway
9/30/46	1	480	ign	B. Paul
9/6/41	2	481	ign	Stanley-Winthrop
12/5/41	2	481	ign	Comet (Oakland)
7/19/46	1	481	ign	Hobby Center Miami
9/9/41	2	482	ign	Plecan
1/30/42	2	482	ign	Skyway
7/30/47	1	482	ign	Burk & Co
3/10/47	1	483	ign	Georgia
9/6/41	2	484	ign	Miami Novelty
9/12/41	2	485	ign	Weeks

1/12/42	2	485	ign	Berkeley
9/5/46	1	485	ign	5th Ave Hobbies
8/25/41	2	486	ign	Comet (Oakland)
12/5/41	2	486	ign	Comet (Oakland)
9/10/46	1	486	ign	Colonel Bob's
8/25/41	2	487	ign	Comet (Oakland)
2/18/47	1	487	ign	Jay Simmons
8/25/41	2	488	ign	Comet (Oakland)
8/25/41	2	488	ign	Comet (Oakland)
1/12/42	2	488	ign	Ballinger
2/18/47	1	488	ign	Hall, Inc
1/12/42	2	489	ign	Berkeley
10/1/46	1	489	ign	Harms
2/18/47	1	490	ign	Jay Simmons
8/25/41	2	491	ign	Comet (Oakland)
1/12/42	2	491	ign	Berkeley
10/1/46	1	491	ign	Bon Marche
8/25/41	2	492	ign	Comet (Oakland)
1/7/42	2	492	ign	Simpson (engine only)
2/20/47	1	492	ign	Colonel Bob's
8/27/46	1	493	ign	Markley's
2/26/42	2	494	ign	Auler
8/26/46	1	494	ign	Tom's Model Center
9/30/46	1	495	ign	B. Paul
9/28/46	1	496	ign	Huffman
2/26/42	2	497	ign	Auler
9/5/46	1	497	ign	Hall, Inc
6/30/47	2	497	ign	Stewart's Sport Shop S/N ?
1/12/42	2	498	ign	Polk's
8/26/46	1	498	ign	Tom's Model Center
10/1/46	1	499	ign	Bon Marche
9/28/46	1	501	ign	Hobby Arts
10/20/39	1	502	ign	Pilcher
12/5/41	2	502	ign	Comet (Oakland)
7/26/46	1	502	ign	Aero Craft
9/28/46	1	503	ign	Hobby Arts
9/10/46	1	504	ign	Colonel Bob's
11/4/39	1	505	ign	Burkhart
2/26/42	2	505	ign	Dallaire
8/12/46	1	506	ign	Huth
1/12/42	2	507	ign	Polk's
9/5/46	1	507	ign	Rouge
12/11/47	1	507	ign	Jay Simmons
1/12/42	2	508	ign	Polk's
2/20/47	1	509	ign	Colonel Bob's
9/11/46	1	510	ign	Air King (Hudspeth)
4/20/50	1	510	ign	All Am Motor Exchange
2/26/42	2	511	ign	Auler
2/18/47	1	511	ign	Hall, Inc
1/12/42	2	512	ign	Berkeley
8/12/46	1	512	ign	Knox Tire Co
9/5/46	1	513	ign	Hillcrest
12/5/41	2	514	ign	Comet (Oakland)
8/12/46	1	514	ign	Knox Tire Co
1/12/42	2	515	ign	Berkeley
8/27/46	1	515	ign	Forster's
2/26/42	2	516	ign	Dallaire
9/5/46	1	516	ign	Wolther
9/28/46	1	517	ign	Fresno Cycle
1/30/42	2	518	ign	Skyway
2/26/47	1	518	ign	Almeida & Veiga
1/12/42	2	519	ign	Polk's
8/27/46	1	519	ign	Forster's
8/27/46	1	520	ign	Forster's
2/26/42	2	521	ign	Dallaire
7/19/46	1	521	ign	Hobby Center Miami
2/8/40	1	522	ign	Aerodiddler
9/10/46	1	522	ign	Colonel Bob's
2/8/40	1	523	ign	Aerodiddler
4/24/40	1	523	ign	R Stab
1/12/42	2	523	ign	Niedyelski
9/5/46	1	523	ign	5th Ave Hobbies
2/26/42	2	524	ign	Auler
8/26/46	1	524	ign	Spokane
8/26/46	1	525	ign	Tom's Model Center
4/24/40	1	526	ign	R Stab
10/1/46	1	526	ign	Bon Marche
4/24/40	1	527	ign	R Stab
1/12/42	2	527	ign	Western Model
10/1/46	1	527	ign	Bon Marche
5/8/40	1	528	ign	American Jr (to Cloud M.A.CO.)
1/12/42	2	528	ign	Western Model
9/30/46	1	528	ign	B. Paul
5/8/40	1	529	ign	American Jr (to Cloud M.A.CO.)

4/18/42	2	529	ign	Comet
7/26/46	1	529	ign	Aero Craft
4/18/42	2	530	ign	Comet
11/19/46	1	530	ign	Cincinatti
7/30/47	1	530	ign	Burk & Co
2/20/40	1	531	ign	Will Gresell (to Amer. Jr) ~~520~~ 531
9/30/46	1	532	ign	B. Paul
4/13/42	2	533	ign	Skyway
2/26/47	1	533	ign	Almeida & Veiga
1/20/40	1	534	ign	Jerguson ~~512~~ 534
3/10/47	1	534	ign	Georgia
6/25/40	1	535	ign	Amer Jr for Nationals
3/10/47	1	535	ign	Georgia
12/1/39	1	536	ign	Western Model Supply Co ~~506~~ 536
3/18/42	2	536	ign	Rothrock
7/19/46	1	536	ign	Hobby Center Miami
4/9/42	2	537	ign	Roth
6/30/47	1	537	ign	Southside
4/18/42	2	538	ign	Comet
9/11/46	1	538	ign	Air King (Hudspeth)
9/28/46	1	539	ign	Jay Simmons
12/7/39	1	540	ign	American Jr ~~508~~ 540
4/18/42	2	540	ign	Comet
7/26/46	1	540	ign	Aero Craft
3/19/42	2	541	ign	Comet
2/18/47	1	541	ign	Burk & Co
1/24/40	1	542	ign	Monmouth Model Supply ~~517~~ 542
8/26/46	1	542	ign	Spokane
9/11/46	1	543	ign	Air King (Hudspeth)
7/3/40	1	544	ign	Meier & Franks
9/4/46	1	544	ign	Osterman
3/19/42	2	545	ign	Comet
3/19/42	2	546	ign	Auler
7/30/47	1	546	ign	Burk & Co
4/13/42	2	547	ign	Skyway
9/30/46	1	547	ign	B. Paul
4/18/42	2	548	ign	Comet
6/17/47	1	548	ign	Southside (Phillips)
1/3/40	1	549	ign	C Van Court ~~PX 49~~ 549
9/5/46	1	549	ign	Hillcrest
8/12/46	1	550	ign	Knox Tire Co
8/12/46	1	552	ign	Knox Tire Co
8/26/46	1	553	ign	Tom's Model Center
12/7/39	1	554	ign	Portland Hobby Supply ~~504~~ 554
1/8/40	1	554	ign	M D Whitman (Aerodiddler) ~~509~~ 554
2/18/47	1	554	ign	Burk & Co
2/26/47	1	555	ign	Almeida & Veiga
5/9/47	4	555	ign	Thomas
3/28/42	2	556	ign	Aircraft (Roseburg)
7/13/40	1	557	ign	Tropical
3/19/42	2	557	ign	Comet
12/1/39	1	558	ign	Eagle Wing ~~507~~ 558
10/19/39	1	559	ign	Lucas ~~503~~ 559
3/19/42	2	559	ign	Ortiz
2/20/47	1	560	ign	Colonel Bob's
3/13/40	1	561	ign	H R Carson
3/23/42	2	561	ign	International
9/5/46	1	561	ign	Rouge Ret'd
9/24/47	1	561	ign	Sims Bros
3/19/42	2	562	ign	Comet
3/12/40	1	563	ign	Phil Foster
3/19/42	2	563	ign	Weeks
4/18/42	2	564	ign	Comet
3/19/42	2	565	ign	Weeks
7/1/42	4	567	ign	Bowden
3/25/40	1	568	ign	E O Boyer
4/18/42	2	568	ign	Comet
3/19/42	2	569	ign	Comet
4/18/42	2	570	ign	Comet
1/3/40	1	571	ign	Herb Meyer (to C VanCourt) ~~510~~ 571
4/25/42	2	571	ign	Rothrock
3/19/42	2	572	ign	Willson Stationery Co
4/18/42	2	574	ign	Skyway
3/23/42	2	577	ign	Keller
3/19/42	2	578	ign	Comet
3/19/42	2	579	ign	Auler
6/20/47	4	583	ign	Howard A. Edwards
7/26/44	2	584	ign	Huffman
8/3/44	2	584	ign	Rose
6/19/44	2	585	ign	Bob Merchant
6/13/44	2	586	ign	Gullikson
6/27/44	2	587	ign	Gee
6/19/44	2	588	ign	Shaw
6/3/44	2	590	ign	Owens

7/26/44	2	591	ign	Huffman
8/15/44	2	593	ign	Huffman
8/15/44	2	594	ign	Huffman
3/12/46	2	595	ign	Vic's
11/16/44	2	596	ign	Vic
9/21/44	2	597	ign	Page
6/14/44	2	598	ign	Wolther
3/15/48	2	598	ign	Wolfe
9/28/44	2	599	ign	Dupree
11/18/44	2	600	ign	Cary
11/2/44	2	601	ign	Glen
11/2/44	2	602	ign	Glen
11/18/44	2	603	ign	Huffman
11/18/44	2	604	ign	Emme
9/19/44	2	605	ign	Roseway
11/18/44	2	606	ign	Huffman
11/29/44	2	607	ign	Rymer
12/1/44	2	608	ign	Cope
11/29/44	2	609	ign	Rymer
11/29/44	2	610	ign	Sims Stolen
12/14/44	2	611	ign	Sims
12/1/44	2	612	ign	Huffman
12/1/44	2	614	ign	Huffman
4/30/45	2	616	ign	Halloway
3/7/45	2	617	ign	Levine
3/7/45	2	618	ign	Truenbach
8/15/44	2	619	ign	Huffman
1/8/46	2	620	ign	Rouge
12/29/45	2	621	ign	Habe's
6/15/46	2	622	ign	Post Exchange
12/28/45	2	623	ign	Vic's
7/12/46	2	623	ign	Douglas Models
8/24/45	2	624	ign	Jeffries
6/15/46	2	624	ign	Model Hobby, Hawaii
12/18/45	2	625	ign	Wachtell
6/15/46	2	625	ign	Model Hobby, Hawaii
12/28/45	2	626	ign	Hayward
6/6/46	2	626	ign	Colonel Bob's
1/8/46	2	627	ign	Rouge
7/12/46	2	627	ign	Hall Inc
12/18/45	2	628	ign	Murphy
7/3/46	2	628	ign	Hobbycraft (Tennessee)
12/29/45	2	629	ign	Wolther
7/3/46	2	629	ign	Hobbycraft (Tennessee)
8/24/45	2	630	ign	Lon
7/12/46	2	630	ign	Hall Inc
8/24/45	2	631	ign	Meyer
7/15/46	2	632	ign	San Diego
12/22/45	2	633	ign	Bob's
7/3/46	2	633	ign	Spokane
1/8/46	2	634	ign	Roth
6/15/46	2	634	ign	Model Hobby, Hawaii
1/8/46	2	635	ign	Model Hobby (Honolulu)
6/24/46	2	635	ign	Tom's
1/8/46	2	636	ign	Roth
7/3/46	2	636	ign	Forster's
6/24/46	2	637	ign	Eagle Wing (Burkhart)
12/18/45	2	638	ign	Broncato
6/24/46	2	638	ign	Tom's
12/22/45	2	639	ign	Glen's
6/24/46	2	639	ign	Harvey
8/30/45	2	640	ign	Johnson
7/12/46	2	640	ign	Douglas Models
12/22/45	2	641	ign	Air King
7/12/46	2	641	ign	Hall Inc
12/18/45	2	642	ign	Caroline
6/24/46	2	642	ign	Harvey
1/15/46	2	643	ign	Hauerwaas
7/15/46	2	643	ign	San Diego
12/29/45	2	644	ign	Maxson S/N ?
6/15/46	2	644	ign	Meccan
1/16/46	2	645	ign	Lt. Regrutto
7/3/46	2	645	ign	Forster's
6/7/46	2	646	ign	Knox
3/15/46	2	647	ign	Huffman
7/13/46	2	647	ign	5th Ave Hobbies S/N ?
6/24/46	2	648	ign	Colonel Bob's
3/12/46	2	649	ign	Bob's
6/15/46	2	649	ign	Meccan
3/15/46	2	650	ign	Fandel
7/3/46	2	650	ign	Hobby Arts (Everett)
6/24/46	2	651	ign	Colonel Bob's
3/13/46	2	652	ign	Busters Super Service
7/13/46	2	652	ign	5th Ave Hobbies

3/12/46	2	653	ign	Glen's
7/2/46	2	653	ign	Kirby
7/8/46	2	654	ign	Enstrum
3/13/46	2	655	ign	Western Hobby
6/26/46	2	656	ign	Bryan's Bicycle Store
12/29/45	2	657	ign	Habe's
2/27/46	2	657	ign	Harry Fosbury
7/8/46	2	657	ign	Portland Hobby Center
7/3/46	2	658	ign	Van's
3/13/46	2	659	ign	Cincinatti
6/15/46	2	659	ign	Edwards
2/26/45	2	660	ign	McClain
7/12/46	2	660	ign	Douglas Models
3/15/46	2	661	ign	Huffman
2/26/46	2	662	ign	Stewart
3/4/46	2	663	ign	Stokes
3/4/46	2	665	ign	Nathan (Cinquanta)
3/15/46	2	666	ign	Mabery
2/27/46	2	668	ign	Crowell
11/19/46	2	668	ign	Huffman
2/26/46	2	671	ign	Thatcher
3/13/46	2	673	ign	Cincinatti
3/12/46	2	674	ign	Air King
5/15/46	2	676	ign	San Diego Model Supply
3/27/46	2	677	ign	Sims
4/6/46	2	678	ign	Meccan
4/16/46	2	679	ign	Harvey Air exp.
4/16/46	2	680	ign	Harvey Air exp.
5/13/46	2	681	ign	Western Hobby
5/8/46	2	683	ign	Walkers
5/8/46	2	684	ign	Walkers
4/6/46	2	685	ign	Meccan
4/6/46	2	686	ign	Kay-Kraft
5/8/46	2	686	ign	Hall
5/15/46	2	687	ign	Hobby Center, Miami
5/15/46	2	689	ign	Hobby Center, Miami
4/18/46	2	690	ign	Everett
3/15/46	2	691	ign	Habe's
5/15/46	2	692	ign	Bert's
5/15/46	2	693	ign	Bert's
3/27/46	2	694	ign	Sweet
5/15/46	2	695	ign	State Hobby
5/1/46	2	696	ign	Stanley-Winthrop
3/27/46	2	697	ign	Sims
4/16/46	2	698	ign	Wolther
5/15/46	2	699	ign	Hillcrest
4/16/46	2	700	ign	Harvey Air exp.
3/20/46	2	701	ign	Guion
4/16/46	2	702	ign	Wolther
5/15/46	2	703	ign	Seattle Hobbyland
5/15/46	2	704	ign	Seattle Hobbyland
3/27/46	2	706	ign	Schweitzer
4/16/46	2	707	ign	Wolther
3/27/46	2	708	ign	Roger's
3/15/46	2	709	ign	Murphy
5/15/46	2	710	ign	Gashoppers
5/8/46	2	711	ign	Hall
5/1/46	2	712	ign	Moscow
3/27/46	2	713	ign	Roy's Hobbyland
3/15/46	2	714	ign	Murphy
4/16/46	2	715	ign	Dick's
4/6/46	2	716	ign	Spokane
3/27/46	2	717	ign	Coast Hobbycraft & Radio
3/27/46	2	718	ign	Roy's Hobbyland
5/15/46	2	719	ign	Hobby Center, Miami
4/18/46	2	720	ign	Van's Carl
4/16/46	2	721	ign	Dick's
8/6/46	2	761	ign	Colonel Bob's
8/8/46	2	762	ign	Hillcrest Hobbies
8/6/46	2	763	ign	Colonel Bob's
8/8/46	2	764	ign	Ray's
8/8/46	2	765	ign	Markley's
8/8/46	2	766	ign	Hillcrest Hobbies
8/8/46	2	767	ign	Ray's
8/6/46	2	768	ign	Ronge
8/6/46	2	769	ign	Ronge
8/8/46	2	770	ign	Hillcrest Hobbies
8/5/46	2	771	ign	Cincinatti
8/5/46	2	773	ign	Hobby Arts
9/12/46	2	774	ign	Gordon Folger
8/5/46	2	775	ign	Cincinatti
8/5/46	2	776	ign	Everett
8/8/46	2	777	ign	M E Henzler(pd for by Carl VanCourt)
8/8/46	2	778	ign	W R Parke Model Co

8/8/46	2	779	ign	Ray's	
8/5/46	2	780	ign	Cincinatti	
8/6/46	2	781	ign	Ace	
8/6/46	2	782	ign	Ace	
8/6/46	2	783	ign	Ace	
1/20/47	2	784	ign	Kevorkian	
7/7/47	2	785	ign	Bernard's Hobby Shop	
10/25/46	2	787	ign	Jay Simmons	
10/26/46	2	788	ign	Miniature Engineering (Harms)	
1/9/47	2	788	ign	Jay Simmons	
11/21/47	4	789	ign	Stewarts Sport Shop	
3/28/47	2	790	ign	Ideal	
9/23/47	2	791	ign	Rouge(exc for 454, 507, 561)	
7/2/47	2	792	ign	Model Air Depot	
2/26/47	2	794	ign	Almeida & Veiga	
6/30/47	2	795	ign	Hall	
7/30/47	2	796	ign	Burk & Co	
5/16/47	2	799	ign	Douglas	
9/15/47	4	800	ign	Jay Simmons	
9/26/47	4	801	ign	Sims Bros	
10/28/46	2	802	ign	Community Hardware	
7/30/47	2	803	ign	Burk & Co	
5/16/47	2	804	ign	Jereb	
6/23/47	2	805	ign	Jay Simmons	
9/23/47	2	807	ign	Rouge(exc for 454, 507, 561)	
12/5/46	2	808	ign	Murfs	
12/16/46	2	808	ign	Donald Jones	
7/18/47	2	808	ign	Hobbies Inc	
7/18/47	2	811	ign	Hobbies Inc	
1/10/47	2	812	ign	Forster's	
9/23/47	4	813	ign	Huffman	
10/11/46	2	814	ign	Donald Jones	
7/23/47	2	815	ign	Southside	
7/30/47	2	816	ign	Burk & Co	
7/23/47	2	817	ign	Root	
1/20/47	2	818	ign	Aero Craft	
1/9/47	2	819	ign	Jay Simmons	
9/24/47	2	820	ign	Sims Bros	
11/19/46	2	821	ign	Huffman	
6/17/47	2	822	ign	Roth	
5/29/47	4	824	ign	Donald Jones	
12/20/46	2	825	ign	Hobby Center	
12/28/46	2	826	ign	Colonel Bob's	
3/9/48	4	827	ign	Cinquanta	
4/9/47	2	828	ign	Bentley's	
4/9/47	2	829	ign	Bentley's	
1/24/47	2	830	ign	Cooley's	
11/19/46	2	831	ign	Cincinatti	
12/20/46	2	832	ign	Hobby Center	
10/25/46	2	833	ign	5th Ave	
1/9/47	2	834	ign	Majic Valley	Unclaimed
2/13/47	2	834	ign	Larsen	
1/10/47	2	835	ign	Forster's	
10/25/46	2	836	ign	Bob's Model Shop	Ret'd
5/16/47	2	837	ign	Don Johnson Jr	
10/26/46	2	838	ign	Douglas	
7/7/47	2	839	ign	Weisman	
10/28/47	4	840	ign	Jim Walker	
7/18/47	2	841	ign	Stewart's	
1/9/47	2	842	ign	Aero Craft	
8/7/47	2	843	ign	Roth	
3/28/47	2	844	ign	Ideal	
12/5/46	2	846	ign	San Diego	Refused
1/6/47	2	846	ign	5th Ave Hobbies	
11/19/46	2	847	ign	Cincinatti	
4/8/47	2	847	ign	Cooley's	
2/26/47	2	849	ign	Almeida & Veiga	
12/28/46	2	850	ign	Colonel Bob's	
8/16/47	2	851	ign	Al Dippert	
2/26/47	2	852	ign	Almeida & Veiga	
10/26/46	2	853	ign	Douglas	
2/26/47	2	854	ign	Almeida & Veiga	
7/23/47	2	855	ign	Jay Simmons	
6/7/46	2	856	ign	Knox	~~631~~ 856
12/6/46	2	857	ign	Girard	
7/30/47	2	858	ign	Burk & Co	
10/26/46	2	859	ign	Douglas	
10/26/46	2	861	ign	Douglas	
12/5/46	2	862	ign	San Diego	Refused
1/6/47	2	862	ign	5th Ave Hobbies	
12/28/46	2	863	ign	Colonel Bob's	
5/16/47	2	864	ign	Douglas	
10/25/46	2	865	ign	Bob's Model Shop	Ret'd
10/26/46	2	866	ign	Douglas	

4/9/47	2	867	ign	Coleman's	
7/2/47	2	869	ign	Model Air Depot	
11/19/46	2	870	ign	Cincinatti	
1/10/47	2	870	ign	Burk & Co	
10/25/46	2	871	ign	Fresno Cycle	
6/20/47	2	872	ign	Steve's	
1/9/47	2	873	ign	Majic Valley	Unclaimed
2/13/47	2	873	ign	Larsen	
10/26/46	2	874	ign	Douglas	
1/10/47	2	875	ign	Burk & Co	
11/19/46	2	876	ign	Aron	
12/16/46	2	877	ign	Donald Jones	
6/23/47	2	878	ign	Jay Simmons	
10/25/46	2	880	ign	5th Ave	
8/7/47	2	881	ign	Brand	
3/28/47	2	882	ign	Ideal	
6/30/47	2	883	ign	Hall	
7/18/47	2	884	ign	Hobbies Inc	
10/24/47	4	885	ign	Hobby Shop	
12/6/46	2	886	ign	Girard	
6/26/47	2	887	ign	Gashoppers	
4/8/48	2	889	ign	Root's	
3/31/48	2	890	ign	Cinquauta	
5/25/48	2	891	ign	Andrews Variety	
5/13/48	2	895	ign	Sims	
6/15/48	2	896	ign	Mult. Doodlebugs	
7/22/49	2	897	ign	Burk & Co	
7/19/48	4	899	ign	Dayne's(fuel instructions given)	
8/23/48	4	900	ign	Colonel Bob's	
7/22/49	2	906	ign	Burk & Co	
8/23/48	4	907	ign	Colonel Bob's	
1/27/49	2	908	ign	Carl Van Court	
8/23/48	4	910	ign	Colonel Bob's	
8/29/49	4	911	ign	Douglas	
1/24/49	4	912	ign	Northwest Hobby Co	
10/19/49	2	913	ign	Scott's Radio Service	
2/15/50	2	914	ign	Mansuro	
8/13/49	4	915	ign	Barger (Pouch)	
10/8/48	4	916	ign	Scott's	
10/27/49	4	918	ign	Colonel Bob's	
10/15/48	4	919	ign	Carl Van Court	
11/18/48	4	920	ign	Buzz & Dong's	
9/1/48	4	921	ign	Jarret	
11/18/48	4	924	ign	Swift	
10/5/49	2	927	ign	Fetsch	
10/12/48	4	929	ign	Fox's	
4/14/49	2	931	ign	Bob's	
3/18/50	4	933	gp	Nat. Sci. Labs	
11/16/49	4	934	ign	Elwell Lutz	
10/8/49	4	935	ign	Northwest Hobby	
6/28/50	4	936	ign	America's Hobby Center	
3/18/50	4	937	ign	Nat. Sci. Labs	
4/24/50	2	938	ign	Pouch	
9/15/48	2	940	ign	George's Model Shop S/N ?	
2/23/50	4	941	ign	Pouch	
3/14/50	2	942	gp	America's Hobby Center	
12/29/49	4	943	ign	Pouch	
9/30/49	4	944	ign	Colonel Bob's	
4/20/50	2	946	ign	America's Hobby Center	
2/15/50	2	947	ign	Mansuro	
12/2/49	4	948	ign	Pouch	
2/20/50	4	950	ign	Beacon Electronics	
11/30/49	4	952	ign	Pouch	
10/27/49	4	953	ign	Pouch	
11/22/49	4	954	ign	Ray's	
12/29/49	4	955	ign	Pouch	
4/5/50	4	956	ign	Pouch	
1/16/50	4	957	ign	Pouch	
1/16/50	4	958	ign	Pouch	
2/20/50	4	959	ign	Rockwell Radio Control	
5/18/50	4	961	ign	Pouch	
1/31/50	4	963	ign	B & E	
4/11/50	2	965	ign	America's Hobby Center (AC PLUGS)	
5/9/49	4	966	ign	Pouch	
5/2/50	4	967	gp	Root's Hobby Hut	
3/14/50	2	968	ign	America's Hobby Center	
4/13/50	4	971	gp	Pouch	
3/14/50	4	973	gp	America's Hobby Center	
3/14/50	4	974	ign	America's Hobby Center	
4/5/50	4	975	gp	Pouch	
1/31/50	4	976	ign	Pouch	
12/2/49	4	977	ign	Pouch	
11/16/49	4	978	ign	Douglas	Ret'd
12/21/49	4	978	ign	Pouch	

12/23/49	4	979	ign	Northwest Hobby	
2/23/50	4	980	ign	Pouch	
2/23/50	4	981	ign	Pouch	
9/12/50	4	983	ign	Bacock (Miniature)	
8/23/50	4	984	ign	Fischers Hobby (Miniature)	
8/23/50	4	985	gp	Wenzel (Miniature)	
4/17/51	6	986	gp	Fosbury	
8/26/50	4	989	ign	Stocker (Miniature)	
9/26/50	4	990	ign	Kirchner (Miniature)	
12/21/51	4	991	gp	O'Neil's (Polk's)	
2/14/51	2	1000	ign	Miniature (Meunier)	
12/13/50	2	1002	ign	Triangle	
11/28/50	2	1004	ign	Miniature	
10/16/50	2	1005	ign	Parr (Miniature)	
12/18/50	2	1005	gp	Miniature	
10/18/50	2	1007	gp	Earl (Miniature)	
1/10/51	2	1008	ign	All American	
12/13/50	2	1009	gp	Triangle	
11/8/50	2	1010	gp	All American	
11/29/50	2	1011	ign	All American	
12/5/50	4	1012	ign	Polk's (Mr. O'Neil)	
2/11/52	4	1013	ign	Fischer's (Miniature)	
3/22/51	4	1014	gp	De Bolt	
3/22/51	4	1015	ign	Pvt Diamond (Miniature)	
2/11/52	4	1016	gp	All American (Hill)	
4/27/51	4	1017	ign	Pryor (Pouch)	
9/10/52	6	1018	gp	Vivell	
5/18/51	4	1019	ign	Canton Key Shop (Miniature)	
11/28/50	4	1020	gp	Miniature	
4/30/51	4	1021	ign	Chesser (Miniature)	
5/16/51	4	1022	ign	Pertone	
3/22/51	4	1023	ign	De Bolt	
5/23/51	4	1024	gp	McKee (Miniature)	
3/30/51	4	1025	ign	All American	
5/4/51	4	1026	ign	Miller Cook Co (Miniature)	
1/24/51	4	1027	ign	Miniature	
7/10/51	4	1028	ign	Bowen Bros (Miniature)	
1/10/52	4	1029	ign	Gonzales (Miniature)	
11/3/54	4	1030	ign	Colonel Bob's	
12/28/51	4	1031	gp	Schmitt's	
11/28/50	4	1032	gp	Miniature	
3/9/51	4	1033	ign	Triangle	
5/23/51	4	1034	gp	Allred's (Miniature)	
1/10/51	4	1035		All American	
4/24/51	4	1036	gp	Pouch (Cote)	
3/8/51	4	1037	gp	America's Hobby Center	
3/22/51	4	1038	ign	Andrews	
3/8/51	4	1039	ign	Ziggy's (Miniature)	
11/22/50	4	1040	IGN	Roach Sons	
12/21/51	4	1041	gp	Pete's	
10/16/50	4	1042	ign	All American	
10/13/54	4	1043	gp	Pouch	
12/28/51	4	1044	gp	Village	
5/28/51	6	1045	gp	Colonel Bob's	
6/19/51	4	1046	ign	Ed Guth (Miniature)	
1/5/51	4	1047	ign	Bartram (Miniature)	
1/10/51	4	1048	ign	America's Hobby Center	
5/28/51	4	1049	ign	Phil Foster	
2/16/51	4	1050	ign	Bowen Bros (Miniature)	
12/13/50	4	1051	ign	America's Hobby Center	
11/4/50	4	1052	ign	All American	
4/27/51	4	1053	ign	May's (Pouch)	
3/11/51	6	1054	gp	J Roach Sims	
3/7/51	4	1055	ign	Harry Fosbury	
3/7/55	4	1055			
2/7/52	2	1056	gp	All American Model Motor EX	
8/3/51	2	1057	gp	Triangle	
4/30/51	2	1058	gp	Triangle	
5/4/51	2	1060	ign	America's Hobby Center	
5/11/51	2	1061	ign	Andrews	
4/30/51	2	1062	gp	Triangle	
4/30/51	2	1063	ign	Petrone (Miniature)	
5/3/51	2	1064	ign	Don's	
3/31/51	2	1065	ign	Robinson (Miniature)	
2/7/52	2	1067	ign	Toy Craft (Miniature)	
5/29/52	2	1068	gp	Toys For Men	Ret'd
2/7/52	2	1070	gp	Western Auto (Miniature)	
8/3/51	2	1071	gp	Triangle	
4/30/51	2	1073	gp	Triangle	
3/26/51	2	1074	ign	Don's	
5/4/51	2	1075	gp	Roy's (Miniature)	
3/31/51	2	1076	ign	Brumbeloe (Miniature)	
5/3/51	2	1077	ign	Gull	
8/3/51	2	1078	ign	Triangle	

7/5/51	2	1079	ign	All American
4/24/51	2	1080	gp	Pouch (Pfc Storer)
5/4/51	2	1081	gp	Cecil Martin (Miniature)
4/30/51	2	1082	gp	Triangle
4/4/51	2	1084	gp	De Bolt
8/3/51	2	1085	gp	Triangle
10/10/52	2	1086	gp	Polk's
12/11/52	2	1087	gp	Modelcraft
3/3/52	2	1088	gp	All American
5/29/52	2	1089	gp	Toys For Men
3/3/52	2	1090	gp	All American
8/15/52	2	1091	gp	Modelcraft
5/26/52	2	1093	ign	All American
12/11/52	2	1094	gp	Thomson
11/14/52	2	1096	gp	Thomson
11/14/52	2	1097	ign	Thomson
4/7/52	2	1098	gp	Polk's
12/11/52	2	1099	gp	Modelcraft
3/4/52	2	1100	gp	Mod Ad
12/11/52	2	1101	gp	
12/11/52	2	1102	gp	Modelcraft
9/11/52	2	1103	ign	Thomson
9/11/52	2	1104	gp	Thomson
4/7/52	2	1105	gp	Thompson
7/30/52	2	1106	gp	
12/9/52	2	1107	gp	Tacoma
8/8/55	2	1107	ign	Fast Engine
5/2/52	2	1108	gp	Modelcraft
10/16/52	2	1109	gp	Modelcraft
5/26/52	2	1110	ign	All American
10/10/52	2	1111	ign	Polk's
12/9/52	2	1112	gp	Tacoma
5/26/52	2	1113	ign	All American
11/14/52	2	1114	gp	Pete's
8/15/52	2	1115	gp	All American
8/15/52	2	1116	gp	All American
2/20/52	2	1117	gp	All American
2/20/52	2	1118	gp	All American
8/15/52	2	1119	gp	Honolulu
4/7/52	2	1120	ign	Polk's
11/14/52	2	1122	gp	Tacoma
5/26/52	2	1123	ign	All American
4/7/52	2	1124	gp	Thompson
5/29/52	2	1126	ign	Toys For Men
3/4/52	2	1127	gp	Bon Marche
11/14/52	2	1128	gp	Tacoma
11/18/52	2	1129	gp	Modelcraft
2/20/52	2	1130	gp	All American
3/4/52	2	1131	gp	Bon Marche
12/11/52	2	1132	gp	Thomson
2/20/52	2	1133	ign	All American
12/11/52	2	1134	ign	Thomson
5/26/52	2	1135	ign	All American
4/7/52	2	1136	ign	Thompson
12/11/52	2	1137	gp	Modelcraft
7/15/52	2	1138	gp	Toys For Men
2/20/52	2	1139	gp	All American
9/24/52	2	1141	gp	Thomson
8/15/52	2	1143	ign	All American
4/7/52	2	1144	gp	Polk's
12/11/52	2	1145	ign	Thomson
8/15/52	2	1146	ign	All American
5/26/52	2	1147	ign	All American
12/9/52	2	1148	ign	Tom Water (Miniature)
5/19/52	2	1149	ign	American's Hobby Center
11/14/52	2	1150	ign	Jack's
4/7/52	2	1151	ign	Miniature
4/15/52	2	1152	gp	All American
12/15/52	4	1153	ign	Two State (Miniature)
3/23/53	4	1154	gp	Webster's (Miniature)
10/16/52	4	1155	ign	Colonel Bob's
10/21/52	4	1156	GP	Polk's
3/23/53	4	1157	gp	All American
2/24/53	4	1158	ign	May's
2/4/53	4	1160	ign	Chamberlain's (Miniature)
2/5/53	4	1161	ign	Bob's
10/21/52	4	1162	GP	Polk's
10/23/52	4	1163	ign	All American
10/23/52	4	1164	gp	All American
3/6/53	4	1165	gp	Modelcraft
3/7/53	4	1166	ign	Miniature
10/21/52	4	1167	GP	Polk's
12/16/52	4	1168	gp	Lil's Candy (Miniature)
2/17/53	4	1169	ign	Zenith
3/23/53	4	1170	ign	Smitty's (Miniature)
11/7/52	4	1171	ign	Schaedla (Miniature)
10/16/52	2	1172	gp	Modelcraft
3/7/53	4	1173	ign	Miniature
3/23/53	4	1174	ign	Edd Alexander's (Miniature)
10/23/52	4	1175	ign	Mariotti (Miniature)
5/8/53	4	1177	ign	Webster's (Miniature)
11/18/52	4	1179	gp	Tacoma
3/23/53	4	1180	gp	Zenith
2/18/53	4	1181	ign	Mod Ad
9/24/52	4	1182	ign	Thomson
2/4/53	4	1183	ign	Chamberlain's (Miniature)
10/16/52	4	1184	ign	Colonel Bob's
5/2/52	4	1185	ign	Mod-Ad
2/17/53	4	1186	gp	Zenith
2/9/53	4	1187	ign	Lucelle Johnson (Miniature)
12/9/52	4	1188	ign	Tom Water (Miniature)
9/19/52	4	1189	gp	All American
8/4/52	4	1190	gp	J C Jersey (Miniature)
5/19/52	4	1191	gp	Bon Marche
8/11/52	4	1192	ign	Polk's
5/29/52	4	1193	ign	Toys For Men
5/2/52	4	1194	gp	Miniature
5/2/52	4	1195	gp	Modelcraft
5/29/52	4	1196	gp	Toys For Men
12/16/52	4	1197	ign	Miniature
9/19/52	4	1198	ign	All American
12/16/52	4	1199	gp	Minacraft (Miniature)
5/12/52	4	1200	gp	All American
4/25/52	4	1201	gp	Miniature
8/4/52	4	1202	ign	Northwest
5/19/52	4	1203	ign	Bon Marche
8/21/52	4	1205	gp	All Amercan
8/5/52	4	1206	ign	All American
8/11/52	4	1207	gp	Allred's
8/5/52	4	1208	gp	All American
5/12/52	4	1209	ign	All American
9/16/52	4	1210	gp	Casto
8/6/52	4	1211	gp	Honolulu
5/26/52	4	1212	gp	Colonel Bob's
5/12/52	4	1213	gp	All American
5/29/52	4	1214	gp	Toys For Men
5/2/52	4	1216	ign	Walker (Miniature)
8/8/52	4	1217	gp	Shelby (Miniature)
5/26/52	4	1218	ign	Colonel Bob's
8/11/52	4	1219	gp	Polk's
5/8/53	2	1220	ign	Mod-Ad
5/8/53	2	1221	ign	El Monte
10/10/53	2	1222	gp	El Monte
3/9/53	2	1223	gp	All American
2/11/53	2	1224	gp	Village
10/10/53	2	1225	gp	El Monte
2/13/53	2	1226	ign	All American
2/11/53	2	1227	gp	Allsman
2/13/53	2	1228	ign	All American
3/9/53	2	1229	gp	Whitman
2/13/53	2	1230	gp	All American
3/9/53	2	1231	ign	All American
2/11/53	2	1232	gp	Whitman
2/10/53	2	1233	ign	Tacoma
5/1/53	2	1234	gp	Colonel Bob's
2/10/53	2	1235	gp	Tacoma
3/9/53	2	1237	ign	All American
2/10/53	2	1238	gp	Graham
2/13/53	2	1238	gp	All American
2/16/53	2	1239	ign	Prain (Miniature)
3/9/53	2	1240	ign	All American
3/9/53	2	1241	gp	All American
2/13/53	2	1242	gp	All American
3/9/53	2	1243	gp	All American
1/15/54	4	1244	gp	Fast Engine
10/14/53	4	1245	gp	Miniature
10/15/53	4	1246	gp	Webster's
11/14/53	4	1247	gp	Thomson
1/7/54	4	1248	gp	Stultz
1/15/54	4	1250	gp	Tulsa
10/14/53	4	1251	hot	Miniature
11/14/53	4	1252	gp	Thomson
3/12/54	4	1253	gp	Fast Engine
10/15/53	4	1254	ign	Mod-Ad
1/7/54	4	1255	ign	Zalonski
1/7/54	4	1256	ign	Miniature
1/7/54	4	1257	ign	Miniature
11/14/53	4	1258	ign	Darnell's (Miniature)

10/14/53	4	1259	ign	Miniature
10/14/53	4	1259	gp	Miniature
1/7/54	4	1261	ign	Miniature
11/14/53	4	1262	ign	Thomson
2/18/54	4	1263	ign	Fast Engine
10/24/53	2	1264	gp	Fast Engine
10/13/54	4	1264	gp	Pouch
1/7/54	4	1265	ign	Miniature
10/7/54	4	1266	ign	Mod-Ad
1/7/54	4	1267	gp	Billett
3/23/54	4	1268	gp	Sikorsky (Miniature)
3/16/54	4	1269	gp	Ray's
3/17/54	4	1270	gp	Fast Engine
3/12/54	4	1271	gp	Fast Engine
11/14/53	4	1273	ign	Colonel Bob's
10/14/53	4	1274	gp	Miniature
3/4/54	4	1275	ign	Colonel Bob's
1/7/54	4	1276	gp	Miniature
1/7/54	4	1277	gp	Modelcraft
10/14/53	4	1278	gp	Miniature
2/9/54	2	1279	gp	Colonel Bob's
11/1/54	2	1281	ign	Mod-Ad
3/16/54	2	1282	gp	Colonel Bob's
4/12/54	2	1283	gp	All American
10/15/54	2	1284	gp	Fast Engine
8/17/54	2	1285	gp	Fast Engine
11/16/53	2	1286	ign	Darnell's (Miniature)
11/1/54	2	1287	gp	Modelcraft
4/29/54	2	1288		Fast Engine
8/12/54	2	1289	gp	Fast Engine
3/16/54	2	1290	gp	Ray's
3/19/54	2	1291	ign	Cpl Kaiser (Miniature)
1/6/55	2	1292	ign	Polk's
3/4/54	2	1293	gp	Fast Engine
10/24/53	2	1294	ign	Santa Ana
2/15/54	2	1295	ign	Fast Eng
11/3/54	2	1296	gp	Colonel Bob's
8/12/54	2	1297	gp	Gib's
2/5/54	2	1300	gp	Tulsa
10/26/53	2	1301	gp	Newman (Fast Engine)
9/21/54	2	1302	gp	Fast Engine
1/7/54	2	1303	gp	Britton (Fast Engine)
2/10/54	2	1304	gp	Thomson
8/12/54	2	1305	gp	Fast Engine
3/19/54	2	1306	gp	Lt. Helmlse (Miniature)
9/1/54	2	1309	gp	Fast Engine
9/15/54	2	1310	gp	Tulsa
2/2/54	2	1312	ign	Hillcrest
10/29/53	2	1313	ign	Modelcraft
11/1/54	2	1314	gp	Modelcraft
2/9/55	2	1315		Tulsa
12/29/54	2	1316	gp	Fast Engine
7/15/54	2	1317	gp	Colonel Bob's
1/6/55	2	1319	gp	Polk's
1/26/54	2	1320	gp	Staffel (Fast Engine)
1/7/54	2	1321	gp	Billett
10/24/53	2	1322	gp	Woodward
9/21/54	2	1323	gp	Fast Engine
11/3/54	2	1326	gp	Colonel Bob's
1/6/55	2	1327	gp	Polk's
11/1/54	2	1328	gp	Mod-Ad
3/16/54	2	1329	gp	Colonel Bob's
1/6/55	2	1330	gp	Polk's
1/28/54	2	1331	gp	Fast Engine
9/1/54	2	1332	gp	Fast Engine
10/26/53	2	1333	gp	Brown
10/13/54	2	1334	gp	Pouch
1/6/55	2	1335	gp	Fast Engine
9/1/54	2	1336	ign	Fast Engine
1/15/54	2	1337	gp	Mod-Ad
5/28/54	2	1338	gp	Tulsa
10/24/53	2	1339	ign	Santa Ana
3/12/54	2	1340	gp	Mod-Ad
3/4/54	4	1341	ign	Colonel Bob's
5/14/54	4	1343	ign	Fast Engine
6/18/54	4	1344	ign	Miniature (Busick)
6/18/54	4	1345	ign	Miniature (Busick)
4/13/54	4	1347	gp	Colonel Bob's
7/15/54	4	1348	gp	Fast Engine
8/25/54	4	1349	ign	Nichols
4/13/54	4	1350	gp	Colonel Bob's
4/29/54	4	1351	gp	Rhoads (Fast Engine)
4/20/54	6	1352	gp	Colonel Bob's
9/21/54	4	1353	gp	Fast Engine

7/15/54	4	1354	gp	Cly Roberts (Miniature)	
9/20/54	6	1355	gp	Franson	
7/13/55	2	1355	ign	Hall, Inc	
9/21/54	4	1356	gp	Fast Engine	
6/30/54	4	1357	gp	Fast Engine	
9/20/54	6	1358	gp	Nichol	
7/15/54	4	1360	gp	Fast Engine	
4/20/54	4	1362	ign	Hillcrest	
9/20/54	6	1363	gp	Colonel Bob's	
5/27/54	4	1364	gp	Schnase	
3/30/54	4	1365	gp	Fast Engine	
6/7/54	4	1366	gp	Modelcraft	
9/20/54	6	1367	gp	Tulsa	
4/12/54	4	1368	ign	Daly City	
8/19/54	4	1370	gp	Tulsa	
7/15/54	4	1371	gp	Fast Engine	
5/28/54	4	1372	ign	Fast Engine	
9/1/54	4	1374	gp	Fast Engine	
9/23/54	4	1375	gp	Tulsa	
7/15/54	4	1376	ign	Fast Engine	
5/28/54	4	1377	gp	Kaiser	
4/20/54	6	1378	gp	Ray's	
3/30/55	4	1380	gp	Polk's	
9/9/55	4	1381	gp	Mod-Ad	
3/24/55	4	1382	gp	Fast Engine	
11/14/55	4	1383	ign	Polk's	
8/26/55	4	1384	ign	Polk's	
4/11/56	4	1386	gp	Mod-Ad	
8/26/55	4	1387	ign	Mod-Ad	
2/1/55	4	1388	gp	Modelcraft	
12/27/55	4	1390	gp	Mod-Ad	
6/14/55	4	1391	gp	Quality Leather	
1/5/55	4	1392	gp	Polk's	
11/22/55	4	1393	ign	Mod-Ad	
6/2/55	4	1394	gp	P.D. Hay's	
6/3/55	4	1395	gp	Jack Dixon, Enterprise, WA	
6/14/55	4	1396	gp	Quality Leather	
1/5/55	4	1397	gp	Miniature	
2/9/55	4	1398	gp	Colonal Bob's	
6/1/55	4	1399	gp	Boy & Boy Towne	
1/5/55	4	1400	gp	Polk's	
5/18/55	4	1402	gp	Cloverleaf	
3/24/55	4	1403	gp	Fast Engine	
4/14/55	4	1404	ign	Burnstein (Miniature)	
3/30/55	4	1405	gp	Polk's	
9/22/55	4	1406	gp	Rocky Mt. (Miniature)	
3/22/55	4	1407	ign	Waldell (Miniature)	
2/17/55	4	1408	ign	Fast Engine	
11/14/55	4	1409	ign	Polk's	
12/29/54	4	1410	gp	Colonel Bob's	
6/13/55	4	1411	gp	Mod-Ad	
2/9/55	4	1412	gp	Sportsman's	
2/9/55	4	1413	gp	Polk's	
3/6/56	4	1414	gp	Root's	
11/22/54	4	1415	ign	Miniature	
9/9/55	4	1417	ign	Southside (Miniature)	
9/22/55	4	1418	gp	Modelcraft	
9/1/55	4	1419	gp	Polk's	
5/23/55	4	1420	gp	Carl's	
2/3/55	4	1421	gp	Sportsman's	
2/1/55	4	1422	gp	Modelcraft	
11/22/55	4	1423	gp	Mod-Ad	
7/13/55	4	1424	gp	Webber's (miniature)	
9/1/55	4	1425	gp	Polk's	
1/5/56	4	1426	gp	Mod-Ad	
6/24/55	4	1427	ign	All American Motor Exchange	
11/22/55	4	1428	ign	Mod-Ad	
4/28/55	4	1429	gp	Fast Engine	
6/14/55	4	1430	gp	Quality Leather	
7/19/55	4	1431	ign	All American Motor Ex	
4/11/56	4	1432	gp	Mod-Ad	
3/12/56	4	1433	gp	Colonel Bob's	
12/29/54	4	1434	gp	Miniature	
9/1/55	4	1435	gp	Polk's	
8/31/55	4	1436	gp	Root's	
2/23/56	4	1437	gp	Wil (All American)	
1/6/55	4	1439	ign	Fast Engine	
5/10/55	4	1440	gp	Fast Engine	
6/14/55	4	1441	gp	Quality Leather	
1/5/55	4	1442	ign	Polk's	
1/29/55	4	1443	ign	Fast Engine	
1/24/56	4	1444	gp	Sawyer (All-American)	Ret'd
4/11/56	4	1444	gp	Mod-Ad	
3/24/55	4	1445	gp	Lee Bishop	

3/22/55	4	1446	ign	John Barry Co (Miniature)
1/5/55	4	1447	ign	Miniature
9/22/55	4	1448	gp	Rocky Mt. (Miniature)
6/14/55	4	1449	gp	Quality Leather
11/14/55	4	1450	ign	Polk's
8/31/55	4	1451	gp	Root's
1/5/56	4	1452	gp	Mod-Ad
1/5/55	4	1454	gp	Miniature
2/9/55	4	1455	gp	Polk's
8/26/55	4	1456	ign	Polk's
9/1/55	4	1457	ign	Root's
1/5/56	4	1458	gp	Mod-Ad
6/14/55	4	1459	gp	Quality Leather
1/24/56	4	1460	gp	P.D.Hays
10/24/55	4	1461	gp	Curio Shop
12/20/55	4	1462	ign	Mod-Ad
4/28/55	4	1463	gp	Fast Engine
4/29/55	4	1464	ign	Mod-Ad
1/29/55	4	1465	ign	Fast Engine
12/29/54	4	1466	ign	Reis
1/24/56	4	1467	gp	P.D.Hays
8/26/55	4	1468	gp	Mod-Ad
6/23/55	4	1469	gp	Modelcraft
12/27/55	4	1470	gp	Mod-Ad
1/5/55	4	1471	ign	Mod-Ad
12/20/55	4	1472	ign	Mod-Ad
11/22/54	4	1473	ign	Miniature
12/27/55	4	1474	gp	Mod-Ad
1/5/55	4	1475	gp	Mod-Ad
4/28/55	4	1476	ign	Fast Engine
8/29/55	2	1477	gp	Polk's
9/1/55	2	1478	gp	Root's
9/9/55	2	1479	ign	Modelcraft
4/29/55	2	1480	gp	Mod-Ad
8/26/55	2	1481	ign	All American
12/14/55	2	1483	gp	Modelcraft
8/24/55	2	1484	gp	Habe's
10/31/55	2	1485	gp	Tulsa
7/26/55	2	1486	gp	Webbers (Miniature)
4/14/55	2	1487	ign	Fast Engine
8/11/55	2	1488	gp	Hall
10/26/55	2	1489	gp	Polk's
9/1/55	2	1491	ign	Polk's
9/27/55	2	1492	ign	Hall
1/24/56	2	1494	gp	P.D.Hays
1/25/55	2	1495	ign	Miniature
8/13/55	2	1496	gp	Pierce
6/20/55	2	1497	gp	Wayne's
10/26/55	2	1497	gp	Hall
7/21/55	2	1498	ign	Salazar
11/14/55	2	1500	ign	Polk's
8/13/55	2	1501	ign	Pierce
2/23/56	2	1502	gp	Mod-Ad
8/24/55	2	1503	ign	Mod-Ad
9/27/55	2	1505	gp	Hall
11/14/55	2	1506	ign	Polk's
8/23/55	2	1508	gp	Walton
2/9/55	2	1509	gp	Polk's
8/24/55	2	1510	gp	Root's
8/1/55	2	1511	gp	Sims
4/21/55	2	1512	gp	Tulsa
3/22/55	2	1513	ign	Ace (Miniature)
2/23/56	2	1514	gp	Mod-Ad
2/23/56	2	1515	gp	Mod-Ad
3/22/55	2	1516	gp	Fast Engine
1/5/56	2	1517	gp	Mod-Ad
2/9/55	2	1518	gp	Sportsman's
4/14/55	2	1519	gp	Simmon's
8/29/55	2	1520	gp	Polk's
9/9/55	2	1521	gp	Woolverton
3/6/56	2	1522	gp	Root's
12/8/55	2	1523	gp	Wunderlock (All American)
7/25/55	2	1524	ign	Beth
12/5/55	2	1526	gp	Danville
4/29/55	2	1527	ign	Mod-Ad
4/14/55	2	1528	gp	Smokey's
10/26/55	2	1529	gp	Polk's
4/14/55	2	1530	gp	Smokey's
10/26/55	2	1531	gp	Polk's
2/7/56	2	1533	gp	Joy
3/22/55	2	1536	gp	Fast Engine
4/29/55	2	1537	gp	Mod-Ad
9/1/55	2	1538	ign	Polk's
9/9/55	2	1539	ign	Modelcraft
10/24/55	2	1540	ign	McKinley
8/24/55	2	1541	gp	Root's
6/8/55	2	1542	gp	Wayne's
4/14/55	2	1543	gp	Tulsa
9/29/55	2	1544	gp	Harris Hobby (Miniature)
10/26/55	2	1545	gp	Polk's
9/20/54	4	1546	gp	Lt. Kenison (Miniature)
8/11/55	2	1546	gp	Hall
10/26/55	2	1547	ign	Hall
7/5/55	2	1548		Danville Hobbies
9/1/55	2	1549	ign	Polk's
10/26/55	2	1550	gp	Polk's
10/26/55	2	1551	gp	Polk's
8/2/55	2	1552	gp	Pierce Hobby
5/25/55	2	1553	gp	Stewart's Sporting Goods
8/24/55	2	1554	gp	Mod-Ad
2/1/55	2	1556	gp	Modelcraft
11/14/55	2	1557	ign	Polk's
8/29/55	2	1558	gp	Polk's
1/5/56	2	1559	gp	Hall
4/14/55	2	1560	gp	Tulsa
8/23/54	4	1561	ign	Morris (Fast Engine)
5/18/55	2	1561	gp	Oceana (Miniature)
4/23/49	1	1562	ign	Stardusters
9/1/55	2	1562	gp	Root's
1/5/56	2	1563	gp	Mod-Ad
1/27/49	1	1564	ign	Sims Bros
9/9/55	2	1564	gp	Modelcraft
8/24/48	1	1565	ign	Stardusters for contest prize
3/21/50	1	1566	gp	Nat. Sci. Labs
9/16/55	2	1567	ign	Danville
8/19/48	1	1568	ign	Carl Van Court
8/23/55	2	1568	ign	Danville
4/14/49	1	1569	ign	Bob's
8/24/55	2	1569	gp	Habe's
2/9/55	2	1571	gp	Polk's
1/31/56	2	1572	gp	Fast Engine
4/14/49	1	1573	ign	Bob's
8/24/55	2	1573	gp	Habe's
3/14/50	1	1577	ign	America's Hobby Center
3/14/50	1	1578	gp	America's Hobby Center
3/26/51	1	5000	ign	Triangle
2/9/51	1	5001	gp	All American
3/26/51	1	5002	ign	Triangle
12/13/50	1	5003	gp	Triangle
8/6/51	1	5004	ign	Bill Crandell
11/2/50	1	5005	ign	Earl
11/8/50	1	5007	gp	All American
2/9/51	1	5007	gp	All American
7/24/51	1	5008	gp	Aerodiddler
7/31/51	1	5009	ign	Bill Crandell
8/26/50	2	4--	gp	Voigtis Hobby (miniature)
1/20/40	1	5--	ign	Whitman (Aerodiddler) ~~515~~ 5--
2/14/51	2	9_6	ign	Miniature (Don's Model Shop)
2/16/51	2	9_7	ign	Brumbeloe (Miniature)
1/10/51	2	9_8	glo	Hobby Center (Miniature)
5/29/40	1		ign	Stoddard
7/1/40	2		ign	Lang's
9/13/40	1		ign	R D Mc Gilora
11/27/40	2		ign	Stanley-Winthrop
11/29/40	2		ign	Meier & Frank
11/29/40	2		ign	Meier & Frank
11/30/40	1		ign	Air-King
12/18/40	1		ign	Lumsden
12/18/40	1		ign	Lumsden
2/21/41	2		ign	Polks
3/29/41	1		ign	Bob Knapp
7/3/41	4		ign	Williams
7/19/41	4		ign	Meyers & Ballman
1/20/42	4		ign	Comet (Los Ang)
1/20/42	4		ign	Comet (Los Ang)
1/20/42	4		ign	Comet (Los Ang)
1/20/42	4		ign	Comet (Los Ang)
3/4/42	4		ign	Laminacao
3/4/42	4		ign	Laminacao
3/4/42	4		ign	Laminacao
3/7/42	4		ign	Skyway
3/7/42	4		ign	Comet
3/7/42	4		ign	Comet
3/7/42	4		ign	Comet
3/9/42	1		ign	Auler
3/9/42	1		ign	Auler
3/9/42	1		ign	Auler
3/9/42	1		ign	Auler

3/9/42	1		ign	Auler
3/16/42	1		ign	Weeks
4/25/42	4		ign	Good's
4/25/42	4		ign	Dallaire
4/25/42	4		ign	Rothrock
4/25/42	4		ign	Comet (L.A.)
4/25/42	4		ign	Comet (L.A.)
4/25/42	4		ign	Comet (L.A.)
4/25/42	4		ign	Comet (L.A.)
4/25/42	4		ign	Comet (L.A.)
5/25/42	1		ign	Aerodiddler
6/27/42	4		ign	Rothrock
6/27/42	4		ign	Rothrock
6/30/42	4		ign	Ortiz
6/30/42	4		ign	Ortiz
6/30/42	4		ign	International
6/30/42	4		ign	Auler
6/30/42	4		ign	Auler
6/30/42	4		ign	Auler
6/30/42	4		ign	Dallaire
6/30/42	4		ign	Dallaire
6/30/42	4		ign	Dallaire
6/30/42	4		ign	Comet
6/30/42	4		ign	Burk & Co.
7/1/42	4		ign	American Jr
7/2/42	4		ign	Persons
7/2/42	4		ign	Nathan
7/2/42	1		ign	Petron
7/29/42	4		ign	Ortiz
7/29/42	4		ign	Ortiz
8/4/42	1		ign	Cash
9/26/42	4		ign	L Lucas
6/13/44	2		ign	Will Grasell
6/27/44	1		ign	Wellington
7/28/44	1		ign	
7/29/44	1		ign	
9/19/44	1		ign	Vic's
9/25/44	1		ign	Vic's
5/11/46	1		ign	Kay-Kraft
5/11/46	1		ign	Kay-Kraft
11/14/49	4		ign	Phil Foster
4/22/52	2		gp	Model Land
10/16/52	4		gp	Modelcraft
7/1/55	2			

App. C NOTES ON MODEL AIRPLANE AND PROPELLER DESIGN

DANIEL CALKIN, B. S., M. S.

INTRODUCTION

These notes contain a brief but practical explanation of the stability of a model airplane and of how stability can be attained, a discussion of the factors affecting the performance of a model, some advice on construction and flying, a method of designing model propellers, and an outline drawing and a description of the main features of several models built and flown by the ELF Co.

This is not a complete and authoritative account of how small gasoline driven models should be built and why they fly as they do but simply an outline of the work that has already been done on such models by the ELF Co. The ELF Co. is glad to get criticisms of and additions to this information. We will appreciate receiving a description of your airplane model.

LONGITUDINAL STABILITY

Most model builders understand well enough the arrangement of forces that keeps an airplane stable longitudinally but we shall review it briefly.

The weight of the airplane may be considered as if it were all concentrated at the point where the airplane balances, the center of gravity.

The whole airplane is supported by the lifting effect of the wing which with reservations may be assumed to act as if concentrated at a point on the wing section called the center of pressure.

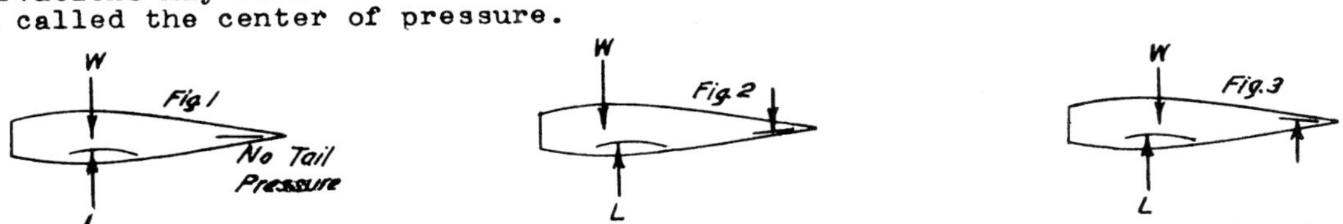

It is easy to see (fig. 1) that if the center of gravity is directly above or below the center of pressure, no pressure is needed on the tail surface to keep the airplane horizontal. If the center of gravity is ahead of the center of pressure (fig. 2), the pressure on the tail must be downward to keep the nose of the ship from dropping. If the center of pressure is ahead (fig. 3), then the tail must lift.

The longitudinal stability is affected by two horizontal forces in addition to the three vertical ones discussed above. They are the resistance of the airplane to its passage through the air and the propelling force. In flight at constant altitude, the propelling force is the propeller thrust (fig. 4). In a glide, the propelling force is a part of the weight of the airplane (fig. 5), just as it is part of the weight of an automobile which makes the automobile roll down hill. In a climb, the propelling force of the propeller must overcome a part of the weight of the ship as well as the resistance of the air (fig. 6).

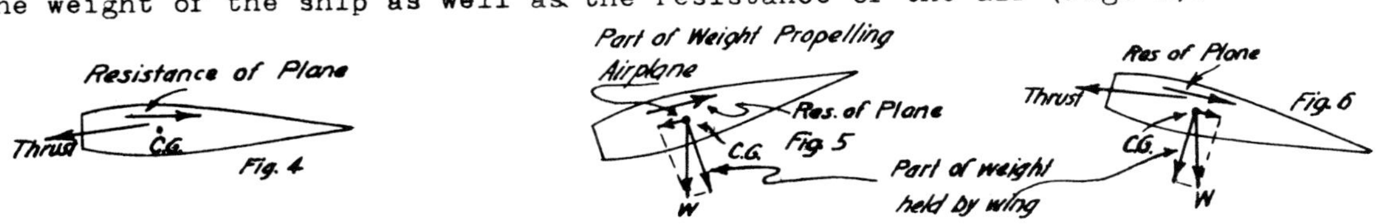

When a model is set to glide properly, its propelling force, which is part of its weight, acts through the center of gravity. In order to make a steady power flight, the propeller must also exert its thrust through the center of gravity, otherwise the arrangement of forces will be upset. This statement must be modified because the line of resistance does not usually pass through the c.g. and because the thrust line is not always parallel to the line of flight. If the propeller thrust is large, it must pass closer to the line of resistance; it will stay close to the center of gravity if it is small. Also, the slip stream of the propeller tends to make the model tail heavy. Therefore in practice the thrust line is always ¼" to ½" above the c.g. The angle of climb of a model is changed by moving or tipping the propeller up or down.

In order for an airplane to fly, these five forces must be arranged so that they tend to return the airplane to its flying position if it is caused to nose up or down. This makes the airplane statically stable. This is not the whole story however.

Put a pencil through a hole in a yardstick, allowing the stick to hang down. If you pull the end of the stick to one side, it will return to its vertical or stable position, showing that it is statically stable. The stick does not stop at its stable position but continues to swing back and forth. It has no dynamic stability. Now fasten a piece of paper on the lower end like the tail of an airplane. You will find that the stick returns to its statically stable position with very little swinging. If you use a bar of iron instead of a wood stick, you will need a much larger tail to keep it from swinging violently, but on the other hand it will take a harder push to start the swinging.

The hanging stick and the fuselage of an airplane behave in the same way. When the airplane is flying, it pivots about its center of gravity. The swing of the fuselage up and down makes the ship climb and dive. It is the resistance of the air acting on the top and bottom of the fuselage and stabilizer that keeps the fuselage from swinging back and forth past its stable position. The heavier the ship (the higher the wing loading) and the more its weight is distributed along the length of the fuselage, the more tail is necessary to keep it from pitching.

In designing a large airplane usually no special consideration is given to dynamic stability. It has been found that if an airplane has enough control surface for good control, it usually has sufficient dynamic stability. As a model has no pilot, special attention must be given to dynamic stability. For dynamic stability, the model should be as light as possible, i.e., the wing loading should be low, the concentrated weights in the model should be as close as possible to the center of gravity, and special attention should be given to making the tail structure and surfaces as light as possible. When these points are taken care of, the only means left of increasing dynamic stability is to increase the area of the stabilizer and its distance from the center of gravity. Apparently no harm other than spoiling the efficiency and appearance of the model comes from having a stabilizer too large.

Designing for Longitudinal Stability

One of the two factors that are important in longitudinal static stability is the position of the center of gravity relative to the wing chord. As every model builder knows, a wing of ordinary section is unstable because of the way that its center of pressure moves when the angle of attack is changed. For instance, when the angle of attack is increased, the center of pressure moves forward, tending to increase the angle of attack still further. Now as the point that the wing turns about, which in the airplane is the center of gravity, is moved further forward, this disturbing effect due to the movement of the center of pressure becomes less and less, and consequently the stabilizer surface required for stability becomes less and less. However, the further forward the c.g. is placed, the more down pressure is required on the stabilizer to keep the airplane from diving. It is apparent that there is a position along the chord of the wing for the c.g. for which the required tail surface is the smallest. For most airfoils this position is between 25% and 33% of the wing chord behind the leading edge and most models perform best with the c.g. between these two points. For highest efficiency, place the c.g. so that it is under the center of pressure at the angle of attack that you want your model to fly with. Then the tail load will be at a minimum (see fig. 1). As long as the c.g. is within 20% of the chord either below or above the wing its vertical position has very little effect on stability. When the c.g. is as much as 40% below the wing it has a marked stabilizing effect. This stabilizing effect necessitates increasing the stabilizer surface in order to gain dynamic stability. The increase in stabilizer area then has more effect on the static stability than the low position of the c.g. does! A model with a low center of gravity is no more stable than one with a high c.g. and it is less efficient. The big advantage of the low c.g. is that it tends automatically to compensate for poor workmanship in the model, such as warped wings, weight to one side etc. Such a model is not so sensitive to adjustment but on the other hand it cannot be adjusted for as good performance as the more efficient low wing models.

The other of the two factors important in longitudinal static stability is the design of the stabilizer. Important points in the design are distance from c.g., area, aspect ratio, and section. The further from the main plane the stabilizer is, the less area it requires. The higher its aspect ratio is, the less its area needs to be. The aspect ratio of a stabilizer runs from 3 to 5; efficiency is gained but slowly with ratios over 5. For a model the only requirement for a stabilizer section is that it be as thin and streamlined as possible consistent with rigidity. For efficiency use a symmetrical section. For ease of construction use a flat stabilizer or a section that is flat on one side.

The relative angle of the stabilizer and main plane has no bearing on stability. This angle is determined by the position of the c.g. along the wing chord. For any given c.g. position there is a small range of stabilizer settings at which the model will fly, ranging from a glide that is too steep and fast to the point at which the model begins to pitch. The c.g. position at which the model will fly with the greatest angle of attack without pitching should be used in order to get the lowest possible landing speed.

The distance from the c.g. to the center of the stabilizer is usually about $2\frac{1}{2}$ times the mean wing chord. The area of a stabilizer usually should be between 20% and 25% of the wing area. In special cases the stabilizer can be as small as 18%, and areas up to 30% can be used at the sacrifice of efficiency. When the stabilizer is very large (33% of the wing), it may be found that the model flies better with the c.g. at the center of the wing section or even behind it. This is not because this c.g. position gives better stability but because the tail is made to carry a large proportion of the weight of the model, thus lowering the wing loading. Such a model is likely to fly backward as well as it does forward and cannot compare with a properly designed model in efficiency.

For ELF powered models the maximum wing loadings for safe flying are 12 oz./ft.^{2} for a three foot span model, 10 oz./ft.^{2} for a four foot span model, 9 oz./ft.^{2} for a five foot model. These figures using an aspect ratio of 6 for the wings will give the model a maximum total weight that an ELF Engine can safely handle. Most models of ordinary balsa construction will have wing loadings of about half these figures. On two engine ships, the wing loading on 5 and 6 foot models can be 12 oz./ft.^{2}. In general, the lighter a model is, the less likely it is to be badly smashed when it hits something and the more stable it is.

LATERAL STABILITY

The weight of an airplane in flight is supported by the lift of its wings. For the purpose of this discussion, the weight of the airplane acts as if it were concentrated at the center of gravity, and the lift of the wing acts as if it were concentrated in a line passing vertically through the center section of the wing, at a point on the section called the center of pressure. Notice that the lift of the wing is not assumed to act as if it were concentrated at a point. This assumption does not seem to be warranted by the facts. An airplane has no "center of lift" analogous to the center of buoyancy of a boat and an airplane does not maintain its upright position in the same way that a boat does.

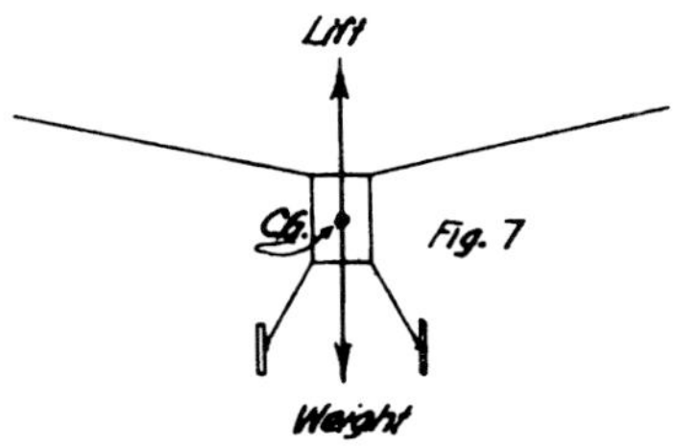

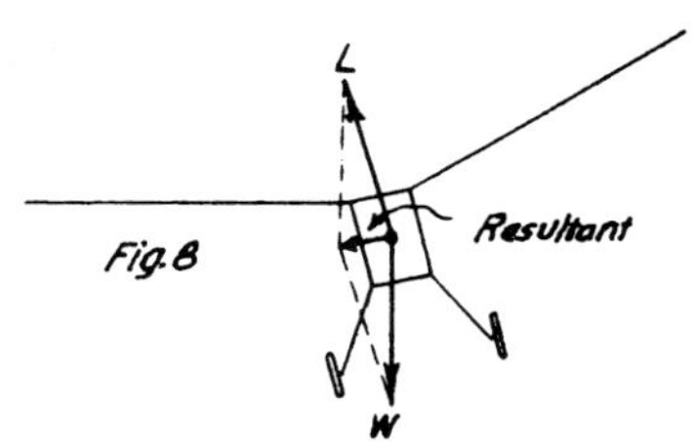

Fig. 7 is a front view of an airplane showing the weight of the airplane acting from the c.g. and the lift of the wing concentrated on a line passing through the center of the wing. This line also passes through the c.g. When the airplane rolls to one side as in fig. 8, these two forces L and W have a resultant R. The force R causes the airplane to add a sideways motion to its original forward motion. This side slip acts on the dihedral of the wing as shown in fig. 9 and 10 increasing the incidence on the low wing and decreasing the incidence on the high wing causing the ship to return to its level position. Fig. 9 represents an airplane in forward flight. The wind moves squarely across the wing from point 1 to point 2. If the airplane has a sideways motion at the same time as in fig. 10 the wind moves across the wing from point 1 to point 3. The dihedral makes point 3 lower than point 2, relative to the fuselage, consequently this wing lifts more. Similarly, the side slip causes the other wing to lift less.

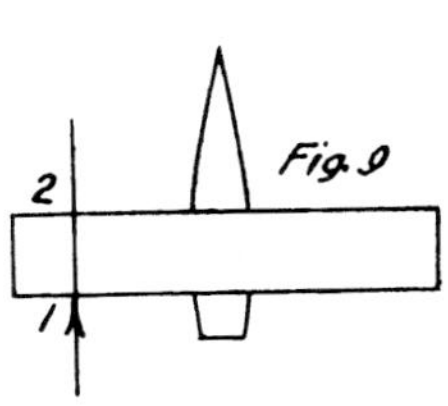

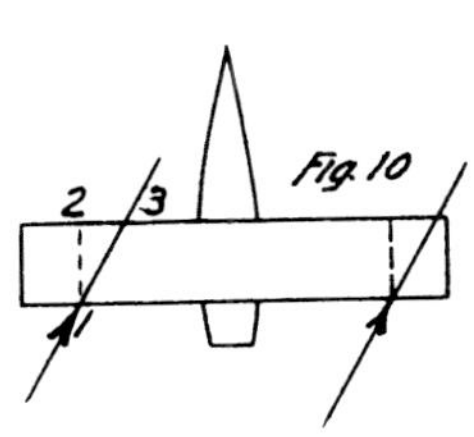

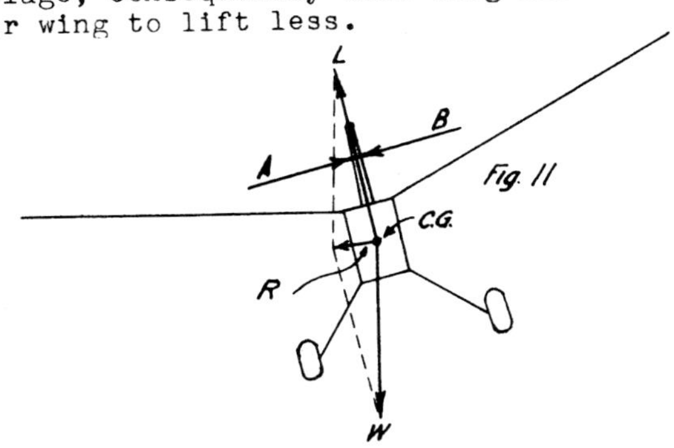

This action of the dihedral angle in side slip is the only important lateral stabilizing effect in a model. Because the side slip is so important, the importance of the fin (vertical tail area) in lateral stability is obvious, as the fin controls the direction of flight. The airplane pictured in fig. 8 has a vertical tail surface as shown in fig. 11. When the resultant R moves the airplane to the left, a side pressure is set up on the fin, as represented by A. This force tends to rotate the ship toward the direction of its side slip, so that by the time the roll is corrected, the plane is no longer side slipping. If the fin is too small, the side slip will not be corrected quickly enough and the ship will roll over the other way and may continue to roll from side to side swinging its tail back and forth (Dutchman's roll). If the fin is too large, the side slip will be removed before the roll is corrected and the ship will circle to the side of its low wing. This circling may set up a pressure on the fin represented by B which is insufficient to correct the turn but which because of its height above the c.g. is sufficient to keep the low wing down. The airplane cannot get out of this position and is in a spiral dive, the first cousin to a tail spin.

It will be noticed in fig. 11 that when the fin is the proper size, the force A which corrects the side slip also tends to correct the roll by virtue of its position above the c.g. It is obvious that all the side area of the airplane that is above the c.g. tends in like manner to correct the roll. However, this effect is small even on parasol wing models with large dihedral and any beneficial effect from it is likely to be more than overshadowed by its ill effect on directional stability.

One factor that has an important effect on lateral stability but is often neglected is the inclination of the rolling axis of the airplane. When an airplane rolls, it does not turn about any random axis running from nose to tail. It always tends to turn about the one that it is easiest for it to turn about. This axis runs through the c.g. and should be higher at the nose than at the tail. In this position it tends to swing the tail to the side of the low wing when the airplane rolls, thus bringing about a correcting side slip more quickly. Its effect is ordinarily small compared to the effect of the fin but in some super-efficient low wing jobs with small tails its stabilizing effect may make the difference between flying and not flying. To get the rolling axis in the proper place, keep weight in the tail of the airplane low and weight forward of the c.g. above the c.g. To show how important this factor may be sometimes, it is possible to make a model stable with the wings sloping down from the body (negative dihedral) by putting weight low forward of the c.g. The pitch of the rolling axis causes a side slip on the side opposite the low wing in a roll and the dihedral acts as before to correct the roll.

Designing for Lateral Stability

In designing for lateral stability, the important points are the dihedral angle, the area of the fin and its height above the c.g. when the airplane is in flying position. For any one dihedral angle there is a range of fin sizes between a Dutchman's roll and a spiral dive for which the airplane is stable. As the angle is increased, the larger the range of sizes is and the larger tail can be used without causing a spiral dive. A practical range for dihedral angles is between 5° and 10°. It is difficult to gain lateral stability by lowering the c.g., for, as we have seen, when the fin area is too high above the c.g. it tends toward a spiral dive. To change the position of the tail relative to the c.g., change the angle of incidence of the wing in relation to the fuselage. The wing should be set on the fuselage so that the axis of the fuselage will be in the line of flight when the wing is at its angle of incidence for the best glide (roughly 8° for a Clark Y). However, since decreasing this angle lowers the tail and promotes lateral stability both by lowering the center of lateral area and the rear end of the rolling axis, most models must set the wing at considerably less incidence than this (from 0° to 5°).

The longitudinal position of the c.g. for best lateral stability is the same as for best longitudinal stability.

Engine torque is a considerable factor in lateral stability in most models. It may be compensated for in two ways, first by setting the propeller thrust off the center line of the model to the proper side, and second by twisting the stabilizer and fin in the proper direction to offset torque. Warping the tail surfaces thus will cause the model to corcle to the right against torque in a glide, but the rudder should be set to make it circle slightly to the left. When the engine is running, the warp in the tail surfaces has more counter torque effect than in a glide because of the velocity of the slip stream. To make sure that the engine will not upset your model, be sure that the tail surfaces have a barely visible warp tending to make the plane circle against the torque, be sure the wings are symmetrical, then set the rudder by trial so that the ship circles slightly with the torque when it glides. Offset the engine to the left from ½ to 1 degree. Light models with a relatively high thrust are sensitive to the alignment and vertical position of the engine, but they can be adjusted so that they circle with the same radius with the power on or off and climb steeply without stalling.

PERFORMANCE

After stability, the first consideration for a model airplane is a low landing speed. The slower it lands, the less likely it is to be damaged in a collision. The speed of an airplane is controlled by the setting of the stabilizer. The greater the angle of the stabilizer, the higher the angle of incidence the wing assumes. The better the stability of the model, the slower it can be set to fly. As high lift wing sections are usually more unstable, it is doubtful whether a slower landing speed can be obtained for a model by using thick section or double camber wings. Increasing the aspect ratio lowers the landing speed by increasing the lift coefficient of the wing. Streamlining or lack of it has no effect on landing speed; it affects only the angle of glide. Landing speed varies inversely with the square root of wing loading.

The next consideration in designing a model is efficiency. You want the model to fly high and far and stay in the air. For a model the term "Efficiency" should be understood to include two related but none the less different things. They are the lift over drag ratio L/D and the power factor $L^{3/2}/D$. In a glide, the airplane with the highest L/D will travel the greatest distance from a given height and the airplane with the greatest $L^{3/2}/D$ will stay in the air the longest time from a given height. For a given model as the angle of the elevator is increased, the glide will become flatter and the airspeed slower until a certain point is reached, then the airspeed will continue to become slower but the glide will get steeper. This point of flattest glide is the elevator setting for which the L/D is greatest for that model. As the airspeed gets slower and the glide gets steeper above this point, it can be seen that there is an elevator setting which will cause the model to take the longest time to reach the ground (slowest sinking speed). This is the setting for which $L^{3/2}/D$ is greatest for that model. Unless the model is properly designed for longitudinal stability, it will begin to pitch before either of these points is reached, as both are close to the stalling speed of the model. To increase the L/D streamline the model as far as possible, reduce the size of wheels and tail surfaces to reduce parasite drag. Increasing the aspect ratio is of help. To increase the $L^{3/2}/D$, reduce the parasite drag and increase the aspect ratio of the plane. Increasing the aspect ratio has a very great effect on keeping the model in the air for a long time. To get the fastest climb, set the elevator by trial for the slowest sinking speed and then set the thrust line by trial for the fastest climb. To get the most speed, set the elevator for the fastest landing that seems safe, then set the thrust line so that the ship does not climb. In a speed model, efficiency is secondary to getting a low minimum drag.

Tapering a wing has the same effect on efficiency as increasing the aspect ratio slightly and has the advantage of making internal bracing easy.

Scale Models

When building an exact scale model, there is no choice of wing section, tail size, aspect ratio, dihedral angle, etc. In order to get a scale model to fly, it is customary to use a larger propeller, a larger tail, increase the dihedral, etc. With an ELF Engine it is possible in many cases to make a flying scale model without any modification. Stability is attained by the arrangements of the concentrated weights in the fuselage and the proper choice of scale.

As the amount you can change the diameter of the propeller of the model engine is limited, pick a scale that makes a propeller that fits the engine the proper scale size for the model. A two bladed propeller can run between 10" and 12" diameter. A three bladed propeller can run from 8" to 10". Diameters outside of these will cause a considerable sacrifice of efficiency.

The wing loading should not be more than 7 oz./ft.^{2} and should be as low as possible ($4\frac{1}{2}$ oz. ideal). The tail surfaces and rear end of the fuselage should be made very, very light. Any weight in the tail should be low. The weights in the model should be kept as close as possible to the c.g. The best longitudinal position for the c.g. should be found.

CONSTRUCTION DETAILS

If the fuselage has longerons equivalent to 4 3/16" squares of medium or hard balsa and is covered with silk, it will be amply strong. The spark coil and battery should be fastened in with strong braces ahead of them to prevent their sliding forward in a head-on collision. Follow the suggestions in the instruction manual for wiring. Tie the coil in with thread, not wire.

The wings should have two strong spars (sheet balsa 1/16" thick and as deep as the section). They will prevent the wing's crumpling in a side slip to the ground. The leading edge should be thin balsa so that no ridge will be formed along the front of the wing. In case of a collision, this leading edge is easily repaired and it protects the spar which is deeper in the wing. The wing tips should be strongly braced to stand the pull of the covering without warping. Movable ailerons are a great convenience in case the wings warp. The wing should not be rigidly attached to the fuselage.

A Clark Y section is used in the stabilizer of the ELF models. The angle between the stabilizer and the wing is therefore easily measured. This section is thick enough to brace internally and its flat bottom makes construction easy. As for efficiency, the minimum drag is nearly as low as any symmetrical section and the more accurate construction possible because of the flat underside probably eliminates this difference. The aspect ratio of the tail is not of great importance but should be fairly large for efficiency. The tail surfaces should be adjustable.

All the ELF models are covered with china silk and doped with two or three coats of airplane dope cut fairly thin with slow drying thinner. For color, pigment is added to the clear dope. Put as little dope on the model as possible. If the model is colored, the dope will weigh as much as the balsa framework unless great care is used to get it on thin.

A landing gear made of piano wire is the most practical as it can be straightened out after a crash and is very springy. The wheels can be solid balsa or light air wheels. The wheels should be placed far forward, up under the engine.

Do not try to make your model so strong that it will not break in a crash. It cannot be done. Make it so that it will come apart in places easy to repair. To facilitate repairs, it is advisable to make all the flying surfaces easily removable from the body.

If the engine is in the front of the plane, mount it on 1/32" soft steel. This material will bend in a crash and prevent injury to the engine and it can be straightened and used again. If the engine is protected, use dural. It is light but it is not safe to straighten it after it has been bent. If your ELF Engine is properly mounted, there is very little risk of damaging it. The hangers for the engine should be fastened to the nose so that they can be easily removed. All wires, pipes, and screws that go through the forward bulkhead should be sealed in with glue to prevent oil working into the fuselage.

The ELF can be cowled easily. Because of its very small size it cools easily and the exhaust is too small in volume to cause trouble. An exhaust pipe can be used but it is not necessary. The pipe should not be fastened too solidly to the cylinder as it may injure the cylinder in a crack-up.

FLYING A MODEL

Get the model to glide properly before trying to fly it under power. Roll it down a steep smooth slope, gradually increasing the negative angle of the stabilizer until it leaves the ground. Continue to change the stabilizer until the glide is as slow as you want it but remember that the slower the plane gets the more unstable it becomes. For the first power flight, have enough negative angle in the propeller so that the plane will not take off. Use the full power of the engine as the plane is less likely to stall if it does take off. Run along beside the plane and if it does not leave the ground, stop it by touching one wing to make it ground loop. This will slow it down so that you can pick it up. Try another run with less negative angle.

PROPELLERS

It is not practical to design a model propeller the way a large propeller is designed. Here is a procedure that gives good results.

1. Estimate the speed of the model in miles per hour, change it to feet per minute. mph (88) = fpm.

2. Divide the speed in fpm by the rpm of the engine in flight, which will be roughly 300 rpm more than its speed on the ground. Use 3800 for an ELF Engine for ordinary use. This will give the distance in feet the airplane moves forward per revolution of the engine. Multiply by twelve to get the distance in inches.

3. Increase this number 50% which will give you the pitch of your propeller.

4. Carve the propeller with constant pitch along its blades and of a diameter that is sure to be too great.

5. Mount the propeller on the engine and try it, trimming off the ends until the engine runs at 3400-3600. This speed can be checked by an electric light as explained in the instruction manual.

1. Do not use a pitch smaller than 6 inches. No thrust is gained even for very slow models. An 8 inch pitch takes about a 10 inch diameter.

2. Plan form.

The plan form has little effect on efficiency. The best plan form is the one that allows the smallest blade deflections with the material used. The layout in fig. 12 is a very good one for wooden propellers where efficiency is required. Any desired plan form can be used with this method of laying out the pitch, however. The aspect ratio of each blade should be about 5 to 1 for a two bladed propeller, 6 to 1 for a three bladed propeller, 7 to 1 for a four bladed propeller.

3. The blade section should approximate a Clark Y section. Propeller sections are not suitable for model propellers because of their low air speeds.

4. Three and four bladed propellers are only slightly less efficient than two bladed propellers and should be used where a small diameter propeller is needed. Fig. 13 shows an easy way to make a three bladed propeller from two standard ELF propellers.

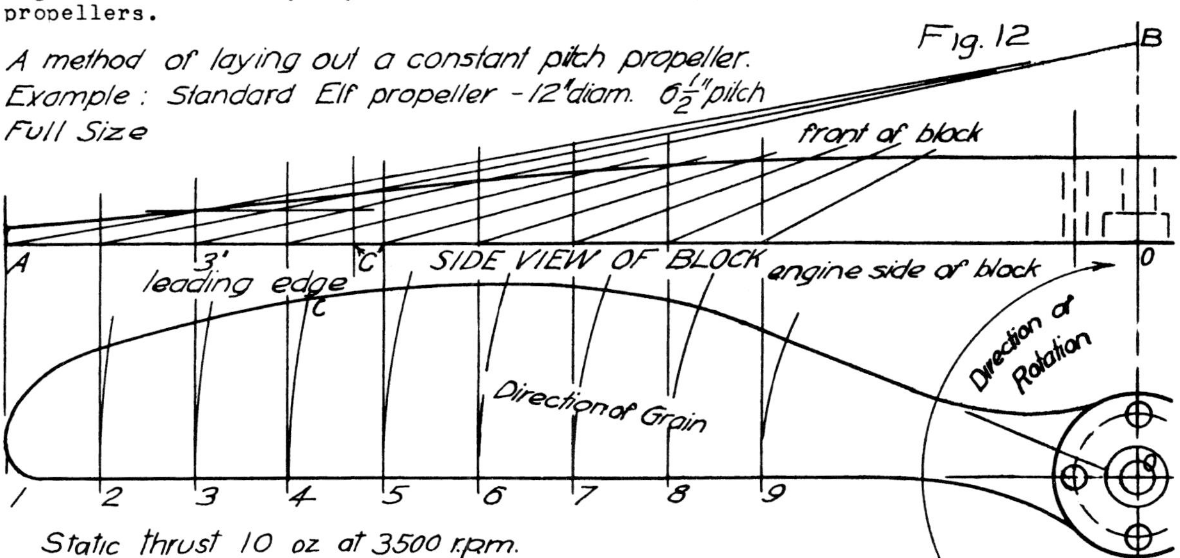

1. Lay out plan form. 2. Mark off convenient intervals as 1,2,3,etc. 3. Swing arcs across plan form from center of prop. as shown.****To get side view of block. 1. Draw line OA; project points 1,2,3,etc. up to it. 2. Find point B thru O.

$$\frac{OB}{pitch} = \frac{OA}{\pi\,diam.}\ ;\quad OB = pitch\left(\frac{\frac{diam.}{2}}{\pi\,diam.}\right) = \frac{pitch}{2\pi} = \frac{6.5}{2\pi} = 1.035''$$

3. Lines drawn from B to stations 1,2,3,etc. on OA give the angle of the blade at each station. 4. To find thickness of block at any section (taking station 3 as an example) lay off chord 3C on OA, as 3'C'. Then the vertical distance from C' to the line 3'B is the thickness of the block at station 3. This construction is repeated at each station and a smooth curve drawn thru the points. 5. The plan form is cut out of a block of uniform thickness, then each end of the blank is thinned to the shape indicated in the side view. 6. Carve back side first making it flat across the blade to the chords of the arc thru the stations. The leading edge has a double curve (curved in two planes) and the trailing edge is straight. 7. Carve the front side to an airfoil section.

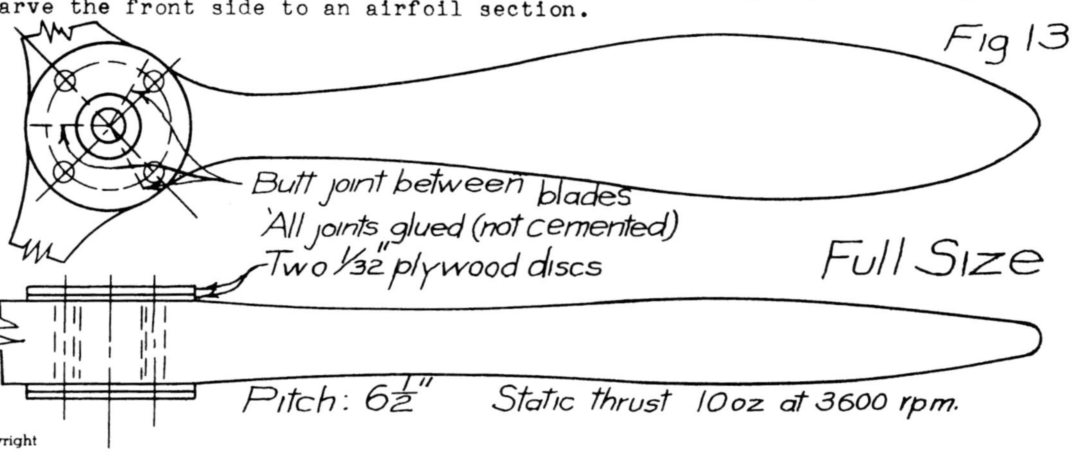

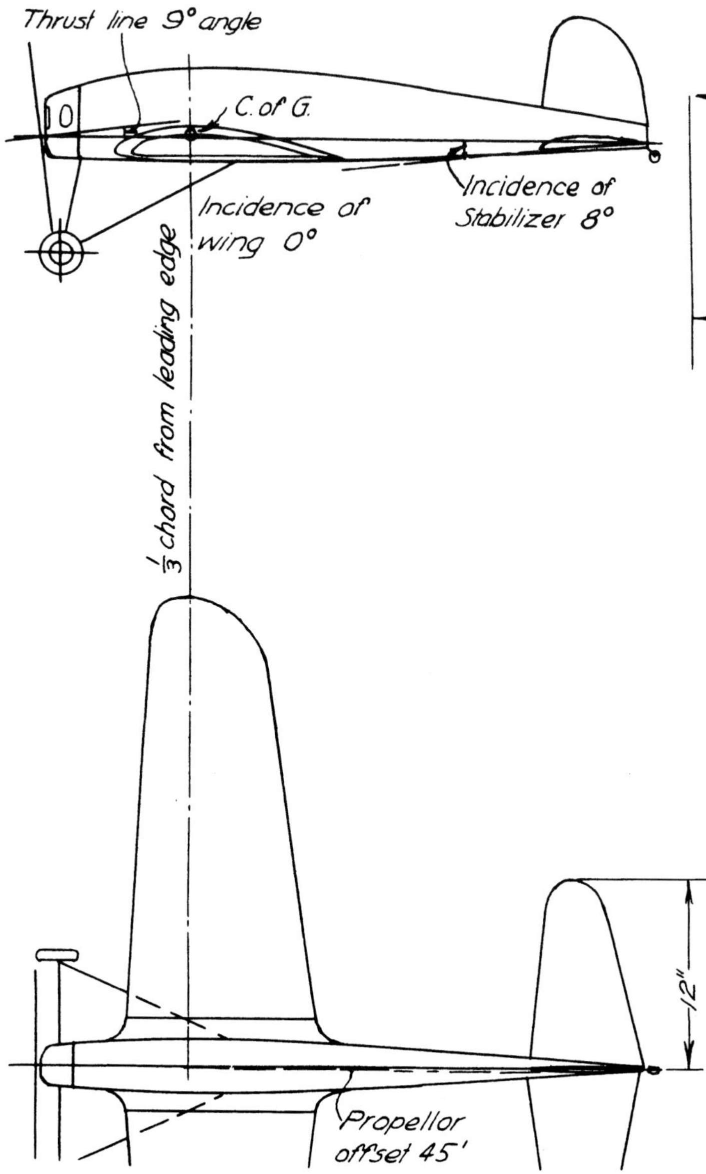

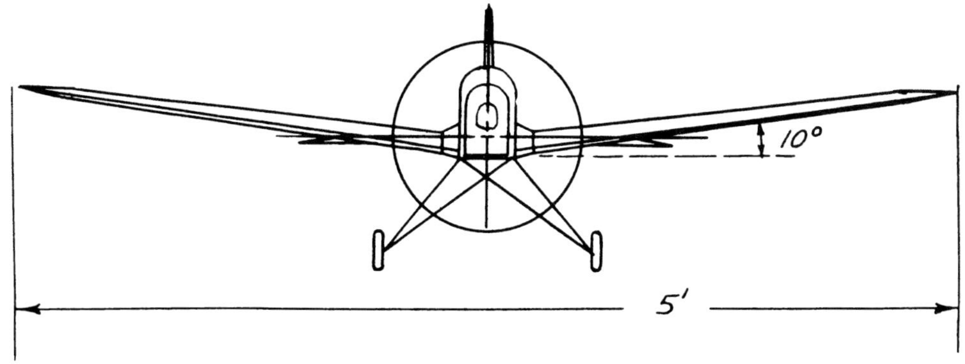

FIVE FOOT WING SPAN AIRPLANE. Scale, 1 inch = 1 foot.

Total Weight	18 oz.
Area of Wing	4.16 sq.ft.
Wing Loading	4.33 oz./sq.ft.
Wing Section	Clark Y throughout
Mean Chord	10 in.
Aspect Ratio	6

Taper--Rib at root 12" long. Rib 20" out 8" long. Points on ribs 1/3 of chord from leading edge in line perpendicular to fuselage.

Section--Clark Y. (stabilizor)

Area of Stabilizer	0.917 sq.ft.
Area of Stabilizer	22% plane area
Section of Stabilizer	
Area of Fin	0.208 sq.ft.
Area of Fin	5% plane area
Symmetrical Section	(fin)
Distance Center of Gravity to Center of Stabilizer	2½ X mean chord
Propeller	12" dia. 6½" pitch

Center of Gravity--1/3 chord from leading edge and even with top of root rib.

Construction Details

Fuselage--4 longerons and cross pieces ¼"x1/8" balsa. Silk covered.
Wings--Ribs every 2 inches 1/16" thick. Two double spars, 1/8"x 3/16" piece top and bottom, no web. End of spars slide into boxes in fuselage, U shaped spring of piano wire inserted from underneath holds ends together.
Landing Gear--1/16" piano wire. Solid balsa wheels 2½" diameter and 1" tread.
Cowling--Of cardboard.
Stabilizer--1 piece, hinged in front, screw adjustment in rear.

This model was originally made with a spruce fuselage and had a total weight of 25 oz. This model was a stable flier and made 40 to 50 flights. The 18 oz. model glides very much slower and is stable over a wider range of stabilizer adjustment. It has made about 20 flights. It is not sensitive to wing or tail adjustment but is very sensitive to engine alignment. With proper adjustment it 8
climbs rapidly in a wide circle, glides down at a speed noticeably much slower.

Appendix D REFERENCES CONSULTED

The following are the sources of information consulted for this book. The author has an original copy or a photocopy of each document listed.

ELF ENGINE COMPANY

ELF Engines shipped from October 20, 1939 to April 11, 1956. Handwritten by Dan Calkin in log book.

Miscellaneous receipts from June 1, 1944 to March 20, 1956. Handwritten log of parts sold and engines repaired.

Direct Material Purchases, April 9, 1945 to September 1, 1955.

Indirect Materials Purchased, November 11, 1941 to November 28, 1955.

Additions to Shop, November 30, 1943 to November 30, 1955.

Additions to Foundry, November 8, 1940 to August 10, 1954.

Advertising, Office expenses, October 31, 1944 to August, 1955.

Radio Control Project, March 17, 1943 to January 1, 1953.

Telephone expenses, January 1945 to September 1951.

Rent records, January 1945 to March 1956.

Water expenses, January 1952 to March 1956.

Electricity expenses, January 1952 to March 1956.

Gas expenses, March 1952 to March 1956.

Fuel Oil expenses, March 1952 to February 1956.

Taxes paid, 1942 to November 1956.

Interest Paid, November 1946 to February 1954.

Contributions, 1942 to 1959.

Insurance, 1947 to 1955.

Labor Costs, 1954 to 1955.

Preliminary drawing for the Model 10 ELF, November 30, 1938.

Design drawings for the ELF four stroke engine. Also patterns, castings and tooling.

Tooling for the various ELF engines.

Josephine Calkin's ELF scrap book. A collection of ELF photos, articles, company literature from earliest days of ELF Engine Co. until 1965.

ELF literature 1935-1956. A collection of ELF brochures, instructions, parts lists, price lists, product announcements, stationary, envelopes. Obtained from many different sources, including the Calkin family; over 60 items.

ELF material borrowed from Josephine Calkin and Cecelia Jackson 1995-1998: Includes original ELF master patterns, match plates, parts, tooling, coils, spark plugs, plans, radio control equipment, literature, photographs, and letters. Also includes about 20 original ELF engines, some of which are prototype engines.

Cecelia Calkin's baby book.

LETTERS

Paramount Publix Corporation to E.D. Calkin, August 7, 1930. Letter about possible filming of engine.

Letter Miniature Aircraft Co to T. J. Thompson, "Elbert Weathers ELF distributor."

E.D. Calkin to Ralph and Etta Calkin, March 13, 1932. Letter written from M.I.T.

F.H. Riddle, Champion Spark Plug Company to E.D. Calkin, March 22, 1935.

F.H. Riddle, Champion Spark Plug Company to E.D. Calkin, April 2, 1925.

E.D. Calkin to Josephine Peck, April 8, 1934.

E.D. Calkin to Josephine Peck, May 6, 1934.

O.C. Rohde, Champion Spark Plug Co. to E.D. Calkin, May 7, 1934.

E.D. Calkin to Josephine Peck, May 18, 1934.

E.D. Calkin to Josephine Peck, May 26, 1934.

E.D. Calkin to Josephine Peck, May 27, 1934.

E.D. Calkin to Josephine Peck, June 5, 1934.

E.D. Calkin to Josephine Peck, June 15, 1934.

E.D. Calkin to Josephine Peck, June 16, 1934.

Etta Calkin to Josephine Peck, June 27, 1934.

E.D. Calkin to Josephine Peck, July 14, 1934.

E.D. Calkin to Josephine Peck, July 16, 1934.

E.D. Calkin to Josephine Peck, July 18, 1934.

E.D. Calkin to Josephine Peck, July 19, 1934.

E.D. Calkin to Josephine Peck, August 3, 1934.

E.D. Calkin to J.P. Young, May 10, 1935.

E.D. Calkin to J.P. Young, May 27, 1935.

E.D. Calkin to J.P. Young, December 16, 1935.

E.D. Calkin to Charles Hyde, July 29, 1936.

E.D. Calkin to Charles Hyde, August 10, 1936.

J.A. Jeffrey, Champion Spark Plug Company, to E.D. Calkin, September 25, 1936.

B.H. Sibley, Champion Spark Plug Company, to E.D. Calkin, July 13, 1937.

J.A. Jeffrey, Champion Spark Plug Company, to E.D. Calkin, October 26, 1937.

E.D. Calkin to J. A. Jeffrey, Champion Spark Plug Company, November 4, 1937.

J.A. Jeffrey, Champion Spark Plug Company, to E.D. Calkin, November 11, 1937.

R.S. Rogers, Champion Spark Plug Company, to E.D. Calkin, January 27, 1936.

E.D. Calkin to Carl VanCourt, August 2, 1944.

E.D. Calkin to Carl VanCourt, August 21, 1944.

E.D. Calkin to William V. Krecek, February 14, 1945.

E.D. Calkin to Carl VanCourt, February 23, 1945.

DeWitt Ross, Jr. to E.D. Calkin, June 26, 1945. (Note: Ross later manufactured the *Wasp Twin* engines).

E.D. Calkin to Carl VanCourt, July 18, 1945.

E.D. Calkin to Carl VanCourt, July 30, 1945.

E.D. Calkin to Carl VanCourt, August 23, 1945.

Carl VanCourt to Jim Walker, August 8, 1945.

E.D. Calkin to Carl VanCourt, September 5, 1945.

E.D. Calkin to Carl VanCourt, October 2, 1945.

E.D. Calkin to Carl VanCourt, November 7, 1945.

E.D. Calkin to Carl VanCourt, December 10, 1945.

Carl VanCourt to E.D. Calkin, November 29, 1945.

E.D. Calkin to "To Whom it May Concern," December 18, 1946.

E.D. Calkin to Gordon S. Light, August 29, 1947.

E.D. Calkin to Carl VanCourt, April 19, 1948.

E.D. Calkin to Carl VanCourt, April 29, 1948.

E.D. Calkin to Carl VanCourt, October 18, 1948.

E.D. Calkin to Carl VanCourt, January 31, 1949.

E.D. Calkin to Carl VanCourt, June 29, 1949.

E.D. Calkin to Carl VanCourt, September 12, 1949.

E.D. Calkin to Carl VanCourt, November 16, 1949.

E.D. Calkin to Carl VanCourt, January, 16, 1950.

E.D. Calkin to Carl VanCourt, August 4, 1950.

E.D. Calkin to Carl VanCourt, August 16, 1950.

E.D. Calkin to Carl VanCourt, November 9, 1950.

E.D. Calkin to Carl VanCourt, January 22, 1951.

E.D. Calkin to Carl VanCourt, January 10, 1952.

E.D. Calkin to Carl VanCourt, October 8, 1952.

E.D. Calkin to Carl VanCourt, May 20, 1953.

E.D. Calkin to Carl VanCourt, July 15, 1953.

E.D. Calkin to David L. McKinley, October 24, 1955.

E.D. Calkin to David L. McKinley, June 15, 1956.

E.D. Calkin to Phil Foster, August 28, 1959.

E.D. Calkin to Phil Foster, December 10, 1960.

E.D. Calkin to Cecelia Calkin Jackson, March 3, 1963.

E. D. Calkin to Cecelia Calkin Jackson, June 23, 1963.

E.D. Calkin to Peter Baylinson, February 5, 1966.

E.D. Calkin to Peter Baylinson, August 5, 1967.

Josephine Calkin to Cecelia Calkin Jackson, June 3, 1972.

E.D. Calkin to Cecelia Calkin Jackson, February 28, 1975.

E.D. Calkin to granddaughter Allanna Jackson, about 1975 or 1976.

Josephine Calkin to Cecelia Calkin Jackson, March 20, 1978.

Josephine Calkin to Cecelia Calkin Jackson, undated, but probably about May 1978.

Plans for ELF three-foot model airplane, 1935-1938. Full size plans, three different versions.

Plans for control line scale model of Culver Cadet, 40" wing span for ELF twin. Published by Magazine Plan Service.

PERSONAL INTERVIEWS

Pete Baylinson, ELF collector, 1997.

Josephine Calkin, widow of Dan Calkin, 1994-1998.

Peter Chinn, engine expert, former engine columnist for *Aeromodeller* magazine, 1997.

Steve Ditta, ELF collector, 1995-1998.

Ted Enticknap, ELF expert and my close assistant through this whole adventure. I can't give him enough credit.

Phil Foster, former employee of Dan Calkin, 1996-1998.

Abe Gallas, former co-worker of Dan Calkin at Douglas Aircraft, 1997

Jim Gerard, ELF collector, 1995-1998.

Alex Imrie, columnist for Aeromodeller magazine, 1997-1999.

Cecelia Calkin Jackson, daughter of Dan Calkin, 1994-1999.

Donald Keeton, former co-worker of Dan Calkin at Douglas Aircraft, 1997

Dale Kirn, control line expert and historian, 1995-1999.

Leonard Lucas, former associate of Dan Calkin, 1998-1999.

Charles Mackey, control line expert, historian and author, 1998.

Frank Macy, model airplane historian, 1997-1999.

Dr. Richard E. Nichol, ELF researcher and collector, 1997-1999.

Dick Rothrock, former employee of Dan Calkin, 1997-1999.

Alan Shively, early ELF collector and researcher, 1970s.

Ron Sporing, former co-worker of Dan Calkin at Douglas Aircraft, 1997-1998.

Carl VanCourt, former associate of Dan Calkin, 1995-1997.

Many other interviews with many people, too numerous to mention, but all important.

SCHOOL DOCUMENTS

E. Daniel Calkin, transcript of school work, 1924-1927, Kansas City, Missouri School District.

E. Daniel Calkin, transcript of college work, 1927-1929, Junior College of Kansas City, Missouri.

E. Daniel Calkin, transcript of college work, 1930-1931, University of Washington at Seattle, Washington.

E. Daniel Calkin, transcript of graduate studies, 1929-1933, Massachusetts Institute of Technology, Cambridge, Massachusetts.

Josephine Peck, transcript of school work, 1924-1927, Kansas City, Missouri School District.

Josephine Peck, transcript of college work, 1927-1929, Junior College of Kansas City, Missouri.

Dan Calkin's Southwest High School Yearbook, 1927, Kansas City, Missouri.

Josephine Peck's Lindenwood College Yearbook, 1930, St. Charles, Missouri.

NEWSPAPERS

Kansas City Times (Kansas City, Missouri), April 22, 1927. Article about Southwestern High School band.

Kansas City Journal-Post (Kansas City, Missouri) April 22, 1928. Article about Calkin boat engine.

Prosser Record (Prosser, Washington) August, 1930. "Calkin Builds Midget Engine."

Seattle Daily Times (Seattle, Washington) August 4, 1930. "Tiniest Gas Engine."

Kansas City Journal -Post (Kansas City, Missouri) August 16, 1930. "Smallest Gas Engine."

University of Washington Daily (Seattle, Washington) October 31, 1930. "Senior Creates Tiny Engine."

University of Washington Daily (Seattle, Washington) April 15, 1931. Semi-annual meeting of ASME, "Design and Construction of Model Gas Engine."

Morning Oregonian (Portland, Oregon) December 5, 1934. Oregon Society of Model Craftsman. Picture and display of Dan Calkin, page 10.

Morning Oregonian (Portland, Oregon) August 15, 1935. "Oregonian Oddities" column, Description of Calkin's Model 1 engine.

Prosser Record-Bulletin (Prosser, Washington) September 10, 1935. "Rotarians hear Rev. Mr. Clearmont Discussing Calkin engine."

Oregonian (Portland, Oregon) September 15, 1935. "Oregonian Oddities" column, ELF Model 1.

Prosser Record-Bulletin (Prosser, Washington) June 4, 1936. "Makes finest small motor in the world."

Sunday Oregonian (Portland, Oregon) November 29, 1936. p. 2, "World's Smallest Plane Motor Built in Portland Factory."

Kansas City Star (Kansas City, Missouri) and many others, syndicated column, December 27, 1936. "Believe it or Not" by Ripley.

Chicago Daily News (Chicago, Illinois) February 5, 1937. "Gas Model show planned."

Chicago Daily News (Chicago, Illinois) February 5, 1937. "Model Planes will feature Loop Air Show"

Chicago Herald and Examiner (Chicago, Illinois) March 9, 1937. "Nifty tiny gas powered model planes."

Chicago Daily Tribune (Chicago, Illinois) March 9, 1937, "Model Makers to show tiniest engine."

Chicago Herald & Examiner (Chicago, Illinois) March 9, 1937. "Fifty tiny gas powered model airplanes."

Chicago Daily News (Chicago, Illinois) March 11, 1937. "Model Planes will feature Loop Air Show."

Chicago Daily News (Chicago, Illinois) March 12, 1937. Central Gas Model Society, "Announcement of show."

Chicago Daily News (Chicago, Illinois) March 12, 1937. "Working plane models bring roaring good times."

Chicago Herald and Examiner (Chicago, Illinois) March 12, 1937. "Tiniest plane."

Chicago Daily News (Chicago, Illinois) March 13, 1937. "Tiny plane really flies."

Kansas City Journal Post (Kansas City, Missouri) March 14, 1937. "Model Plane's Motor Small as Watch."

Seattle Post Intelligencer (Seattle, Washington) March 18, 1937. "Pee Wee gas motor."

Oregonian (Portland, Oregon) July 17, 1939. Contest report.

Sunday Oregonian (Portland, Oregon) August 27, 1939. Dan Calkin featured in article.

Daily Oregonian (Portland, Oregon) August 28, 1939. Contest report.

Oregonian (Portland, Oregon) June 5, 1940. VanCourt's winning model.

Oregonian (Portland, Oregon) July 15, 1940. Contest article, photo of Calkin, Foster, Mitchell.

Tribune-Sun (Place of publication unknown) August 29, 1941. Contest article.

Sunday Oregonian (Portland, Oregon) August 30, 1942. Classified ad for operator.

Oregonian (Portland, Oregon) August 6, 1945. p. 3 Contest photos.

Sunday Oregonian (Portland, Oregon) January 5, 1947. "Tiny Plane Motor".

Lindenwood College Bulletin (St. Charles, Missouri) July 20, 1947. Vol. 20, No. 11. Article about Josephine Peck Calkin and the work she did at ELF.

Milwaukie Review (Milwaukie, Oregon) September 28, 1951. "Building rises for Airplane Engine Works."

Enterprize-Courier (Milwaukie, Oregon) September 19, 1952. "Dan Calkin, Owner of the ELF Engine Company."

Enterprize-Courier (Milwaukie, Oregon) September 20, 1952. Article about ELF operation.

Enterprize-Courier (Milwaukie, Oregon) October 25, 1953. p. 3, "One-man plus makes engine."

Unknown Newspaper, undated, "About possible hazards of gas powered model airplanes."

PERIODICALS

Everyday Engineering, Yates, R. F. Editor, NY, July 1918.

The Scientific Youth, Kansas City, Missouri, November 28, 1925. Michael Faraday Article by Dan Calkin pp. 2, 3.

Popular Mechanics, June 1930. "Tiny Gas Engine Drives 40" Boat."

Science and Invention, October 1930. "Senior Creates Tiny Engine."

Popular Mechanics, June, 1931 "Tiny Gas Engine drives 40" boat."

Model Airplane News, June 1935. Charles H. Grant's Column, "Airways." Photo #2 "Description of Calkin 3 Foot airplane." pp. 22, 23, 36, 37, 47, 48.

Model Airplane News, July 1935, Airways column, Photo #6 of Calkin 3 Foot airplane with ELF engine.

Model Airplane News, February 1936. Gas Lines Column, Discusses need for smaller engines.

Model Airplane News, March 1936. Gas Lines column, "Elbert Weathers ELF powered airplane," photo #2.

Popular Aviation, March 1936. "Dan Calkin trying to build them smaller."

Model Airplane News, April 1936, Gas Lines column, "Carl VanCourt member of IGMAA."

Model Airplane News, May 1936. Gas Lines column, "Elbert Weathers ELF powered airplane."

Aeromodeller, June 1936. "Petrol Model Aeroplane Topics," C E Bowden, Mention of ELF engine on order.

Aeromodeller, August 1936. "Petrol Model Aeroplane Topics," Description of new ELF (MODEL 1) engine.

Model Airplane News, September 1936. Gas Lines column. p. 16."Elbert Weathers Photo #1 Weathers "Tubby" and #7". Tubby.

Model Airplane News, September 1936. Gas Lines column,

Aeromodeller, December 1936. "Elbert Weather's ELF powered Tubby."

Popular Aviation, January 1937. "ELF engine for gas models."

Popular Aviation, February 1937. pp 36, 37, 54.

Model Airplane News, February 1937. Gas Lines column, "photo #10, Ken Willard airplane."

Model Airplane News, April 1937. photo #6, 9, Gas Lines column.

Flying Aces, July 1937. pp. 58, 59, "Gas show big business"

Model Airplane News, July 1937. Gas Lines column, photo #11.

Model Craftsman, July 1937. p. 30, picture and text of ELF engine.

Flying Aces, August 1937. pp 52, 53 "Logging the Motor Market," ELF review.

Model Airplane News, November 1937. pp. 33-36, 54-56 "How to build a Pee Wee gas model."

Model Airplane News, January 1938. p. 51 First ELF (Model 1) advertisement.

Model Airplane News, February 1938. p. 40 ELF (Model 1) engine advertisement.

Model Airplane News, March 1938. p. 38, ELF (Model 1) advertisement.

Model Airplane News, April 1938, pp. 16-19, 38-40, "Building the CAVU" by Ken Willard.

Air Trails, May 1938. pp. 50-53, 76-78. "Selecting your engine"

Flying Aces, September 1938. pp. 37, 38, "Logging the Motor Mart."

Model Airplane News, December 1937. Photo #15 Carl VanCourt airplane.

Air Trials, September 1940. VanCourt record and photo of new Series 1 ELF Twin.

Air Trails, November 1940. Miniature Aircraft includes advertisement for ELF Single and Twin.

Model Plane Annual, January 1941. pp 151, 151, 170. ELF Single, Twin and Four.

Air Trails, April 1941, pp. 28-32. Gas Engine Census.

Flying Aces, April 1941. "Logging the Motor Mart," ELF Twin.

Model Airplane News, June 1941. p. 46, "The New ELF Twin"

Model Craftsman, June 1941. p. 37, 38, ELF.

Model Craftsman, June 1941. "A Swell Job", by John Ulman.

Aeromodeller, July 1941, p. 29, photo of ELF Model 40.

Air Trails, September 1941. p 56, Ad for Model 10, 20 and 40.

Flying Aces, December 1941. "Logging the Motor Mart", Model 10 ELF.

Flying Aces, April 1942, "Logging the Motor Mart", ELF Model 10.

Flying Aces, October 1942. "Logging the Motor Mart", ELF Four.

Air Trails Annual, January 1943. pp. 36, 37. ELF Single and Twin.

Air Trials, May, 1944. "Model Airplane Engine Performance." p. 22, 23, 44, 46. 48.

Air Trails, October 1944. pp. 48-50, 92, 93, "Engines Return."

Model Airplane News, June 1945. pp. 14, 15, 59-64, "Construction Features of Model Engines."

National Geographic, December 1946. "Oregon Finds New Riches."

Air Trails, March 1949. p. 108. Reader wants address of ELF.

West Coast Model News, September 1953. Earl Vivell and his ELF Six powered R/C plane.

Model Airplane News, November 1953. "Picture of ELF 6 in an airplane."

Model Airplane News, August 1955. "These engines were different."

Engine Collectors' Journal, August 1963. pp, 4,7. Several ELF Engines.

The Engine Collectors' Journal, May 1965. p. 7. Photo of Dan Calkin and Mel Anderson.

The Engine Collectors' Journal, July 1965. p. 11, reed valve ELF Single.

The Engine Collectors' Journal, April 1966. "Odds and Ends" from Doc Nichol, about ELF spark plugs.

The Engine Collectors' Journal, November 1966. Odds and Ends from Doc Nichol, More about ELF spark plugs.

Radio Control Modeler, July 1973. Dick Tichenor's Photo Album, featuring Alan Shively's ELF collection.

Model Builder, November 1976. "Plug Sparks" Column by John Pond includes drawing and discussion of the Model 1 ELF.

Cloud 9, AMA, March 1988. p. 3. "Next Month, More about the ELF by Carl VanCourt."

Aeromodeller, March 1989. pp 160-163. Alex Imrie's "Vintage Corner" column. Article about Imrie's reproductions of a Model 1 ELF and Elbert Weathers' Miss San Diego.

Cloud 9, AMA, May 1989. p. 1. "Cloud 9 in Florida with Carl VanCourt."

Aeromodeller, April, 1991. "Vintage Corner" by Alex Imrie.

Model Engines World, February 1995. p. 16, readers home-built engines, How to build a Model 1 ELF.

CATALOGS

Berkeley, New York, 1941. p. 40, Ad for Single, Twin and Four.

Miniature Aircraft of New York, January 1943. Ad for ELF Single and Twin.

Portland's Hobby Center flyer, Portland, Oregon, March 1947. "Calkin Pioneers Model Plane Motors Here."

Miniature Aircraft Company of New York, January 1950. until 1956. Ads for all current ELF models. ELF ads appeared in many of their annual catalogs.

MISCELLANEOUS

Portland Chamber of Commerce Publication September 19, 1936. "Smallest engine developed here."

1937 California Championship Contest official program. ELF ad on rear cover and on an inside page.

Notes on Model Airplane and Propeller Design by Daniel Calkin 1937, 8 pages, ELF Engine Company, Portland, Oregon.

ELF ADVERTISEMENTS

Model Airplane News, January 1938, p. 51

Model Airplane News, February 1938, p. 40

Model Airplane News, March 1938, p. 46

Model Airplane News, December 1938, p. 42

Air Trails, November 1940, p. 57

Air Trails, December 1940, Miniature Aircraft Corp. ad.

Model Airplane News, December 8, 1940. Also ad copy for ELF Twin for this advertisement.

Model Airplane News, January 1941. p 55, Ad "Super Power for Class A ELF Twin.

Air Trails, April 1941. p. 48 "Super Power for Class A"

Model Airplane News, April 8, 1941. Ad Copy "Dan Announces the New ELF Four."

Air Trails, April 1941. p. 57. Ad announcing the ELF 4, also Twin and Single.

Model Airplane News, May 1941, p. 57

Air Trails, May 1941, p. 57.

Air Trails, June 1941, p. 52, Ad for Model 10, 20 and 40.

Model Airplane News, June, 1941, p. 46.

Air Trails, September 1941, p. 56, Ad for Model 10, 20 and 40.

BOOKS

Bowden, C. E. *Petrol-Engined Model Aircraft*, (Percival Marshall, London, 1937) p. 40, 88, 89, 139, 153.

Bowden, C. E. *Petrol-Engined Model Aircraft*, (Percival Marshall, London, circa 1945) p. 24, 27, 45, 149, 170.

Camm, F. J. *Model Aeroplane Handbook*, (George Newnes, London, 1949) p. 143,

Cooke, David, C. and Davidson, Jesse, *The Model Plane Annual 1942*, (McBride, NY, NY) p. 148, 151-152, 170.

Clanford, Mike, *A Pictorial A to Z of Vintage and Classic Model Airplane Engines*, (Clan Enterprises, Mitchum, Surrey, England, 1987) Note: The Model 1 ELF on page 60 is a reproduction engine.

Fisher, O.F.W. *Collector's Guide to Model Aero Engines.* (Model And Allied Publications, 1977).

Forster, J. F. P. *Model Aircraft Petrol Engines,* Cover title *Petrol Engines For Model Aircraft*, (Aircraft Publications Ltd, circa 1943).

Garami, Louis, and McEntee, Howard, *Model Motor Manual*, (Air Age, NY, NY).

Gierke, C. David *2-Stroke Glow Engines for R/C Aircraft*, (Air Age, Inc, Wilton, CT, 1994).

Graham, Graham P. and Cleveland, Reginald M., *The Model Plane Annual 1941-1942*, (McBride, NY, NY, 1941) pp. 137, 148, 151-152, 170.

Hertz, Louis H. *The Complete Book of Model Aircraft Spacecraft and Rockets*, (Crown, NY, NY. 1967) p. 264.

Mackey, Charles, *Pioneers of Control Line Flying* (Precision Aerobatic Model Pilots' Association, Anniston, AL, 1995) pp. 19-25.

Sheelds, Danny, *Antique and Old Timer Model Aircraft*, (Free State Press, Annapolis, MD, 1980). pp. 46, 54.

Winston, Bernard, *Model Gas Engine Handbook*, (Winston Pub., NY, NY, 1944) pp. 52,53, 66, 88-92.

Pond, John, *Vintage Engines of Yesteryear Vol. 1*, (Pond Associates, 1984) 104 p. Article and drawings, pp. 47, 48.

Suzor, Gems, *Petits Moteurs pour Modeles Reduits 1938.* Paris, 1938 121p. Model 1 ELF on front cover and title page. A description and photo appear on pp. 64-65, ELF spark plug p. 78.

Thornberg, Dave, *Do You Speak Model Airplane*, (Pony XPress, Albuquerque, NM, 1992).

Warring, Ron, *Miniature Aero Motors* (Ian Allen, Ltd, England, 1949) pp. 13, 147, 154-157.

Winters, William, *Engines for Models* (Berkeley Models, NY, NY, 1953) 94p. ELF engines on front cover and p. 48.

Yates, Raymond F. *Model Making*, (Henley, NY, NY, 1919, 1925, 1929) pp. 335-345.

Zaic, Frank, *1938 Model Aeronautics Yearbook*, (Model Aeronautics Publications, NY, NY, 1938.) p. 61-63, 67, 71.

Appendix E ELF INSIDERS

This book is the result of a cooperative effort by many people. Whatever contribution this book makes is because of the invaluable help provided. The information obtained has exceeded my wildest dreams. Everyone listed contributed something. Of course, some contributed more than others and I have tried to acknowledge the magnitude of all the contributions in a crude way. Please accept my sincerest apology if your contribution is overlooked or undervalued.

INVALUABLE CONTRIBUTION

Josephine Jane Peck Calkin*
Cecelia Jane Calkin Jackson
Ted Enticknap
Philip Foster
Carl VanCourt*
Jean R. Brown
Richard Rothrock*
Dan Lutz
Dr. Richard E. Nichol
Leonard "Lucky" Lucas
Alex Imrie
Ron Sporing
Abe Gallas
Dick Tichenor*
Tim Dannels
Irwin Ohlsson*

MAJOR CONTRIBUTION

Jim Adams
Ted Belcher*
John J. Brown, Jr.
Randy Chase
Peter Chinn
Rich Clymer
Fred Collins
Steve Ditta
Don Edmunds
Charles Enstrum
Chuck Entler
Deno Gaganis
George Genevro
Jim Gerard
Walt Grigg
Bill Hannan
Dale Kirn
William V. Krecek
John Lorence
Joe Macias
Charles Mackey
Mike Patience
Miles Patience
Phil Riner
Allan R. Shively*
Chuck Welman

IMPORTANT CONTRIBUTION

George Abbott
Jim Alaback
Robert Arthur
Buzz Averill
Woody Bartelt
Chuck Bartunek
Mike Bassett
Bruce Batch
Peter Baylinson
Ross Beale
Don Belote
Bill Bickel
Toby Blizzard
Ron Boots
Bob Bowen
Larry Boyer
Jack Brown
Charlie Bruce
Bill Caldwell
John Camp
John Carlson
Donald Chapton
Ray Chalker
Bruce Chandler
Mike Clanford
Gordon Codding
George Cook
Tom Cope
Gil Coughlin
Bob Cowles
Tom Crompton
Doug Dahlke
Richard Dalby
Bill Daniels

Tom Dean
Walter DeGenring
Art Dekalb
Leslie Deline
Dave Demilt
Vic Didelot
Anthony Dominick
Mike Drozda*
Ken Dullea
Robert Eireman
Dave Evans
Ronald Ford
Clark Foster
Joe B. Frey
Red Garlow
Donald Galurneou
Jim Gerard
Dave Gierke
John Goodall
Dick Gooding
John Gossman
Gale Grant
Jerry Greaves
Bill Grove
Steve Guardino
Pete Gutierrez
Jim Gwaltney
Ward Hallenburg
Bob Hartman, Jr.
George Hellmer
Bob Hess
Hank Hilscher
Basil Hiotakis
Robert Hopper
Bill Ives
David Jessee
Fred Kaiser
Joan Kaiser
Herb Keener
Ji Kennerman
Frank Kidd
Bill Kincheloe
Guy Kirkwood
John Knepper
Richard Kohl
Dick Lalley
Bill Lane
Jim Large
Bob Laybourne
Clarence Lee
Don Long
Frank Macy
William Mahlstadt
Bob McClelland
John McDonald
Bob McKinley
George Milano
Bob Murphy
Jerry Nielson
Ron Olsen
Robert H. Osenga
Jean Pailet
Jack Passey
Roger Paul
Bob Peterson
Arthur Polson
Ivan Prior
Floyd Reck
Steve Remmington
Kevin Richard
Willie Ritzheimer
John Roberts
Don Saunders
Aubrey Schaefer
Dan Schaefer
Norman Schehl
Ron Schorr
Charles Schultz
Danny Sheelds
Dick Sherman
Gary Sherman
Bill Simpson
Francis Smith
Fred Stewart
William Stokes
Les Stone
Art Swift
Bill Thompson
Halcom Tilley
John Trinder
Gordon Truax
Joe C. Walker
Gene Wallock
Bill Warner
Robert Watkins
Doug Wendt
Lester Weise
Earl Vincent
Joe Wagner
Charles Yost

*** Deceased**

INDEX

D

E

F

G

H

I

J

K

Q

R

S

T

U

V

W

X

Y

Z